BIOTECHNOLOGY AND THE CONSUMER

Biotechnology and the Consumer

A research project sponsored by the Office of Consumer Affairs of Industry Canada

Edited by

Bartha M. Knoppers and Alan D. Mathios

Partly reprinted from the *Journal of Consumer Policy*, Volume 21, No. 4 (1998)

SPRINGER SCIENCE+BUSINESS MEDIA, B.V.

Library of Congress Cataloging-in-Publication Data

Biotechnology and the consumer / edited by Bartha Maria Knoppers and Alan Mathios.
p. cm.
ISBN 978-1-4613-7423-7 ISBN 978-1-4615-5311-3 (eBook)
DOI 10.1007/978-1-4615-5311-3
1. Biotechnology. 2. Consumers. 3. Biotechnology--Canada. 4. Biotechnology--Public opinion. 5. Biotechnology industries. 6. Biotechnology--Moral and ethical aspects. I. Knoppers, Bartha Maria. II. Mathios, Alan D.
TP248.2.B55128 1999
338.4'76606--dc21 98-55183

ISBN 978-1-4613-7423-7

Printed on acid-free paper

Contents

Introduction [vii–xiv]

Articles

HEATHER SHEEHY in collaboration with MARC LEGAULT and DEREK IRELAND / Consumers and Biotechnology: A Synopsis of Survey and Focus Group Research* [1]

JENNIFER B. WOHL / Consumers' Decision-Making and Risk Perceptions Regarding Foods Produced with Biotechnology* [29]

EDNA F. EINSIEDEL / The Market for Credible Information in Biotechnology* [47]

THÉRÈSE LEROUX, MARIE HIRTLE and LOUIS-NICOLAS FORTIN / An Overview of Public Consultation Mechanisms Developed to Address the Ethical and Social Issues Raised by Biotechnology* [87]

TIMOTHY CAULFIELD / The Commercialization of Human Genetics: A Discussion of Issues Relevant to the Canadian Consumer* [125]

ZHIQI CHEN and ALISON MCDERMOTT / International Comparisons of Biotechnology Policies* [169]

GILLIAN K. HADFIELD and DAVID THOMSON / An Information-Based Approach to Labeling Biotechnology Consumer Products* [193]

ALAN D. MATHIOS / Economic Perspectives on the Dissemination of Science-Based Information to Consumers [221]

* Chapters indicated with an asterisk are reprinted from the *Journal of Consumer Policy* volume 21, No. 4 (1998).

ANDREW N. KLEIT / Using Advertising to Generate Information and Signals for Product Quality: Lessons for Biotechnology Markets in Canada from Pharmaceutical Markets in the United States [257]

STEPHEN STRAUSS / Biotechnology and the Media [277]

G. BRUCE DOERN and HEATHER SHEEHY / The Federal Biotechnology Regulatory System: A Commentary on an Institutional Work in Progress [309]

CHRISTOPHER GREEN / The Industrial Economics of Biotechnology [345]

CLAUDE CRAMPES and ABRAHAM HOLLANDER / Biotechnological Innovation and Industrial Performance [377]

JEFFREY G. MACINTOSH and DOUGLAS J. CUMMING / Consumer Controversy and the Funding of Biotechnology Research [411]

Appendices

MARC LEGAULT and ALISON MCDERMOTT in collaboration with BARTHA M. KNOPPERS and DEREK IRELAND / Integration Document – Biotechnology, the Consumer, and the Canadian Marketplace [457]

BRYNE PURCHASE / Rapporteur's Remarks – Symposium on Biotechnology and the Consumer [499]

Bartha M. Knoppers and Alan D. Mathios
Introduction

Biotechnology is a rapidly developing sector of the economy for countries throughout the world. This rapid development has led to heated debate over its risks and benefits. Advocates of biotechnology point to the potential benefits offered by products that promise to eliminate disease, provide for more efficient diagnostic techniques, treatments and drugs, yield increased food production, and so forth. Others fear that the rapid developments of this technology have occurred without appropriate consideration having been given to the ethical ramifications, the potential health risks and long-term environmental impacts, implications for income distribution, and potential for abuse. Consumers and producers share concern for the future of biotechnology: the realities and even the perceptions, informed or otherwise.

This book is the outcome of a research project on *Biotechnology and the Consumer* sponsored by the Office of Consumer Affairs of Industry Canada. The project was designed to foster informed public policy on biotechnology and in particular, to contribute to and inform the Canadian government's development of a Canadian Biotechnology Strategy. The Office funded a group of authors to prepare a series of analytical papers on a range of consumer and informational issues related to biotechnology. This project also involved an interim workshop in which the authors presented their papers, and culminated in a symposium on *Biotechnology and the Consumer Interest*, held on September 24–25, 1997, in Ottawa, Canada. At this symposium the ideas of the papers and their implications were discussed among a wider group of stakeholders, consisting of academics, government representatives, and representatives of advocacy groups, industry associations, and consumer organizations. Two documents integrate the themes of the research papers. Appendix A includes the integration paper prepared by Marc Legault and Alison McDermott (in collaboration with Bartha M. Knoppers and Derek Ireland). This appendix organizes the research into three themes: the consumer perspective on biotechnology, producers of biotechnology, and the

Journal of Consumer Policy **21**: vii–xiv, 1998.
Bartha M. Knoppers & Alan D. Mathios (eds), Biotechnology and the Consumer.

regulation of biotechnology. Appendix B contains the Remarks on the Symposium by the Rapporteur, Bryne Purchase.

The editors of this book are grateful to the large number of individuals who reviewed these papers through each stage of preparation. Appendix A provides a description of this review process. Heiner Imkamp and Folke Ölander provided timely and insightful technical editing of each of the papers.

The term biotechnology is typically used in a variety of contexts and means different things to different individuals. The Canadian Environmental Protection Act defines biotechnology as: *The application of science and engineering in the direct use of living organisms or parts or products of living organisms in their natural or modified forms*. Biotechnology applications are vast and diverse and include: genetically altered tomatoes for increased shelf life and enhanced flavor, the introduction of human genes into bacteria to produce medicines and vaccines, genetic testing of human embryos to determine predisposition for disease, injection of human genetic material into animals to produce organs for transplant, and increased production of milk from cows. Just as the applications are diverse, so are the stakeholders in any policies regarding biotechnology. The main stakeholders can be categorized as consumers of biotechnology, the broad citizen base as represented by government, and producers of biotechnology. Though each chapter tends to focus on one of these three stakeholder groups, it will become obvious to the reader that government policy, producer decisions, and consumer perceptions of biotechnology are fundamentally interrelated with each other.

The first chapter, written by Heather Sheehy (in collaboration with Marc Legault and Derek Ireland) provides a summary of survey and focus group research on consumer perceptions of biotechnology. This synopsis was distributed at the outset to each of the authors and served as a common empirical base. Overall, the survey and focus group research indicates that consumers are cautiously optimistic towards biotechnology. Consumers have limited knowledge of the broad range of biotechnology issues, although they are most familiar with some of the more traditional applications in food and health. Consumers are found to have some anxieties related to possible negative long-term consequences of biotechnology applications and make frequent mention of past examples of negative outcomes associated with other technologies. Consumer perceptions of biotechnology are found to be very complex and involve an array of non-price issues, including environmental concerns and moral beliefs. Most survey respondents admit that government could play an important role in providing information and in ensuring consumer safety, although they also suggest that a consortium of representatives from government, academia, and the private sector would provide a more balanced perspective.

How do consumers form these perceptions, make decisions, and assess the risks associated with biotechnology products? There are a number of papers that address these issues. The first, by Jennifer B. Wohl, develops an analytical model of how consumers make choices about foods produced with biotechnology. An expected-value framework is used to evaluate potential circumstances under which consumers would choose a biotechnology product over a conventional food product. The model also examines choice in the absence of conventional products. Consumers are said to choose biotechnology products when the perceived benefits of these products exceed the perceived costs associated with such a choice. As these costs are a function not only of the perceived probability of an adverse outcome, but also the magnitude of dread associated with such an outcome, Wohl demonstrates that consumers may choose to avoid biotechnology products even when the perceived probability of an adverse outcome is extremely small. Perceived costs may also include ethical and moral implications and other motivations underlying the development of biotechnology, safeguards and controls, and unintended impacts of biotechnology applications. The paper also discusses how consumers form their risk estimates.

Edna F. Einsiedel also examines how consumers form perceptions of biotechnology products. The paper starts from the premise that evolving technologies typically undergo public and citizen scrutiny at early stages of their development. This phenomena reinforces the importance of understanding how consumers build trust of technology, what sources of information are credible to consumers, and how likely consumers are to accept certain technologies. It is argued that consumer perceptions of biotechnology products incorporate factors well beyond simple estimates of utilities (or benefits) and costs, and include factors such as consumer views on moral responsibility. Comparisons with other industries (energy and chemical industries) facing similar public scrutiny are made, and the author discusses how these industries interfaced with consumers about risk. The study includes a discussion of international approaches toward public communication about biotechnology and concludes with a discussion of industry initiatives.

Each of the first three papers addresses, in some form, the role that ethical considerations play in the formation of consumer perceptions of biotechnology. Ethical considerations are directly addressed in the paper written by Thérèse Leroux, Marie Hirtle, and Louis-Nicholas Fortin. This study includes a comparative study of mechanisms developed for examining the ethical and social issues in biotechnology. The study seeks to present the variety of mechanisms established by certain countries regarding these issues. It includes discussion of approaches taken in Australia, Denmark and other Nordic countries, the United States, and the United Kingdom. A variety of approaches have been developed to examine the ethical and social implications of biotechnology. The paper includes a detailed discussion of the information collection strategies and the information dissemination approaches utilized by these institutes and other agencies.

Timothy Caulfield reviews many of the concerns, including ethical considerations, which have been articulated with respect to the commercialization of human genetics, examining them from the perspective of the consumer. The study begins by reviewing the state of the biotechnology industry and finding that human genetics is undeniably "big business," and that encouraging the growth of the biotech sector is viewed as good social policy. However, there also exists a fear of the impact of the commercialization process. Frequently raised issues surrounding the commercialization process include: non-disclosure agreements and other patenting issues, premature entrance of genetic services into the marketplace, "eugenic" pressures that skew consumers' view of disease and disability, marketing and advertising, altering the direction of university-based research, liability concerns, and finally costs to the health care system.

In addition to developing approaches to examine the ethical implications of biotechnology, governments have also devised policies for the direct regulation of biotechnology products. The study by Zhiqi Chen and Alison McDermott provides an international comparison of biotechnology policies. It compares the policies of the United States, Japan, the European Union, and three of its member countries (France, Germany, and the United Kingdom). Several types of government intervention in the marketplace are addressed including industrial policies, regulatory frameworks, and protection of intellectual property.

Within industrial policies attention is focused on government assistance in research and development and government assistance in capital investment. The study includes a comparative analysis of regulatory frameworks and intellectual property protection and examines recent international policy initiatives.

Regulation of biotechnology has also led to heated debate about which types of information can best inform consumers about the risks and benefits of biotechnology.

The study by Gillian K. Hadfield and David Thomson considers labeling as a mechanism to provide consumers with relevant risk information. The paper begins from the premise that the central issue to be examined from a consumer protection standpoint, is a basic information problem: Are consumers getting what they intend and expect to get from their transactions in the marketplace? The chapter analyzes the role of labeling in assisting consumers to get what they expect. The analysis reflects a deference to a consumer-centered approach over paternalism. This is an important distinction in the case of biotechnology products, where significant differences exist between scientific and consumer assessments of risk. It is noted that identifying a consumer protection problem solely with respect to scientific assessments of risk ignores the reality that scientific assessments are uncertain themselves. The study focuses on the potential for labeling to serve the basic function of promoting choice, by allowing consumers who wish to avoid these products a way of doing so. The alert label is discussed for its role in structuring a mechanism whereby information is mediated through the marketplace rather than disseminated by government.

Alan D. Mathios examines, from an economics perspective, the dissemination of science-based information to the consumer and regulatory policies to ensure the truthfulness of the information. The benefits and limitations of government-provided information are discussed and this information source is compared with marketing and advertising by firms. In the mid 1980s, the United States reversed a prohibition on the advertising, by food manufacturers, of the link between diet and health. This policy reversal provided a useful period for evaluating whether the onset of producer-provided claims led to improved consumer choices or deceived consumers and led to less healthy food choices. A variety of data are presented to examine this issue. The study concludes with suggestions of how the results of this analysis can be applied to biotechnology and a review of the institutions that regulate truthfulness of advertising in the United States.

Andrew N. Kleit also reviews the role of advertising in communicating information to consumers. Brand recognition as a potential signal of quality is especially important in situations where quality cannot be judged easily by the consumer. Economic theories of advertising are reviewed; these range from advertising in the role of a performance bond, to advertising as a barrier to entry into an industry. Since many biotechnology products are anticipated to be promoted by large pharmaceutical companies, the chapter reviews how pharmaceutical advertising is regulated in the United States. Arguments for and against direct-to-consumer advertising are reviewed and the chapter concludes with suggestions for the regulation of advertising for biotechnology products.

Stephen Strauss considers the role of the media in providing information about biotechnology to consumers. The study examines the general principles of media operations and then turns to the issues unique to the reporting of biotechnology. The focus is on daily newspapers, in part, because of the practical difficulties of examining television coverage, retrospectively. Some of the unique features of reporting biotechnology issues are discussed: The technology is too new to be understood, its non-dramatic advances tend to be ignored, and it is still largely a business story. The study also discusses the types of biases that might exist among experts who are quoted in the media.

How do governments set up institutions to address the variety of consumer protection issues raised in the previous chapters while preserving the benefits of biotechnology? The next study, written by G. Bruce Doern and Heather Sheehy, provides a detailed description of the Canadian federal biotechnology regulatory framework. Selected features of the emerging regulatory framework are examined and the commentary is based on regulatory institutional analysis. Six principles underlying the regulatory framework are described. These principles are intended to reflect a reasonable and balanced tradeoff between securing the practical benefits of biotechnology and protecting individuals with respect to environmental, human, and animal health and safety. The three core functions of the biotechnology regulatory system are discussed. These include legislative mandates, policy and regulation making, the case handling and approval processes, and compliance and enforcement processes.

The final three chapters of the book examine the producers of biotechnology. Unlike many industries, the business of biotechnology cannot be categorized into a standard industrial classification on the basis of product or process characteristics. The potential applications of biotechnology span such a diverse set of industry sectors that it is difficult to apply standard industrial organization principles to analyze the impact of this economic activity.

The study by Christopher Green examines the industrial economics of biotechnology. The chapter focuses on identifying the producers of biotechnology, the links between various producers, and the structure, behavior, and performance of new biotechnology firms (NBF). The NBF component of the biotechnology sector is found to consist solely of small firms, although there are a few relatively large NBF's. In addition, biotechnology activities take place within the divisions of large pharmaceutical companies and chemical firms. Issues of economics of scale in production, vertical integration, and entry are analyzed. Particular attention is devoted to the static and dynamic vertical links between new biotechnology firms, university-based researchers, and large established pharmaceutical firms. The chapter concludes with three case studies of biotechnology firms.

Claude Crampes and David Hollander examine biotechnology innovation and industrial performance. The chapter reviews developments in biotechnology and examines their impacts on user industries. Partnership agreements between small biotechnology firms and large industrial firms are identified as important elements for proceeding from the research phase to the product development stage. Additional ties to research universities also appear to be important factors in product development. The chapter examines the impact of biotechnology on the agri-food sector. Probable impacts on production and trade balances, on prices, on employment, and on international trade are some of the key measures considered. The impact of biotechnology on health industries is also considered. The necessity for regulation and establishment of product protection law is discussed.

Many of the previous studies in this book suggest that consumer perceptions of biotechnology may be one of the major obstacles facing producers of biotechnology. The final chapter of the book, written by Jeffrey G. McIntosh and Douglas Cumming, focuses on the interaction between the consumer controversy surrounding biotechnology and the funding of research and development (R&D). The chapter highlights some potential controversies associated with food technologies and genetic manipulations including damage to local ecosystems, enhanced use of herbicides, effects on local plant species, the creation of super-pests, and gene jumps. Attention is then focused on factors that mediate consumer reactions to biotechnology. The chapter concludes with a detailed empirical analysis of the interplay between consumer controversy and the funding of biotechnology R&D. The principal questions addressed by this empirical section are: (a) In deciding whether to fund particular R&D projects, do firms take into account current or potential consumer controversies in their due diligence review? (b) What are the financial and other characteristics of firms that address consumer controversies in their R&D expenditures decisions? (c) In which sectors of the biotechnology industry is the funding of R&D most influenced by consumer controversies? (d) Would greater consumer acceptance of biotechnology lead to more funds being directed towards research and development? (e) How do current and potential consumer controversies compare to other factors that influence the funding of R&D in Canadian biotechnology firms?

In conclusion, biotechnology, covering as it does, all "living organisms" presents a dual challenge to consumers, producers, and regulators: (i) informed collaboration and mutual respect, and (ii) common stewardship in the public interest both now and for the future. Hopefully, this project of the Office of Consumer Affairs of Industry Canada begins to meet this challenge in providing an improved analytical framework to guide the ongoing debate on biotechnology and the consumer.

Heather Sheehy in collaboration with
Marc Legault and Derek Ireland

Consumers and Biotechnology: A Synopsis of Survey and Focus Group Research

ABSTRACT. The Office of Consumer Affairs of Industry Canada, in conjunction with other federal departments, has undertaken preliminary work to identify concerns and highlight consumer issues with respect to biotechnology and its applications. This has included conducting a number of surveys and focus groups to identify what consumers know about biotechnology; where there are information gaps; and how information impacts upon consumers' purchasing decisions.

This paper is structured into five sections. First, it provides an overview of some background information on biotechnology. The second section summarises consumers' knowledge of, and reactions to, biotechnology. Next, the paper highlights factors that influence consumers' attitudes towards biotechnology. The fourth section examines two ways of overcoming marketplace barriers contrasting increased consumer awareness with more traditional legislative mechanisms. The concluding section outlines questions that need to be answered in order to determine how the marketplace can best address issues related to biotechnological products.

The April 2nd, 1996 edition of the *Charlottetown Guardian* carried an article reporting the introduction of a new variety of potato, genetically engineered to resist infection by the Colorado potato beetle. In a province known for its "love of spuds," the NewLeaf potato could end the need to spray crops with chemical pesticides in the fight against the devastating impact of beetles.

Trying innovative ways to fight pests is not something new for farmers. However, unlike other, more traditional methods, the NewLeaf potato deflects beetles from the "inside-out" through the addition of an extra gene that produces bacteria that are safe for human consumption, but deadly to beetles. This type of genetic engineering is commonly referred to as biotechnology, a broader term describing the use of parts of living organisms to create new products with innovative characteristics.

Although initial reaction to genetic engineering may evoke images of *Attack of the Killer Tomato*, in fact, there are many traditional applications of biotechnology. The use of yeast to make beer and the selective breeding of animals to enhance desired characteristics are

Journal of Consumer Policy 21: 359–386, 1998.
Bartha M. Knoppers & Alan D. Mathios (eds), Biotechnology and the Consumer.

examples of how living organisms have been used to create consumer products in the past.

The Office of Consumer Affairs (OCA) of Industry Canada, in conjunction with other federal departments, has, over the past two years, completed preliminary work to identify concerns and to highlight consumer issues with respect to biotechnology and its applications. This has included conducting a number of surveys and focus groups to expand our understanding of how much consumers already know about biotechnology, where there are gaps in their information, and how information impacts upon their perceptions and purchasing decisions.

Not surprisingly, these studies found that consumers proceed very cautiously with respect to biotechnology and its applications. They expressed concern over a broad range of unknowns. For example, while beetle-resistant potatoes may appear innocuous enough, the introduction of a broad range of genetically engineered products onto Canada's markets raises a number of consumer, social, and ethical issues. Questions that were raised include: Who will ensure that the products are safe? When I go shopping, will I be able to choose between biotech and traditional products? What does biotechnology entail – are we talking about playing God?

Consumers indicated that although they want to be able to choose between products using biotech and traditional products, at present, many feel ill-prepared to make informed decisions. As a new, complex, and sometimes alarming topic, biotechnology can leave consumers confused. Thus, not surprisingly, rather than only using traditional marketplace indicators such as price and quality to decide whether or not to purchase a biotech product, such as price and quality, consumers rely heavily on personal experiences and values.

The purpose of this paper is threefold. First, it will take an inventory of some of the information that the Office of Consumer Affairs has gathered with respect to biotechnology. Second, it will provide a common reference-point for the authors of a series of expert analytical papers being developed in conjunction with the Office of Consumer Affairs as part of the National Biotechnology Strategy. Third, the paper will investigate whether the nature of biotechnology itself prevents it from operating within the traditional marketplace structure.

In order to achieve these three goals, the paper is structured into five sections. The first gives some background information on biotechnology, framing it as a new and complex issue. Using the survey and focus group results, the second section summarises consumers' knowledge of, and reactions to, biotechnology. The third section highlights factors that influence consumers' attitudes toward biotechnology including those mentioned within the focus groups. In particular, it focuses upon those factors that impact upon consumers' ability to analyse information. The fourth section looks at two ways that marketplace barriers can be overcome, contrasting increased consumer awareness with more traditional legislative mechanisms. The concluding section will outline a number of questions that need to be answered in order to determine which of the two options provides a better solution to overcome marketplace barriers.

BIOTECHNOLOGY: WHAT IS CONFUSING THE ISSUE?

Biotechnology can be confusing for consumers. Not only does biotechnology appear to have emerged as an issue relatively recently, it also involves complexities that are not necessarily evident to the casual observer. This paper will show that these qualities can have repercussions on the marketplace behaviour of consumers. First, however, it is important to understand what biotechnology is, and what it is not.

Biotechnology Is . . . Not New

In fact, biotechnology is very old. The *Canadian Environmental Protection Act* (CEPA), Section 3(1), defines biotechnology as "the application of science and engineering in the direct use of living organisms or parts or products of living organisms in their natural or modified forms." Using this broad definition, biotechnology includes a number of familiar processes such as using yeast (which is a unicellular fungus) to make bread and beer or using rennin (an enzyme taken from the lining of calves' stomachs) to make cheese. Given that there are many of these traditional applications, why is biotechnology often associated with recent scientific advances?

Most likely it is because much of the recent attention surrounding biotechnology concerns the issue of genetic engineering, which is, indeed, a new technology. Developed in the early 1970's, it was not until a decade later that the first genetically engineered products reached the market.

The process of genetic engineering basically involves scientifically altering an organism's genetic code. DNA (or the chemical, deoxyribonucleic acid) is found in the nucleus of all living cells and consists of genes that control the production of proteins. Each living organism has a unique arrangement of DNA and genes that create its own individual characteristics. For a more complete discussion of DNA, please refer to Consumers Association of Canada (1995, pp. 4–5). As depicted in Figure 1, scientists can alter the naturally-occurring characteristics of an organism by cutting and pasting genes. In some cases, this involves taking a gene from one organism and transplanting it to another. In other cases, this might involve re-arranging the order of genes within the same organism. The result however, is an infinite number of genetic combinations.

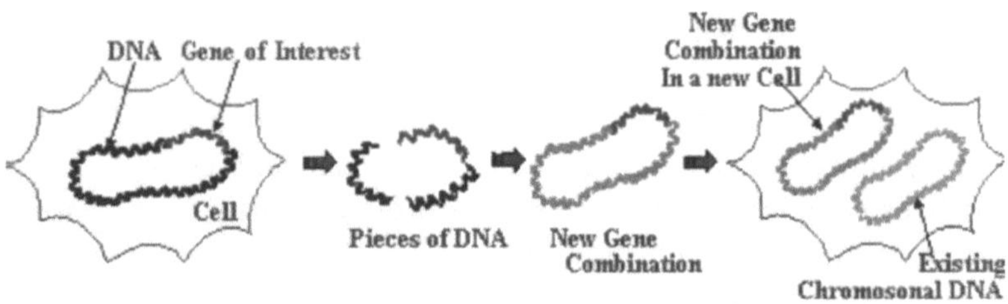

Figure 1. How are new gene combinations made?

Biotechnology Is . . . Hidden to the Observer

Although a highly complex process, biotechnological change is largely hidden from the observer. This is especially true of genetic engineering which is not only a hidden process, but has, to date, made largely small, unremarkable alterations to organisms' characteristics, rather than fundamental changes to the nature of a species. For example, returning to the NewLeaf potato, despite its beetle-resistant qualities, the casual observer would be unable to differentiate it from other "non-biotech" potatoes. As a result, consumers must seek out this information from specialists and others.

Biotechnology Can . . . Raise Alarm

For generations, people's imaginations have created fantastic stories that parallel genetic engineering. Movies such as Jurassic Park, where genetically-engineered dinosaurs create havoc, encourage the public to think of terrible outcomes that could result from re-arranging genes. Nevertheless, to date, genetic engineering has been used in small, largely inconspicuous ways and none of the negative events found in films, the popular literature, and the press, or forecasted by those alarmed by the technology, have come to pass. In fact, the scientific evidence up to now does not suggest that genetic engineering by its very nature would raise more health and safety alarms than other more traditional technologies. The Third Report of the House of Commons Standing Committee on Environment and Sustainable Development looked at whether biotechnology merits specific process-oriented legislation to ensure consumer/environmental protection. Its recommendation to proceed with a system of providing protection under current legislation, with the *Canadian Environmental Protection Act* (CEPA) acting as a "safety-net" for processes not covered elsewhere, supports the view that biotech products do not by definition pose an additional safety concern (Biotechnology regulation in Canada, 1996).

Biotechnology Can Be . . . Complex to the Layperson

Biotechnology, and in particular, genetic engineering, has been made possible through advanced research in science and technology. Generally, the technical complexity of the technology prevents the public from understanding details of the product and the specific attributes of biotechnology applications. The result is an unequal relationship between the producer and the consumer which can leave the consumer without the necessary tools by which to analyse the costs and benefits of a product. The consumer is left to speculate on his/her own about possible consequences that may or may not be reasonable. This situation is compounded by the rapid pace of change and the information overload of consumers.

Rapid rate of change. Technology is evolving at such a rapid pace that what is "new" one year is obsolete the next. Just as consumers become comfortable with one technology, another, new process emerges. As noted in *The Economist* (The hitchhiker's guide to cybernomics, 1996), "during the past two decades the global network of computers, telephones and televisions has increased its information-carrying capacity a million times over." As a result of this rapid rate of change, more and more time is being demanded of people who want to "keep up."

This acceleration of technological change can be daunting, even to those specialising in a particular field. While the science of genetic engineering itself is not accelerating quickly, technology has promoted the rapid expansion of biotechnology applications. It would be presumptuous to assume that consumers have the time to keep up with all the changes to a level that would allow them to appropriately evaluate the pros and cons of biotech products. Moreover, relatively few consumers are early adopters of the products of new technologies. More of us are followers. As cited in Ireland (1996),

> the 1995 CROP Environics survey shows that perhaps fewer than one in ten Canadians are eager to embrace the goods and services provided by new technologies, while about 50% or so bring varying degrees of enthusiasm to purchases of high technology goods and services. At the other end of the spectrum, about one in ten Canadians continue to resist and distrust technology, while the remaining 30% display varying degrees of reluctance to become full participants in the high-tech market and economy (p. 20).

Information overload. Even if consumers undertook a significant effort to try to keep up, the sheer amount of information available is huge and growing. Driving this growth in information are the rapid expansion in information technologies and the wider opportunities for the accumulation, dissemination, and storage of information. Pigeon (1996) notes that the growth in information is being driven by two factors: deregulation in the telecommunications and broadcasting sectors, and the emergence of new information technologies such as the Internet.

Increasingly, new information distribution channels such as 1-800 numbers, the Internet, and cable television are becoming more accessible to consumers in addition to more traditional newspapers and mail. However, although an expanded access to information could lead to greater consumer knowledge, wider choice, debate, and consultation, in reality, the volume of information readily available to the consumer has never presented such a challenge! Even with the help of rapidly evolving "search engines" on the Internet to guide one through what Varian (1996) has termed the "glut" of information, consumers are simply unable to access and digest the breadth and complexity of information now becoming available on biotechnology as well as all other subject areas.

Biotechnology's appearance as a new technology, its inherently hidden attributes, and its complex nature compound the challenges which this technology must overcome to gain the public's attention and confidence. We will return to this momentarily, but first, the next section summarises what consumers have said about biotechnology in the focus groups and surveys.

SURVEY AND FOCUS GROUPS: WHAT WE HAVE LEARNT SO FAR

The Office of Consumer Affairs has conducted three separate studies designed to better understand consumers' attitudes and opinions with respect to biotechnology.

The first, *Understanding the Consumer Interest in the New Biotechnology Industry*, was completed by Optima Consultants in November 1994 (hereafter referred to as Optima Consultants, 1994). It has two sections: The first, conducted in early 1994, consisted of a series of pre-survey focus group workshops in Montreal and Toronto used to pre-test the questionnaire; the second, reports on the results of a national telephone survey that took place mid-April, 1994. It should be noted that the Optima survey had a larger than normal proportion of participants, who, once they knew the topic of the survey, declined to participate. As well, a much larger than normal proportion (10% versus 2%) of respondents who agreed to take part, discontinued the interview, after it had begun. This may have affected some of the results that were received from participants who completed the interview. It should also be noted that at the time, very few biotechnology applications had actually made their way onto the Canadian market. As such, many of the opinions expressed have been described as "behaviour intention" rather than past behaviour.

The other two studies, *Environmental Applications of Biotechnology* (Creative Research, 1996) and *Focus Groups on Agri-food Applications of Biotechnology* (Ekos Research, 1996) used a focus-group format to gauge consumer reactions to food and environmental applications of biotechnology. The final reports of these studies were released in March, 1996. By using a focus group format, these reports provided anecdotal stories that offer insight into consumers' particular opinions and concerns. In addition, the focus-group format served to examine how much information, whether complete or not, impacts upon consumers' perceptions of biotechnology. Focus groups are designed to establish hypotheses and explore the range of opinions and views which may exist, whether they are based on complete information or not. Because of the qualitative nature of the study design, the reader is cautioned to view the findings as hypotheses rather than as definitive conclusions. Although consistencies and logic lend confidence to the analysis and interpretations, there is no way of determining the degree to which the opinions expressed are reflective of the study population at large.

In addition to these three studies, the Office of Consumer Affairs included two questions in CROP's annual survey of socio-cultural change in Canada (CROP, 1996) and contracted Ekos Research Associates Inc. to undertake multivariate analysis of the Optima survey data to help develop a research strategy (Ekos Research, 1995). The following section looks at the results of all of this work. In particular, it highlights consumers' knowledge of, and reactions to, biotechnology.

Consumer Awareness

The studies asked consumers about their current knowledge of biotechnology. The results were similar across all four projects: Consumers in the Ekos focus groups on agri-food applications of biotechnology indicated that they had little or no *knowledge* of biotechnology, even those who were more highly educated. This compares with 70% of consumers who consider themselves either very well informed or somewhat well informed of current issues in general (Optima Consultants, 1994, p. 13).

However, *awareness* of biotechnology, defined as having "heard of" the term, was substantially higher, even amongst those who had previously indicated little or no knowledge. When presented with examples of specific applications, consumer awareness was found to be even higher. For example, Figure 2 shows that more than 50% of the Optima respondents recognised biotechnology applications, even those more obscure products such as bacteria that clean up oil slicks. As well, half of the respondents had heard of at least five applications and an additional 38% had heard of three or four.

General applications specific to food and health were the most easily recognisable, especially more traditional biotechnology applications such as the use of bacteria to make yogurt or issues widely reported in the media such as giving cows hormones to increase milk production. Environmental applications of biotechnology were less well known, with knowledge limited to very general applications such as composting and biologically-produced fuels. The Annex outlines a number of applications that have resulted from biotechnology.

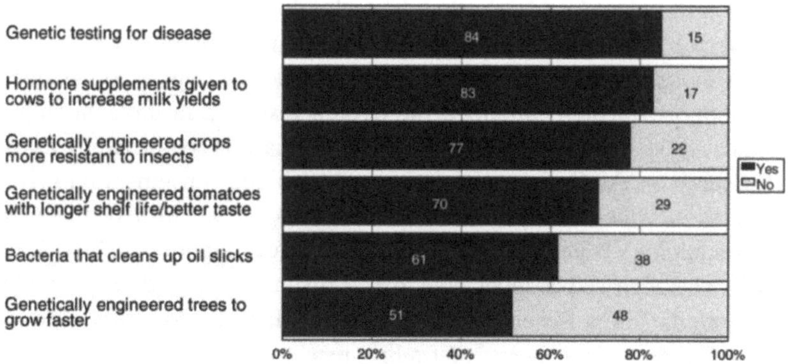

Figure 2. Respondents' familiarity with various biotechnical products.

Two factors explain the low knowledge/high awareness dichotomy. First, as mentioned in the previous section, biotechnology – particularly genetic engineering – is new, complex, and not evident to the casual observer. For example, changes to the genetic mapping of products can significantly affect a product's characteristics but are generally well-concealed from the consumer. As such, there may be awareness of a biotechnology application, but no association made between it and the genetic engineering process that created the novel characteristics. Second, although some consumers may be familiar with the products themselves, and understand that they are genetically altered, they may not associate them with the *term* biotechnology.

It is interesting to note how participants became aware of biotechnology. Although there was no consistency between focus groups, magazines (National Geographic), David Suzuki, government, and local industries were all cited as sources of information. In particular, knowledge of alternative fuels was high in Vancouver and Saskatoon where some gas stations have been marketing a "gasohol" product.

Consumer Reaction to Biotechnology Applications

While some respondents welcomed the positive benefits associated with biotechnology, others focused more on the possible unknown side-effects often associated with new technologies. Like Thalidomide, silicone implants, and asbestos, biotechnology is regarded as a very powerful technology with potentially alarming side-effects.

Although there were variations between focus groups as to the level of support for biotechnology, the overall response to biotechnology was cautiously optimistic. The multivariate analysis conducted by Ekos Research on the Optima data attempted to divide consumers into five groups of which 44%, in all, were characterised as positive towards biotechnology (true believers, 21%, and fearful supporters, 23%), 32% were characterised as indecisive, 6% as disinterested, and 18% as avid opponents (Ekos Research, 1995, p. 26). Many respondents viewed biotechnology as a means of finding new solutions to old problems. As could be expected, familiar, traditional processes, such as composting, were more widely accepted than other less well-known applications.

Balancing risks against benefits was viewed as extremely important. Respondents seemed willing to accept products provided that the benefits outweighed the risks. It is important to note that consumers' risk/value assessment is based upon perceived risk rather than objective risk (Ekos Research, 1995, p. 16). As such, acceptance of biotechnology would appear to require that products be perceived as having added value above what was currently available to them on the market. As well, they were generally more accepting of biotechnology applications that responded to broad societal needs (such as new drugs to fight against disease) than products that offered only personal benefit. Respondents were particularly hesitant to endorse new products where improvements were either very small or potential harm was very large. The feeling that "all it takes is one mistake" was cited frequently.

The example of two similar, genetically-altered vegetables illustrates consumers' unwillingness to accept biotech products that are perceived to provide little additional value beyond what is currently available in other more "traditional" products. When presented with a genetically-engineered tomato featuring longer shelf life and improved taste, consumer reaction ranged from neutral to negative. Although respondents did not reject genetically-altered food outright, many respondents questioned the need for a new variety of tomato when the tomatoes that were currently available were "perfectly good." In comparison, the response was generally more favourable to the genetically-engineered potatoes that reduce the need for environmentally harmful pesticides.

Although reaction to specific biotech products varied widely, environmental and health applications were generally better received than food products. Consumers were more willing to accept biotechnology risks associated with altruistic purposes such as protecting the environment and fighting diseases, than for purposes where the benefits are perceived as more trivial, such as improving the taste of food. More work is needed on whether these differences in risk/benefit tolerances are truly product specific or are related more to the current state of public knowledge where health and environmental applications are associated with major technological breakthroughs and food applications are associated with marginal improvements. If the second situation prevails, major breakthroughs in the food area could substantially alter public perceptions and consumer acceptance of agri-biotechnology applications.

More generally, biotechnology's association with science and technology raised concern among some of the survey group participants. For instance, a question in the Optima survey measured respondents' attitudes toward science and technology. More than 51% of respondents agreed with the statement that a high technology society is important for improving our quality of life (Optima Consultants, 1994); see Figure 3. Nevertheless, the surveys also indicated that many people have difficulty adapting to, what can be, unfamiliar territory. Others expressed frustration with the rapid rate of change and higher levels of stress often associated with science and technology.

The studies also highlighted the impact of education on people's comfort with science and technology generally, and biotechnology in particular. They concluded that education was positively related to one's comfort with biotechnology. Two explanations of why people with more education have an easier time accepting biotechnology were offered. A greater ability to analyse the information on biotechnology and thus more reassurance that the product was safe was suggested as one explanation.

Anecdotal comments made by participants provide another explanation. These comments suggest that those with more education often have had more positive experiences with science and technology than those with less education. For example, one respondent in the Environment focus group attributed job-loss to the introduction of new technology in the workplace. He/she stated "I used to work in offices. Now it's all computers and I'm in the dark" (Ekos Research, 1996, p. 15). In this group, there seems to be resignation to the fact that

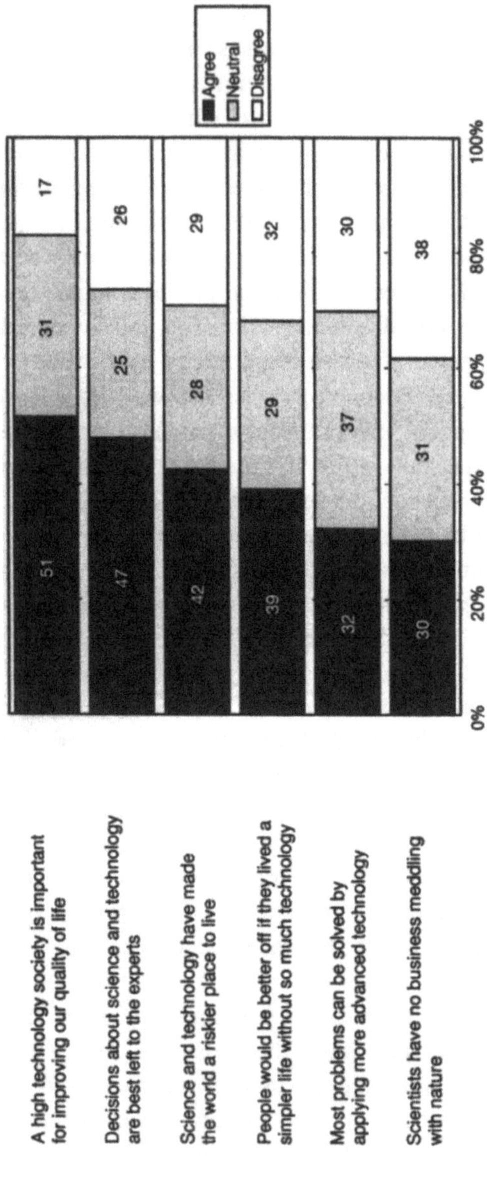

Figure 3. Respondents' attitudes towards science and technology. *Source*: Optima Consultants, 1994.

the rate of change makes it impossible to keep up with the issues. Although such experiences are only tenuously related to biotechnology, they can influence, and, at the extreme, perhaps undermine, the willingness of some consumers to accept new technologies such as biotech products.

The CROP survey also made some interesting observations on consumers' typical reactions to biotechnology (CROP, 1996, p. 13). They found that only a minority of Canadian consumers are tempted to buy biotech products. According to their research, people who are "very likely" to buy biotech products are typically individuals "with great faith in their personal potential, who like to experiment with and explore all aspects of their lives, including what they buy." This group would also tend to be more conscious of social status and to place a high value on success. Those respondents unlikely to purchase biotech products were characterised as being less secure individuals who perceive biotech products as threatening to the environment, their autonomy, and their personal potential. CROP (1996, pp. 20, 29) also pointed out that insofar as this latter point of view is widely held, it represents a major impediment to the commercialisation of biotech products.

From these results, it is evident that a number of factors influence consumers' perspectives with respect to biotechnology. The following section highlights some of their concerns.

CONSUMERS' CONCERNS

Traditional marketplace theory tells us that consumers are most influenced by the price of a product and its perceived quality. However, as previously mentioned, the nature of biotechnology as a new, complex topic, accompanied by ever-increasing amounts of information and rising levels of consumer awareness and education, raises the number of concerns that consumers bring to their marketplace decisions. In their work on antitrust and consumer protection law, Averitt and Lande (1997) suggest that both external and internal factors can prevent consumers from making rational decisions which can, in turn, inhibit the smooth functioning of the marketplace.

External factors, or those outside the direct influence of the consumer, prevent the market from providing sufficient marketplace options and thereby cause competition problems. Examples of external factors are too few producers, differentiated products, and monopolistic conditions that prevent the market from providing sufficient marketplace options and thereby cause competition problems. Internal factors are those that inhibit consumers from distinguishing between what is real and what is unreal. Averitt and Lande (1997) suggest that there are five primary types of internal factors: coercion; vulnerability; incorrect information; incomplete information; and information that is unduly hard to process.

The complexity involved with biotechnology makes it particularly vulnerable to some of these marketplace barriers that are internal to the consumer (i.e., literally inside the consumer's "head"). Ireland (1998) suggests that consumers are applying social, political, "lifestyle," and other broader concerns to their purchasing decisions which in turn poses new challenges to both private sector marketers and government policy makers, including those responsible for competition policy and intellectual property. The following section will briefly outline some of the non-traditional internal factors that consumers take into consideration with respect to biotechnology which might lie beyond traditional marketplace theory. It highlights the concerns that consumers expressed in the focus groups and surveys. In particular, the environment, the economy, safety, and ethical considerations are examined.

The Environment

There has been an increased awareness, in recent years, in the consumer's consciousness of the environment as seen in the rise in popularity of environmentally-friendly products, recycling, and composting. The Optima survey posed several questions to measure respondents' attitudes toward the environment in order to gauge how this affects their opinions towards biotechnology (Figure 4). It found that respondents were very anxious about the safety of the environment with 81% agreeing that land, air, and water are more contaminated than ever before, and 63% of respondents indicating that they regularly purchase environmentally-friendly products.

This increased awareness of the environment is reflected in the responses heard in the focus groups with respect to biotechnology. Biotechnology applications that were positively received were those that offered an alternative to current, environmentally-unfriendly practices or those that offered a solution to current environmental

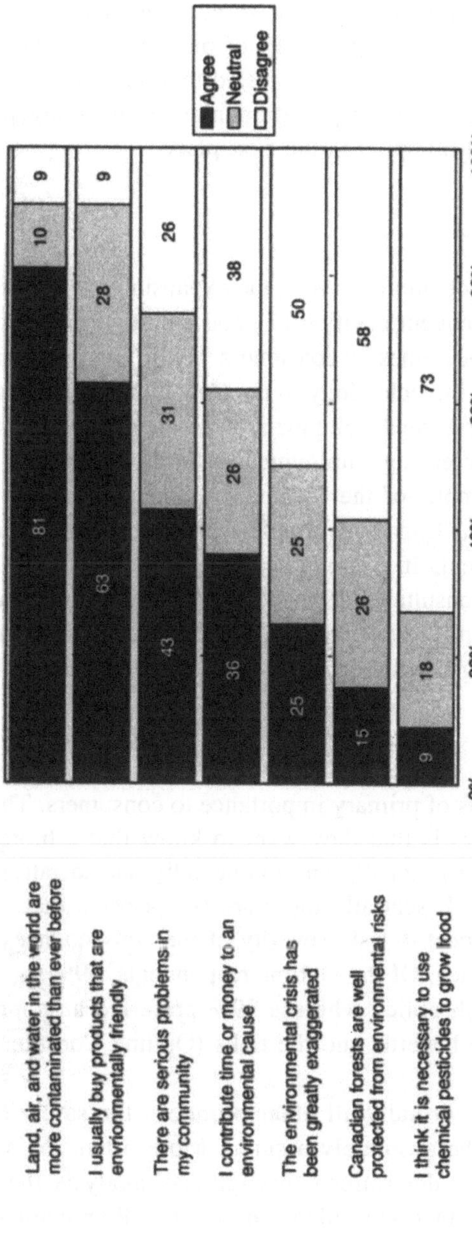

Figure 4. Respondents' environmental attitudes. *Source:* Optima Consultants, 1994.

problems. Focus group respondents were able to readily identify the benefits from these applications to the environment.

However, respondents also had some concerns with respect to even seemingly positive applications of biotechnology. For example, some focus group participants suggested that using bacteria to clean up oil spills might discourage companies from taking preventive measures to avoid oil spills in the first place.

The Economy

The perceived association between the expansion of high-tech industries and the subsequent creation of good employment opportunities, was noted numerous times by consumers. While consumers mentioned the importance of biotechnology to the Canadian economy, they were unwilling to endorse the development of unsafe or dangerous products for the sake of increasing employment or improving Gross Domestic Product. For example, of the Optima respondents, only 28% agreed with the statement, "Canadians should accept some risks from biotechnology developments if it strengthens the economy" while 41% disagreed (Optima Consultants, 1994, p. 27). Some consumers also raised the concern that only certain segments of the population would benefit from biotechnology, while others would be left to absorb the risks.

Safety Concerns

Product safety was of primary importance to consumers. The bottom-line for consumers is that they want to know that a biotechnology application is safe (personally, environmentally, and socially) both now and in the future. Essentially there are two aspects that consumers consider with respect to risk: severity of the consequences and likelihood of occurrence. Of the Optima respondents, 69% said that they preferred a no-risk policy whereas 29% preferred an approach that weighed both the benefits and the risks (Optima Consultants, 1994, p. 30).

However, like virtually all other products, the safety of biotech products cannot be absolutely assured. More often, the safety of a biotech product is determined through risk analysis that is based upon the best information available at the time. Regulation of biotech products for health and safety is the responsibility of a number of departments within the federal government: Health Canada,

Agriculture and Agri-Food Canada, Fisheries and Oceans Canada, and Environment Canada. They work closely together to ensure that products are properly assessed before being introduced into the Canadian market. Together, they use science-based risk assessments to determine the safety of a product. The Canadian risk assessment process to determine health, safety, and efficacy compares favourably with those in other industrialised countries (Consumers' Association of Canada, 1995, p. 72; KPMG, 1995). Of course, there is always the danger (however unlikely) that other factors will become known in the future that will undermine the safety of a biotechnology application.

Ethical Considerations

Ethical considerations were also frequently mentioned in the focus groups. Discomfort with genetic engineering led to discussions over whether it is desirable for humans to "play God." For example, whether or not scientists should be mixing genes between different species raised a number of ethical debates. Although some respondents suggested that curiosity and progress were instinctively human traits and should be supported, others questioned at what point the line should be drawn.

As noted earlier, respondents appear ready to set aside their ethical concerns if genetic engineering would lead to a major break-through of benefit to all of society, i.e., a cure for cancer or AIDS. In addition, there appears to be a hierarchical ranking of biotechnology in terms of acceptability. Whereas genetic engineering of fruit and vegetables seems to be relatively innocuous to most people (regardless of whether or not they would eat these products themselves), the alteration of animals' genetic composition is more controversial, while processes that tamper with human tissues raise even greater concerns. Perhaps the most contentious processes are those that involve copying genes originally derived from humans in conjunction with other living organisms. For example, although switching tomato and potato genes might not be highly controversial for many consumers, mixing human genes with those of other animals would be more likely to raise ethical debates.

Ethical considerations also extend into whether firms should be allowed to have ownership of a living organism. The Optima survey found that 90% accepted the general principle of patents. However this support dropped to 24% (a difference of 66%) when it came to the patenting of genetically-engineered higher life forms. The 24% figure represents participants' responses to the question, "Do you agree with 'patents on any form of life developed through biotechnology'?" Support for patents on higher life forms increased to 57% when used to develop cures for diseases (Optima Consultants, 1994, p. 19). See also Figure 5.

For example, if a firm in the biotechnology sector created a genetically-altered prototype of an animal, should it be allowed to protect its investment through patent protection? In August 1995, patent protection was denied to researchers of Harvard University for a genetically-engineered mouse which they had developed to aid in scientific research. The Commissioner of Patents ruled that a genetically-altered mammal was now under the control of nature and thus could not be patented under Canadian law. Specifically, the Commissioner of Patents ruled that a gene that had been altered through recombinant DNA techniques and introduced into a cell that was in turn introduced into a mammal was now under the control of the laws of nature so as not to give rise to a patentable invention for that mammal per se. The Commissioner's decision was appealed by the President and Fellows of Harvard College to the Federal Court, Trial Division. This appeal was rejected in an April 1998 ruling. This decision is now being appealed to the Federal Court of Canada, Appeal Division.

OVERCOMING "INTERNAL" MARKETPLACE BARRIERS

The previous section looked at a number of concerns that affect consumers' decisions in the biotech market that could lead to "internal" marketplace failure (i.e., based on factors internal to the consumers' thought processes): the environment, economy, safety, and ethical concerns. This raises the question of what can be done to promote balanced decisions with respect to biotechnology. In their work on marketplace forces, Hadfield, Howse, and Trebilcock (1996, 1998) identify a number of ways that consumer protection can be assured. These include: setting minimum standards for the provision of products and services, preapproval of products, product bans, provision of dispute resolution mechanisms, providing consumer information, and labelling.

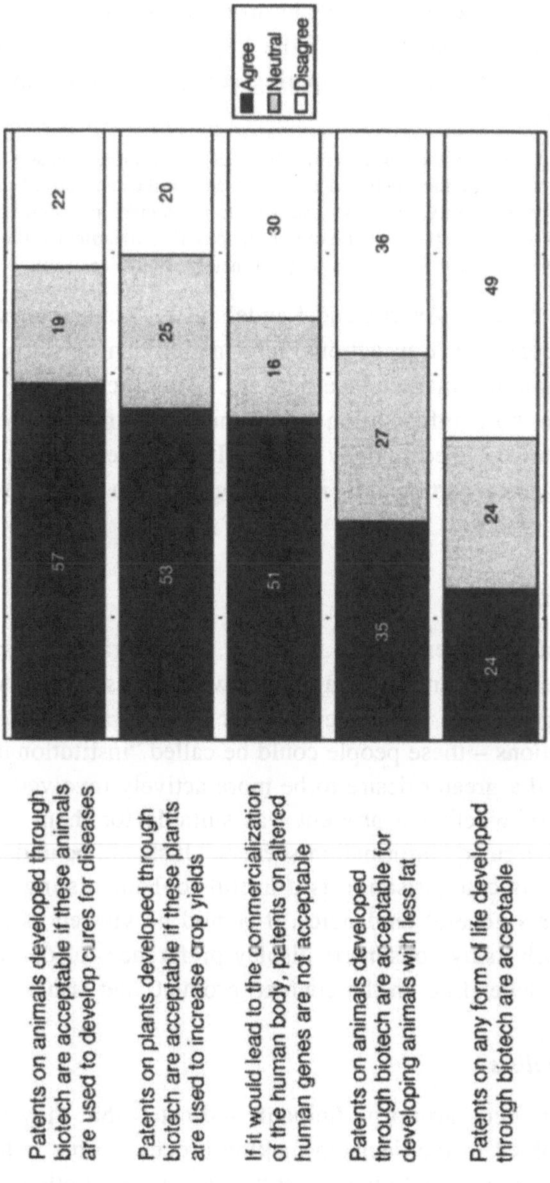

Figure 5. Respondents who agree with patenting life forms created through biotechnology. *Source*: Optima Consultants, 1994.

In deciding upon the most appropriate methods, it is important to consider who will be the recipients/audience of the consumer protection mechanism and how it will be used. Hadfield et al. (1998) suggest that information is one of the key elements to providing modern consumer protection. Comparing consumer and competition policy they state,

> Information lies at the heart of the essential problem of consumer protection as a distinct policy area: Whereas competition policy engages a focus on the structure of markets and the options – price, quality, quantity– available to consumers, consumer protection policy engages a focus on the structure of transactions and, in particular, the match between what consumers expect and what they ultimately receive (p. 150).

Thus, to go back to Averitt and Lande's (1997) characterisation of barriers to consumer transactions, information can be introduced to overcome "internal marketplace barriers." This section will focus specifically on how information and other consumer protection mechanisms can be used and by whom. It will then take a brief glance at focus group participants' impressions on who should provide information.

Who's to Decide?

Almost all consumers expressed the need for assurance of safety. However, some people indicated a greater willingness than others to trust outside agencies to make decisions for them regarding biotechnology applications – these people could be called "institutionalists." Others indicated a greater desire to be more actively involved in the determination of whether a product was suitable for them – these people could be termed "information-seekers." It should be underlined that although this categorisation is a useful tool for distinguishing between people's general tendencies, it should be viewed as a continuum on which many consumers display preference for one or the other category depending on the context, product, and issue.

The Institutionalists

Although most focus group participants indicated that they would like to be involved in the decision-making process, some indicated that they would be more willing to trust the judgement of others. Most consumers appear to be institutionalists when it comes to safety issues. Slightly more than 51% of respondents of the Optima survey indicated that decisions about the *safety* of biotechnology should be left to the experts (Optima Consultants, p. 27). For others, this could be extended to include ethical, environmental, and/or economic considerations of whether a product should be made available on the market.

The desire to have others make decisions with respect to whether or not a product should be available in the marketplace should not be confused with disinterest. Rather, the institutionalists feel that the complexity of biotechnology and genetic information undermines their ability to make informed personal decisions. They would prefer mechanisms that would draw on the experience of experts to make decisions, rather than rely on their own incomplete knowledge to make these types of decisions. For example, when participants of the Environmental Applications focus group were asked about their information needs, responses that were given included, "I'd prefer to have confidence in experts than trying to make up my own opinion," and "I wouldn't have the capacity to understand it all" (Creative Research, 1996, p. 53).

Although one might assume that this group consists of people with less formal education, Hadfield et al. (1998) suggest that even those with advanced education can, at times, feel that they are ill-equipped to make good decisions. They state that "in some cases, consumers may need substantive protection, not because they are members of a vulnerable group, such as the poorly educated, but because, as a class, consumers are systematically unable to adequately process the information, they need to make good decisions" (p. 145). This is particularly true as consumers are faced with increasing amounts of information. As such, they make the case that there is some rationale for outside intervention that protects the consumer from the more serious consequences of information asymmetries. As well, some people may have institutionalist tendencies, not because they are willing to let experts decide on their behalf, but because they see the institutions as upholding their ideals or opinions.

Consumer protection instruments that institutionalists would most likely support would be those mechanisms that rely less heavily on consumer input such as regulation, setting minimum standards for the provision of services, preapproval of products, and product bans.

The Information Seekers

While the majority of focus group participants supported an "institutionalist" approach to safety standards, the surveys suggest that when it comes to other consumer issues such as exercising their ethical preferences, they have a strong desire to have information provided to them on which to make their own personal choices in the marketplace. Optima Consultants (1994, p. 17) note that 68% of respondents felt that choice was an important issue, even if they were uncomfortable with the products themselves. For information seekers, products that offer individual and societal benefits are more likely to be well-received than those that pose risks. Information seekers also want to be able to make a risk assessment based upon their individual beliefs and preferences. They are more likely to have higher levels of formal education than the institutionalists and generally have greater confidence in their ability to ascertain what products/applications of biotechnology they are willing to endorse.

In order to undertake meaningful analysis, however, these consumers need access to information that is complete, accurate, and easily understood. Marketplace analysis tells us that information acts as a key dimension in consumer decision-making. Used as "bargaining power," information is the tool by which consumers are able to decide how to allocate their resources:

> Bargaining models routinely identify private information as the source of "bargaining power" and critical determinant of the costs of bargaining and the likelihood of breakdown . . . information is a key ingredient in the operation of reputation and repeat play mechanisms that discipline the exercise of bargaining power that may arise from a variety of sources (Hadfield et al., 1998, p. 140).

Information asymmetries resulting from the unavailability of information to either or both parties to a transaction, may lead a consumer to make choices that he/she might not otherwise make.

The environmental focus group participants identified three reasons to support labelling: to protect consumer choice; to provide information on what a product is made of for health reasons, e.g., allergies; and to encourage companies to provide safer products by having disclosure requirements. Not surprisingly, the analysis of Optima data conducted by Ekos Research (1995) showed that those who advocated labelling of biotechnology applications also displayed a higher risk aversion; in other words, they were less likely to accept the perceived risks of biotech products, even if there were associated benefits. For these consumers, labelling products provides them with a means to undertake risk analysis for themselves.

The demand for credible, quality information however, in itself raises a key question: Who can be trusted to provide reliable, accurate, non-biased information?

Information

The Creative Research Focus Group participants suggested that an ideal information source would (Creative Research, 1996, p. 55):

- not be afraid to tell the truth
- be neutral
- have experience
- not have a vested interest
- not have a profit motive.

In particular, respondents suggested a number of institutions that would provide credible information to the public. Four of the most commonly cited sources (government, universities, private sector, and a consortium of the three) are discussed below.

Government. Almost all focus group and survey results supported a role for government with respect to biotechnology. Despite some misgivings about the efficiency of government, generally it was acknowledged that regulation and safety protection were two legitimate areas where government has made significant contributions in the past. Specifically, 62% of Optima survey respondents indicated that government should increase its regulation of biotechnology and 50% would hold government responsible if something were to go wrong (Optima Consultants, 1994, p. 28). Nevertheless, some focus group participants pointed to the lack of appropriate precautions taken with respect to Canada's blood supply, suggesting a need for outside sources to monitor government's safety precautions.

While information is generally seen as a public or mixed public/private good (in the sense that many can "own" the same information at the same time and that "consumption" by one individual does not detract from the ability of others to use that information), much of the information available in the marketplace is provided by private industry. However, once information has been shared, it can be rapidly distributed with no compensation to the original provider. As such, private industry will tend to provide only that information which serves its marketing needs. Thus, there is a natural tendency for consumers to expect not-for-profit organisations and governments to play a role in disseminating information.

University. Some participants suggested that universities, as the source of much of the research and development for genetic engineering and biotechnology, would be well-placed to provide a safety-assurance role. Concern was raised, however, that in the context of private-sector funding, impartiality is not guaranteed.

Private sector. As with the university sector, some focus group participants suggested self-regulation by the private sector as a possible source of safety regulation. Generally, however, participants were most distrustful of an industry-regulated safety system. They felt that firms that had invested research and development resources into a product would be less likely to stop a product from entering the market over questions of safety.

Consortium/multi-stakeholder approach. A number of respondents suggested that the best solution might be a board consisting of representatives from the government, university, and private sectors. This, they concluded, would overcome some of the difficulties associated with individual sectors, while drawing on the strengths of each. These suggestions are consistent with the work that was conducted by OCA and the Treasury Board Secretariat on the advantages and drawbacks of voluntary arrangements which have multi-stakeholder involvement and support (Office of Consumer Affairs, 1996).

SUMMARY AND IMPLICATIONS FOR FUTURE WORK PLAN

Biotechnology has the potential to contribute new and valuable products to the marketplace. As has been noted, genetic engineering has already led to the introduction of a number of new and exciting applications. However, despite some enthusiasm for the benefits that biotech products might offer, consumers are clearly worried about the associated risks. As was pointed out, no consensus has been reached on how best to respond to these risks. While some consumers want increased legislation, others are demanding more information that will enable them to make their own decisions.

Each of these approaches raises a number of issues. For example, advocates of increased regulation would like answers to questions like:

1. Which regulatory mechanisms are most suitable to biotechnology? What are the strengths and weaknesses of the existing regulatory framework?

2. How appropriate is it, for example, to have regulatory and promotional activities carried out by the same government agency?
3. Is there a need for additional health and safety testing of biotechnology processes (as distinct from products)?
4. To what extent can government regulation be supplemented and extended by voluntary codes of practice and other industry-led arrangements?
5. What are the roles of market and non-market mechanisms in addressing the socio-economic and ethical issues raised by biotechnology?

Consumers demanding greater decision-making power have their own set of concerns such as:

1. Given the technical nature of biotechnology, is the general public capable of analysing the data?
2. How desirable is choice, and to what extent is it provided and facilitated by the market?
3. Who should provide information and should the information providers be held responsible for ensuring that it is accurately and reasonably disseminated to the public?
4. What are the benefits and drawbacks of product labelling, as compared with other forms of information dissemination?
5. How does information influence consumer attitudes and perceptions about risk?
6. How do prior levels of risk aversion influence consumers' perceptions of risk?
7. What are the advantages and drawbacks of information dissemination by electronic means compared with more traditional methods (i.e., reports, brochures, pamphlets in hard copy)?

This paper does not try to answer these questions. Rather, its purpose is to create a common understanding of what we know about how consumers view biotechnology. Certainly, biotechnology, which has been with us for centuries, and genetic engineering, which offers such interesting prospects, are unlikely to disappear. What we need is an understanding of how the marketplace will respond and where and how the marketplace needs to be supplemented by non-market mechanisms. Using a marketplace framework, the Office of Consumer Affairs, working together with other government departments and our partners in the academic and consulting communities, are investigating these questions by implementing a work plan on biotechnology, the consumer, and the marketplace. We hope, that in this way, the Office will contribute to broadening the discussion of where markets are clearly sufficient and where intervention by government, or by multi-stakeholder partnerships, should be considered.

ANNEX: EXAMPLES OF BIOTECHNOLOGY APPLICATIONS

Taken from Chemical and Bio-Industries Branch (1996)

FOOD APPLICATIONS

Flavr Savr Tomato

This tomato has been genetically engineered to slow the rate at which it softens, meaning that it can be left to ripen on the vine longer than other tomatoes so that its flavour can develop more fully before it is picked.

Canola

Canola, a type of rapeseed, is found in numerous products, including salad oils, mayonnaise, coffee creamers, cosmetics, printing inks, suntan oil, and livestock feed. Unfortunately, Canola is very often ravaged by weeds. To combat these weeds, improved canola varieties have been specially designed to tolerate glyphosate, the active ingredient in a brand name herbicide that is used to destroy the harmful weeds. To combat the effects of the herbicide, a gene is removed from a plant or bacterium and inserted into the genetic structure of a canola plant to become herbicide resistant. As a result, Canola is able to withstand herbicides and allowed to grow normally.

HEALTH APPLICATIONS

Insulin

Insulin is used in the treatment of diabetes and was originally extracted from animal sources, not always available in adequate supply. Through the use of genetic engineering techniques, scientists can produce large quantities of human insulin by placing the human insulin gene into bacteria. Because this insulin has the same composition as human insulin, diabetics are less likely to have allergic reactions to it than to the original animal insulin.

Human Growth Hormone (HGH)

HGH has been produced using similar methods as insulin and is used to treat children who are diagnosed with "dwarfism," in other words, children who naturally lack an adequate supply of this growth hormone. The natural hormone was previously produced from cadavers. Supplies of the hormone were limited, and they carried the risk of transmitting disease.

Vaccines

Vaccines have been one of the greatest successes in medicine. They allow us to control some of the world's most deadly bacterial and viral diseases such as polio, tetanus, and diphtheria. Smallpox has been eradicated because of intensive world wide vaccination efforts. A vaccine contains components of disease-causing micro-organisms or their products. These components stimulate the body's immune system so that when we are exposed to a live disease-causing micro-organism, we will be protected from it. Through genetic engineering, vaccines have been produced that are safer and less expensive than traditional vaccines, because they are less likely to cause side-effects. Vaccines for hepatitis B and influenza are produced by genetic engineering.

FORESTRY APPLICATIONS

Bacterial Insecticides

Since the 1960s, scientists have used a naturally occurring bacterium, Bacillus thuringiensis (B.t.) to help manage populations of forest pests, such as the spruce budworm and gypsy moth, and agricultural pests such as caterpillars and potato beetles.

Fast Growing Trees

Although trees normally take years to reproduce, trees with superior growth characteristics can now be grown much more quickly. Tree "plantlings" are produced using tissue cultures which allow production of a large number of similar plants.

ENVIRONMENTAL APPLICATIONS

Clean-Up of Contaminated Environments

An important application of biotechnology is the harnessing of naturally occurring microorganisms and plants to help clean up contaminated environments. Micro-organisms can feed on toxic chemicals, breaking them down into non-toxic substances. They can be used to biodegrade sewage, oil spills, and a variety of toxic chemicals found in waste. In the pulp and paper industry, enzymes produced by micro-organisms are used to reduce the need for chlorine bleaching. Currently, researchers are genetically engineering micro-organisms to biodegrade toxic chemicals like PCBs more easily than naturally occurring micro-organisms can.

REFERENCES

Averitt, N. W. & Lande, R. H. (1997). Consumer sovereignty: A unified theory of antitrust and consumer protection law. *Antitrust Law Journal, 65*, 713–756.

Biotechnology regulation in Canada: A matter of public confidence (1996). Third Report to the Standing Committee on Environment and Sustainable Development. Ottawa: House of Commons.

Chemical and Bio-Industries Branch (1996). *A layperson's guide to biotechnology*. Draft. Ottawa: Industry Canada.

Consumers' Association of Canada (1995). *Background paper on food biotechnology in Canada*. Prepared for the National Round Table on the Environment and Economy (NRTEE). Ottawa: Consumers' Association of Canada.

Creative Research International (1996). *Environmental applications of biotechnology: Focus groups, final report.* Ottawa: Industry Canada.

CROP Inc. (1996). *CROP Annual study: Analysis of Industry Canada's omnibus questions included with the 1995 3SC Survey.* Prepared for the Office of Consumer Affairs. Ottawa: Industry Canada.

Ekos Research Associates Inc. (1995). *Biotechnology research design. Final Report.* Prepared for the Office of Consumer Affairs. Ottawa: Industry Canada.

Ekos Research Associates Inc. (1996). *Focus groups on agri-food applications of biotechnology. Summary Report.* Prepared for the Office of Consumer Affairs. Ottawa: Industry Canada.

Hadfield, G., Howse, R., & Trebilcock, M. J. (1996). *Rethinking consumer protection policy.* Toronto: University of Toronto, Faculty of Law, Centre for the Study of State and Market. Working Paper.

Hadfield, G., Howse, R., & Trebilcock, M. J. (1998). Information-based principles for rethinking consumer protection policy. *Journal of Consumer Policy, 21*, 131–169.

The hitchhiker's guide to cybernomics (1996). *The Economist, 340* (September 28), pp. 3–7.

Ireland, D. (1998). Competition policy, intellectual property, and the consumer. In: R. B. Anderson & N. T. Gallini (Eds.), *Competition policy and intellectual property rights in the knowledge-based economy*, pp. 293–338. Ottawa: Industry Canada Research Series, Vol. 9.

KPMG (1995). *Improving Canadian biotechnology regulation – A study of the U.S. experience.* Prepared for the National Biotechnology Advisory Committee and the Chemicals and Bio-Industries Branch. Ottawa: Industry Canada.

Office of Consumer Affairs (1996). Voluntary codes and the consumer interest. *Consumer Quarterly, 1*(4), 1–4. Ottawa: Industry Canada.

Optima Consultants (1994). *Understanding the consumer interest in the new biotechnology industry.* Prepared for the Office of Consumer Affairs. Ottawa: Industry Canada.

Pigeon, A. (1996). *Consumers and the marketplace for information.* Draft. Prepared for the Office of Consumer Affairs. Ottawa: Industry Canada.

Varian, H. R. (1996). *Economic issues facing the internet.* *ftp://alfred.sims.berkeley.edu/pub/Papers/econ-issues-internet.html*

THE AUTHORS

Heather Sheehy is a policy analyst at Health Canada, P.L. 3007A, Holland Cross, Tower A, Suite 709, Ottawa, Ontario, K1A 0K9, Canada. E-mail: heather_sheehy@hc-sc.gc.ca. At the time of writing she was with the Office of Consumer Affairs, Industry Canada. Marc Legault is a Senior Policy Analyst, International Investment and Services Policy, Industry Canada, 235 Queen Street, Room 502B, Ottawa, Ontario, K1A 0H5, Canada. At the time of writing he was with the Office of Consumer Affairs, Industry Canada. Derek Ireland was previously Director of Consumer Research and Analysis.

Jennifer B. Wohl

Consumers' Decision-Making and Risk Perceptions Regarding Foods Produced with Biotechnology

ABSTRACT. This paper presents a basic framework for how people make decisions about products that may have been produced with biotechnology and examines how decision makers form their quantitative risk estimates. It also reviews several studies which argue that the acceptability to consumers of foods produced with the use of biotechnology is influenced not only by their perceptions about the magnitude of the risks involved, but also by the qualitative characteristics of the potential risks. The paper concludes with a suggested approach to enhancing communication between risk communicators and the public regarding the role of biotechnology in the food and other sectors.

Proponents of the use of biotechnology in the agri-food sector claim that biotechnology can be used to produce food that has several desirable qualities. These include, but are not limited to, longer shelf life, less risk of contamination from bacteria, better taste, appearance, nutritional value, and other desirable qualities (ADA Reports, 1995; Industry Canada, 1996; Ontario Ministry of Agriculture, 1996). Many of these beneficial attributes, proponents argue, come with low risk to human health and low prices. Proponents would likely argue further that the potential for safely feeding more people for less money, which benefits producers and consumers alike, should be clear enough reason to support the development of biotechnology in food. However, several surveys and focus groups conducted in Canada and the U.S. in the last several years indicate that consumers have ambivalent feelings about the use of biotechnology in the food system (e.g., Ekos Research, 1996; Grobe & Douthitt, 1995; Optima Consultants, 1994; Sparks & Shepherd, 1994; van Ravenswaay, Wohl, & Hoehn, 1992). These surveys show that while consumers recognize the benefits from biotechnology, they also have several concerns. These include the potential human health risks posed, the moral and ethical implications of the use of biotechnology, and other non-health implications of biotechnology such as the possibility of an inequitable distribu-

Journal of Consumer Policy **21**: 387–404, 1998.
Bartha M. Knoppers & Alan D. Mathios (eds), Biotechnology and the Consumer.

tion of its benefits (Ekos Research, 1996; Misra, Fletcher, & Huang, 1995; Optima Consultants, 1994).

This paper examines the literature on how consumers form their perceptions about biotechnology in food, and presents a framework of consumer decision making under uncertainty about the safety and other attributes of the food they consume. This information will facilitate further dialogue between risk communicators and the public regarding the role of biotechnology in our economy. Assuring the public that scientists believe some biotechnology is safe is not likely to suffice. Consumers may ask "what if the scientists are wrong, as they have been in the past?" Recent public-health scares such as the uncertainty about the safety of the growth hormone, *Alar*, the mad-cow disease scare, and consumers' mistaken associations between irradiation of food and radiation all raise questions in consumers' minds about how much they should trust products produced with biotechnology. This is especially true when consumers may not perceive any problems with "conventional" foods. The paper focuses primarily on biotechnology in the food system with the recognition that the potential reach of biotechnology extends to other sectors as well (the health care industry, for example). However, the framework is general enough to be extended to other sectors.

The next section presents a basic framework for understanding how consumers make choices about foods produced with biotechnology. As in a traditional choice model, a consumer weighs the additional benefits derived from choosing the good against any additional costs. He/she will then choose the product if the former outweigh the later. Because costs from biotechnology can be evaluated only in a probabilistic sense, costs must be considered in an "expected value" framework. The section discusses the difficulties that arise in using the expected value framework when evaluating choices that involve biotechnology in food. It then discusses how these difficulties can be used to explain some seemingly "irrational" behaviour on the part of consumers (e.g., consumers rejecting foods produced with biotechnology even when consumers believe and scientific evidence suggests that the human health risks are negligible, or even non-existent).

The section following discusses some of the literature on how consumers form their quantitative risk perceptions and explains why consumers' risk perceptions may diverge substantially from those of experts. This explanation may help risk communicators direct risk information more effectively.

The fourth section of the paper is a review of the literature on the characteristics of risk that influence choice. Psychologist Paul Slovic and his colleagues have speculated that there is more to risk perceptions than simply the probability of an adverse outcome (Slovic, Fischhoff, & Lichtenstein, 1980, 1990). Characteristics of risk such as the severity of the outcome should it occur, the degree of control one feels over the likelihood of the outcome, and whether or not a risk is faced voluntarily all influence the "acceptability" of risk. Throughout this and other sections, data from surveys conducted in Canada and the U.S. on consumers' attitudes about biotechnology and general food safety issues are used to illustrate the concepts.

The final section makes some recommendations about strategies for risk management and communication regarding biotechnology in food. It should be noted that this paper focuses primarily on consumer decision making and does not address the wider question of the effect of biotechnology on the financial viability of the agricultural or other sectors.

CONSUMER CHOICE ABOUT BIOTECHNOLOGY

There is a large segment of the population in North America that is concerned about the potential health and other risks from biotechnology (see survey results in Ekos Research, 1996; Grobe & Douthitt, 1995; Optima Consultants, 1994; Sparks & Shepherd, 1994; van Ravenswaay et al., 1992). It is thus worth exploring how consumers make choices about foods produced using biotechnology when they have little experience on which to rely for guidance.

A Basic Framework for Consumer Choice

In the surveys cited above, consumers were wary of biotechnology even when they felt the associated human health risks were small. This section presents a simple framework for how consumers make choices about foods produced with biotechnology. There are two scenarios to consider: (a) the consumer has a choice between a product produced with biotechnology, and is labelled as such, and some other "conventional" product; and (b) the consumer has a choice between consuming a product produced with biotechnology and not consuming at all (i.e., either the product is not labelled and there is thus no way to distinguish goods produced with biotechnology from goods produced conventionally, or all of that product is produced with biotechnology – novel or new goods, for example).

Scenario 1: Consumers Can Choose

Assume first that the consumer can choose to consume the conventional product at some price, or he/she can choose to consume the product that has been produced using biotechnology and is labelled as such. The price of the biotechnology good may or may not differ from that of the conventional product. Standard economic theory predicts that the consumer would choose the product if the gain in benefits outweighs what must be forfeited with the consumption of the new product. The important consideration is the change in the benefits relative to the change in the costs when the product produced with biotechnology is chosen instead of the conventional product. The consumer will choose the biotechnology product if he/she perceives that

$$\Delta EB > \Delta EC \quad (1)$$

where:

ΔEB = the change in benefits from choosing the good produced with biotechnology instead of choosing the conventional product

ΔEC = the change in costs associated with choosing the good produced with biotechnology instead of choosing the conventional product.

The change in benefits when switching from a "conventional" product to one produced using biotechnology depends on the product. The benefits may be limited to financial savings to the consumer and the consequent ability to consume either more of that good, or more of other goods. Or, the benefits could include a longer shelf life, lower risk of bacterial contamination, and cosmetic and other positive physical attributes. Consider the case of the NewLeaf Russet Burbank potato, which was genetically engineered to be resistant to the Colorado Potato Beetle. The protein in the NewLeaf potato that kills the beetle when it eats the potato is not harmful to humans, and is predicted to reduce the use of chemical insecticides for insect control (Mitchler & Murphy, 1995). The potato also has the same composition, taste, and nutritional quality as a conventionally produced potato

(Mitchler & Murphy, 1995). Assume that choosing the NewLeaf potato rather than a conventional potato benefits the consumer only to the extent that it results in lower expenditures on potatoes. The product itself is indistinguishable from a conventional potato.

Continuing with the example of the NewLeaf potato, consider the cost side of Equation (1). If the consumer does not know for certain whether there are any long-term health consequences from the consumption of the NewLeaf potato, he/she can consider the consequences of consuming the NewLeaf potato only in a probabilistic sense. In other words, the consumer may feel there is some probability that there will be an adverse health outcome resulting from the consumption of the product, and some complementary probability that there will be no health outcome different from what would have resulted from the consumption of the conventional product. The health outcome considered in this case is the one that results specifically from the characteristics of the biotechnology good, not from the characteristics of the good in general. In the case of the NewLeaf potato, for example, the consideration is some health outcome that results because of the protein in the potato that kills the beetle, not from consuming potatoes in general. The consumer can thus evaluate the expected value of this "lottery" as follows:

$$C = \alpha(HO) + (1 - \alpha)(NHO) \tag{2}$$

where:

C = costs (as in Equation (1))
HO = some adverse health outcome
NHO = no health outcome different from that resulting from consumption of the conventional product
α = the consumer's subjective probability that some adverse health outcome will result from the consumption of the product produced with biotechnology
$1 - \alpha$ = the probability that there will be a health outcome no different from that which might result from consuming the conventional potato.

Consider the case where α in Equation (2) is so low that it can be considered negligible, or effectively, zero. This might be the case in which consumers' views coincide with scientists' and other experts' views that there is no health risk posed by consuming the NewLeaf potato. If we normalize the "no health outcome" to zero, the "health outcome" would assume some non-zero cost, and the total cost of consuming the NewLeaf potato would be zero. Given this framework, as long as there is some benefit to the consumer, regardless of how small, the consumer will choose to consume the NewLeaf potato.

The possibility of some adverse health outcome resulting from the consumption of the good produced with biotechnology is only one aspect of the decision-making process. Consumers may also be concerned about the animal welfare and general ethical implications of biotechnology. Focus-group research by Ekos Research (1996) identified several concerns on the part of participants regarding the future of biotechnology. They included ethical and moral implications, motivations underlying the development of biotechnology, safeguards and controls, and unintended impacts. The presence of these concerns should be included as possible "outcomes" on the right-hand side of Equation (2). Equation (2) is thus modified as follows:

$$C = \alpha(HO) + (1-\alpha)(NHO) + \delta(NonHO) + (1-\delta)(NNonHO) \quad (3)$$

where:

C	=	costs (as in Equation (1))
HO	=	some adverse health outcome
NHO	=	no health outcome different from that resulting from consumption of the conventional product
α	=	the consumer's subjective probability that some adverse health outcome will result from the consumption of the product produced with biotechnology
$1-\alpha$	=	the probability that there will be a health outcome no different from that which might result from consuming the conventional potato
NonHO	=	some adverse non-health outcome
NnonHO	=	no non-health outcome different from that resulting from consumption of the conventional product
δ	=	the consumer's subjective probability that some adverse non-health outcome will result from the consumption of the product produced with biotechnology
$(1-\delta)$	=	the probability that there will be a non-health outcome no different from that which might result from consuming the conventional potato.

In this formulation, even if the consumer believes that there is no chance of an adverse health outcome resulting from consuming the

NewLeaf potato (i.e., α is believed to be zero), and the personal health cost is thus zero, the other factors mentioned above may be considered "outcomes" to the consumer. That is, there may be "outcomes" to the consumer in the form of negative feelings about potatoes being genetically engineered, feelings that large corporations and corporate farms are "profit grabbing," or feelings that the use of this technology may contribute to the demise of the family farm. As suggested by Equation (3), these feelings would increase the "costs" on the right hand side of Equation (1) if the consumer attaches some positive probability to their occurrence. The presence of these "costs" may help explain why consumers would opt to not purchase some product produced with biotechnology, even if he/she believes that consuming the product posed no personal health risks. This expanded view of "costs" to include costs to personal integrity helps bridge the gap between predictions from economic theory (in which "costs" are narrowly defined) and consumer behaviour.

Now consider the case (in Equations (2) and (3)) where the consumer believes there is some positive probability that an adverse health outcome will result from consuming the product ($\alpha > 0$). In particular, consider the case in which the consumer thinks the probability is very small (but not zero). The decision criterion depends on not only the magnitude of the probability, but also on the consumer's perception of the severity of the outcome, should it occur. Assume the consumer believes an outcome as severe as death could result from consuming the product. What is the "cost" to the consumer of death? It is likely to be infinitely high. In this case, regardless of how small the probability of the adverse outcome occurring, it would be infinitely bad if it were to occur and the consumer will avoid it, especially if the only benefit to him/her is a few cents less spent on the food bill.

A further consideration is how people evaluate risks. Imagine a decision maker contemplating a game of roulette where a ball will spin on a wheel with a million slots. If the ball falls in the player's slot, that player will be killed. Every player will receive some small payment for playing. On average, a player will be killed in one out of a million rounds of the game. But if a player (or consumer) can avoid playing at all, simply by doing what he/she was doing already (consuming the conventional good), why would he/she choose to participate (consume the biotechnology good)? Furthermore, although the risk may be as low as one in a million, or even smaller, the consumer does not have a million lives to live, only one. He/she may ask "what if that one in a million happens first, in this lifetime, not in some other?" The motivation to "play" (buy the product) may then be missing.

Scenario 2: Consumers Have Limited Choice

A different case is the one in which the only choice the consumer has is between consuming the biotechnology product and not consuming at all. That is, there is no information available about whether or not the product was produced with biotechnology, or the product is a new product produced using biotechnology for which there is no substitute. The choice is then to consume the product, and potentially expose oneself to harm, or to exit (or never enter) the market for that product. The comparison is then not between the conventional product and the one produced with biotechnology, but between consuming the one produced with biotechnology and not consuming at all. One would choose the product if the health risks posed by consuming are not sufficiently high to warrant exposing yourself to the risks of *not* consuming. Consider the case of a drug produced using recombinant biotechnology. If the choice is between taking the drug and not taking the drug at all (or finding a substitute), one would have to evaluate the risks posed by consuming the drug relative to the risks posed by not taking the drug at all. This choice could leave one feeling that the choice is between two evils. Choose the drug and expose yourself to its potential health risks, or choose to forgo the drug and assume the risks associated with no treatment.

Other products produced with biotechnology offer more personal health, taste, and quality benefits. Improved food taste, texture, shelf life, and less exposure to pesticide residues and microbial contamination are just some of the possible characteristics of goods produced with biotechnology. There is some possibility, then, that consuming products produced with biotechnology can improve one's health and enjoyment of food. These benefits must be weighed against any potential health risks and other costs on the right hand side of Equation (1). The choice calculus becomes even more complex than before because in this scenario there are more unknowns (i.e., there may be some probability that the product will improve health, and some offsetting probability that it will be harmful to health). The focus-group results from Ekos Research (1996) reveal that consumers acknowledge the health and other benefits to be gained from goods produced with biotechnology, but are concerned about the associated long-term health risks.

THE FORMATION OF RISK PERCEPTIONS

In the discussion thus far, we have assumed that both the potential outcome and the probability of that outcome were known with certainty. In this section, we consider how risk perceptions are formed as well as some of the qualitative aspects of risk that may influence choice.

Smith, Desvourges, Johnson, and Fisher (1990) postulate that people use a Bayesian updating process to form their risk perceptions. That is, decision makers have some notion, however derived, of the risk of some activity. When new information becomes available, they update their original estimate. The new information does not replace the old, but is added to the bank of information the decision maker already has. More formally, the updating process can be represented by the following:

$$(4) \qquad P_U = \phi(P_O) + \beta(P_R)$$

where:

P_U = the "updated," subjective estimate of the probability of an adverse outcome

ϕ, β = parameters

P_O = the probability of an adverse outcome the decision maker holds prior to the availability of any additional information

P_R = some "reported" probability.

The magnitudes of ϕ and β reflect the weight the decision maker gives to old and new information, respectively. These values depend on how much the individual trusts the two sources of information. The media may report, for example, that based on scientific evidence, there is no risk posed by consuming some new product. If the decision maker does not trust this source of information (either because he/she does not trust the medium where the information was presented, or because he/she does not trust experts' opinions), then β will be zero, and the "updated" risk estimate will not differ from the original. At the other extreme, if the consumer has no preconceived notions of the risks of some activity, and completely trusts the source of information, then β will be unity; the decision maker's "updated" estimate of the risk will be the same as the experts.

There are several conditions under which β will be small. One is when the decision maker has no trust that the scientific community, or anyone else, knows what the risk is. It may be that an individual trusts both the scientific community and the media to accurately represent what is currently known about risks. However, many people understand the limits of deriving estimates of human health risks by conducting experiments on laboratory animals and extrapolating laboratory results to the human population. In other words, individuals know that science is not infallible and that the risks posed by certain technologies are difficult to assess. This concern was expressed by focus-group participants in Ekos Research (1996). The main consideration for the decision maker in this case is the question "what if the scientists are wrong? What if the probability is not one in a million but one in a hundred?" This uncertainty about risks is called "ambiguity" and has been shown to have implications for choice that are distinct from risk alone (Wohl, 1994). Nevertheless, technical risk assessments continue to treat ambiguity as unimportant. A new technology that has an estimated risk of one fatality per year is considered to be no more risky than an old technology for which actuarial data predict there will be one fatality per year (Lopes, 1992).

The other case in which β would be close to zero is when a decision maker believes that the truth is misrepresented in information sources. It may be the authorities who are in charge of regulation, the industry involved, the scientists studying the issue, or the media who are misrepresenting the truth. The public is cognizant of past risk cover-ups by parties with interest in withholding information. Those who benefit from the sale of cigarettes, silicone breast implants, thalidomide, and other products have been less than forthcoming about the true risks, often with dire consequences. Consider the *Challenger* space shuttle, which exploded in March, 1986. While serving on the committee investigating the causes of the explosion, Nobel-prize winning physicist, Richard Feynman, discovered that although the risk of explosion was assessed by technicians to be as low as one in two hundred, the "official" estimate of the risk was one in one hundred thousand (Lopes, 1992). Why the discrepancy between the technicians' estimate and the "official" estimate? According to Feynman (1988), it was because of the political expediency of having the shuttle launched on the day for which it was scheduled (the day of the U.S. President's State of the Union Address). There was a massive cover up of the hazards of sending the shuttle off in the cold weather, and several lives were lost as a result. Tobacco companies, it is now understood, knew about the risks from cigarette smoking long before the Surgeon General's 1964 warning. Yet these risks were not revealed to the public. The rationale in this case was not political expediency but potential profits.

At the other end of the spectrum, special-interest groups often have something to gain by exaggerating the dangers of new technologies. Francis (1995) cites several examples of cases in which statistics were misused for special-cause purposes. The Natural Resources Defense Council (NRDC) reported that 200 million tons of pesticides were applied in California in 1989. The actual tonnage, according to Francis, was 42,000. The risks of cancer from Alar were overblown by the NRDC, again according to Francis, as were the risks of cancer from fish from Lake Michigan estimated by the National Wildlife Foundation.

The magnitude of β, then, has much to do with the level of trust an individual has in the science that produces risk estimates and the trust he/she has in the relevant institutions to be truthful about what is understood about the risks. The survey by Optima Consultants (1994) did not ask respondents specifically about which organizations they trusted to provide information about the safety of biotechnology. However, they did ask respondents how they would feel if they learned that rbST was approved by Health Canada (this question was asked of the 86% of respondents who said they were concerned or unsure of the safety of using supplemental rbST in the production of milk). Only 17% of the respondents said they would be "a lot less concerned" (46% felt they would be "a little less concerned," 33% said their "opinion would not change").

In a 1990 survey of over 1,500 Canadian adults, the most "trustworthy" sources of information about food safety were "health professionals," "the Consumers' Association of Canada," and "family and friends." Thirty two percent of respondents said they had a lot of trust in government (Consumers' Association of Canada, 1990).

Focus-group research conducted by Creative Research (1996) found that the groups with the least credibility on biotechnology issues include religious leaders, corporations, and industry associations. Consumer associations appear to have the highest credibility rating.

It should be noted that these findings are from small groups (eight focus groups consisting of nine or ten individuals each) and may not be applicable to a more general population.

Although U.S. consumers are likely different in their attitudes toward various institutions, it is worth noting, for the sake of comparison, some results from surveys done in the U.S. In a study of Michigan households, van Ravenswaay et al. (1992) found that when asked about organizations that would be most effective in conducting tests and issuing certification for levels of pesticide residues in food, the most popular choice was "a well-known consumers' group." The U.S. federal government and state government were each chosen by fewer than 20% of respondents. The same study found that almost 50% of survey respondents either "somewhat disagreed" or "strongly disagreed" with the statement "the scientific community can be trusted to be truthful about what they know about the health risks from pesticide residues" and with the statement "the health risks associated with current levels of pesticide residues in food are well known and understood by the scientific community."

Misra et al. (1995) asked respondents to express the extent of their confidence in comments on food safety issues by different interest groups. They found the most trustworthy group was university scientists: 58% of respondents indicated that they had a high level of confidence in this group. Independent laboratories and consumer groups were also trusted: 56% and 53%, respectively, indicated that they had a high level of confidence in these groups. Chemical manufacturers, government agencies, supermarket, and media personalities, all had less than 23% of respondents expressing a high degree of confidence in those groups.

Focus-group research in the U.S. by Chipman, Kendall, Auld, Slater, and Keefe (1995) found that the most reliable sources of information about food and food safety were nutrition professors, extension professionals, and nutritionists/dieticians.

In both Canada and the U.S., consumer groups and university scientists rank high in trustworthiness, whereas the government in both countries is not trusted by a substantial segment of survey respondents.

The results of these studies suggest that respondents might not "update" their probability estimates significantly in the presence of new information if that information does not come from a trustworthy source.

QUALITATIVE ASPECTS OF RISK

This section explores how people qualitatively evaluate risks. Consider two scenarios in which the probability of an adverse outcome is the same, the magnitude of the adverse outcome, should it occur, is the same, and the perceived benefits are the same. Why might consumers react differently to these two scenarios when the benefits and the magnitude of the risks are identical? Slovic et al. (1980) have postulated that there are several qualitative features of risk that are not captured in a quantitative risk estimate, but which nonetheless influence peoples' acceptance of risk. These characteristics include the amount of "dread" associated with the outcome should it occur (cancer is dreaded more than emphysema, for example), whether or not the risk is taken voluntarily (voluntary risks are more acceptable than non-voluntary), how much control one has over the incidence of the risk, and how much is known about the magnitude of the risk.

Sparks and Shepherd (1994) surveyed 216 people about their perceptions of the risk characteristics of potential hazards associated with various aspects of food production and consumption. The authors explored two characteristics of food-related potential hazards: "severity" (which encompassed "greater concern," "more serious for future generations," "threat of disastrous consequences," "degree of dread," and "becoming more serious") and "unknown" (which encompassed "risks now known to those exposed," "risks not known to science," and "less accurate risk assessment"). High in both "severity" and "unknown" were genetic manipulation in animals, food irradiation, BST, and veterinary drug residues. Hazards considered low in both of these qualities included alcohol consumption, vitamin C deficiency, and nutritional deficiencies in general. These findings corroborate Slovic's theories with data on perceptions about risks in the food area. They also help explain why the risks posed by biotechnology may be less acceptable than the risks from other activities, such as skiing, even though on a pure magnitude of risk basis, skiing may be more risky.

Busch (1991) discusses three aspects other than the magnitude of risk that are important to consumers: consent, knowledge, and fairness. Consent is similar to the idea of voluntary risk taking. Busch suspects that the lack of consumers' consent may be a large obstacle in furthering the use of biotechnology. As he says "people want to have a choice when confronting risk. . . . It is also worth noting that people are willing to accept much greater risks when the risk appears to have been imposed not by corporate decisions, but by nature itself" (p. 96). Harlander (1991) also argues that although people are willing to accept risks that are imposed by nature, they become "outraged" when they are forced to take risks involuntarily.

Knowledge, in Busch's framework, is similar to the degree to which the risks are understood. As he says, "as uncertainty increases, many of us become increasingly uncomfortable" (p. 96). The public is aware, as has been mentioned earlier in this report, that risks are often not very well understood, even by experts.

The final consideration in Busch's discussion is fairness. Although a benefit-cost analysis may lead us to the conclusion that the benefits of some new technology outweigh the costs, the analysis says nothing about the distribution of the benefits and costs. Consumers may feel that in the case of biotechnology in food, the distribution of benefits and costs is unfair, thus limiting their acceptability of the new technology. Harlander (1991) argues the same point, claiming "if it is perceived that the primary beneficiaries of the technology are corporations, while consumers bear a disproportionate amount of risk, it will not be acceptable to the public" (p. 158).

O'Riordan (1991) discusses three types of uncertainty: uncertainty that results from inadequate data (the solution: gather more information), inadequacies of understanding rather than knowledge (solution: conduct more interdisciplinary work), and events that are inherently indeterminate. This last type of uncertainty cannot be eliminated and there is therefore no "solution." O'Riordan suggests that in cases where this final type of uncertainty is present and where the consequences of being wrong are high, we need a more "vernacular science." By this he means "a special mixture of scientific analysis, communication, and participation that both beckons and unites" (p. 150).

West, Sandman, and Greenberg (1995) have another way of representing the "characteristics" approach to risk. They theorize that risk is a function of both hazard, or the magnitude of the risk, and outrage, or how "upsetting, frightening, or enraging" (p. 6) is the situation to those who must endure it (i.e., how significant are the dimensions of risk discussed by Slovic et al., 1980, 1990). The hazard to a population of some activity is a function of how bad the outcome is when it happens, how likely it is to happen, and the number of people exposed. Scientists and other experts are interested in the hazard, and seem to have a hard time accepting that consumers do not see things the same way. Decision makers, scientists say, overestimate the small probabilities and underestimate the large risks. If this were so, it would be easy to argue that illuminating consumers about the magnitude of risks could remove the obstacle to consumers' acceptance of biotechnology.

Several researchers have shown, however, that consumers' misweighing of probabilities is quite rational. Kahneman and Tversky (1979) argue that the systematic overweighting of small risks and underweighting of large risks does not demonstrate that people are irrational, but rather that our standard model for consumer choice about risky prospects (expected utility theory) is faulty. They develop a theory called "prospect theory" that allows the decision maker to "weigh" a given probability of an outcome by factors such as ambiguity or dread. West et al. (1995) argue that reporters, like consumers, are more concerned with "outrage" than with hazard. The authors believe the media are correct to focus on these "outrage" issues because of the legitimacy of outrage as a concern for consumers. John Stossel (1997), an ABC correspondent, points out that "Tylenol poisonings were a huge story in 1982 – weeks of headlines and breathless news reports. Yet the poisonings killed only seven people, while cars kill more than 100 Americans every day. Most car crashes aren't 'news'" (p. D6). Why not? The relative risks of Tylenol poisonings and car accidents is not the issue that warrants concern, it is the fact that in the Tylenol case, people may be exposed to risks without their consent. Further, there is no way to minimize the risks from Tylenol poisonings (other than to not use the product), as there is with driving a car (drive more slowly, pay more attention to the road, abstain from drinking if driving, etc.). The outrage factor in the Tylenol case is high; in the car accident case it is low.

RECOMMENDATIONS AND CONCLUSIONS

This paper has presented a basic framework for how people make decisions about products that may have been produced with biotechnology. It also examined how decision makers form their quantitative risk estimates and reviewed several studies that argue that qualitative factors of risk, other than the magnitude of the probability of an adverse outcome, influence the acceptability to consumers of foods produced using biotechnology. This section discusses some possible approaches to making biotechnology in food more acceptable to consumers.

There is no doubt that informing the public about the use of biotechnology is critical. Several of the studies mentioned thus far indicated a low level of familiarity with biotechnology in the marketplace. However, this approach is necessary but not sufficient. Familiarity with products may allay some fears about biotechnology, but will not address the "dread" factor, the appreciation by consumers that many of the risks are to date not known with certainty, and the moral and ethical concerns of consumers. Education is not a panacea. It is based on the notion that if consumers only knew how small the risks really were and how much we all have to gain from biotechnology, they would see the rationality of biotechnology. This attitude is not likely to ease most of the concerns of the public (Douthitt, 1995). One conclusion of this paper is that goods produced with biotechnology may not be chosen even in an environment of full disclosure of information because "ambiguity" and non-risk concerns remain.

The framework presented in this paper suggests a two-pronged approach to communication about biotechnology. The first prong requires filling any informational gap that may exist between what is currently known by experts about the health risks (human, animal, and environmental) and what is currently understood by consumers. This prong addresses the situation in which consumers' estimation of the magnitude of risks and the severity of potential outcomes are made on incomplete information. It refers to a situation in which consumers are not well informed about the actual risks of a product or are not well informed about risk thresholds. Results from a survey of 1,500 Canadian adults in 1992 (Gregory, Slovic, & Flynn, 1996) indicate that a large percentage of respondents believe that risks are dichotomous rather than a continuum.

The second prong addresses non-informational concerns by opening a dialogue between consumers and policy makers and risk communicators regarding the qualitative aspects of biotechnology that may lead to its rejection by consumers. These aspects include the ethical, social, and religious implications of biotechnology. This prong acknowledges that these consumer concerns are warranted and significant and may persist even when consumers are well-informed. It has less to do with "correcting" consumers' misperceptions and more to do with creating a forum for dialogue about these issues.

REFERENCES

ADA Reports (1995). Position of the American Dietetic Association: Biotechnology and the future of food. *Journal of the American Dietetic Association, 95*, 1429–1432.

Busch, L. (1991). Biotechnology: Consumer concerns about risks and values. *Food Technology, 45*, 96–101.

Chipman, H., Kendall, P., Auld, G., Slater, M., & Keefe, T. (1995). Consumer reaction to a risk/benefit/option message about agricultural chemicals in the food supply. *Journal of Consumer Affairs, 29*, 144–163.

Consumers' Association of Canada (1990). *Food safety in Canada: A survey of consumer attitudes and opinions*. Ottawa: Consumers' Association of Canada.

Creative Research International (1996). *Environmental applications of biotechnology: Focus groups*. Final report prepared for Environment Canada and Industry Canada.

Douthitt, R. A. (1995). Consumer risk perception and recombinant bovine growth hormone: The case of labelling dairy products made from untreated herd milk. *Journal of Public Policy and Marketing, 14*, 328–330.

Ekos Research Associates Inc. (1996). *Focus groups on agri-food applications of biotechnology*. Final report submitted to Office of Consumer Affairs. Ottawa: Industry Canada.

Feynman, R. P. (1988). *What do you care what other people think?* New York: Norton.

Francis, F. J. (1995). Food safety: Inferences for the scientific community. *Journal of Agribusiness, 13*(2), 77–83.

Gregory, R., Slovic, P., & Flynn, J. (1996). Risk perceptions, stigma, and health policy. *Health & Place, 2*(4), 213–220.

Grobe, D., & Douthitt, R. (1995). Consumer acceptance of recombinant bovine growth hormone: Interplay between beliefs and perceived risks. *Journal of Consumer Affairs, 29*, 128–143.

Harlander, S. (1991). Social, moral, and ethical issues in food biotechnology. *Food Technology, 45*, 152–346.

Industry Canada (1996). Biotechnology: What is it all about? *Consumer Quarterly, 1*(3), 1–4.

Kahneman, D., & Tversky, A. (1979). Prospect theory: An analysis of decision under risk. *Econometrica, 47*, 263–291.

Lopes, L. L. (1992). Risk perception and the perceived public. In: D. W. Bromley & K. Segerson (Eds.), *The social response to environmental risk: Policy formulation in an age of uncertainty*, pp. 57–74. Boston: Kluwer Academic Publishers.

Misra, S. K., Fletcher, S. M., & Huang, C. L. (1995). Irradiation and food safety: Consumer attitudes and awareness. In: J. A. Caswell (Ed.), *Valuing food safety and nutrition*, pp. 435–455. Boulder: Westview Press.

Mitchler, C., & Murphy, L. A. (1995). *Background paper on food biotechnology in Canada*. Paper prepared for the National Round Table on the Environment and Economy (NRTEE).

Ontario Ministry of Agriculture, Food and Rural Affairs (1996). *Agri-Food Research, 19*(2).

Optima Consultants (1994). *Understanding the consumer interest in the new biotechnology industry: Study findings*. Ottawa: Office of Consumer Affairs.

O'Riordan, T. (1991). Towards a vernacular science of environmental change. In: L. Roberts & A. Weale (Eds.), *Innovation and environmental risk*, pp. 149–162. London: Belhaven Press.

Slovic, P., Fischhoff, B., & Lichtenstein, S. (1980). Facts and fears: Understanding perceived risk. In: R. C. Schwing & W. A. Albers Jr. (Eds.), *Societal risk assessment: How safe is safe enough?*, pp. 181–214. New York: Plenum Press.

Slovic, P., Fischhoff, B., & Lichtenstein, S. (1990). Rating the risks. In: T. S. Glickman & M. Gough (Eds.), *Readings in risk*, pp. 61–75. Washington, D.C.: Resources for the Future.

Smith, V. K., Desvousges, W. H., Johnson, F. R., & Fisher, A. (1990). Can public information programs affect risk perceptions? *Journal of Policy Analysis and Management, 9*(1), 41–59.

Sparks, P., & Shepherd, R. (1994). Public perceptions of the potential hazards associated with food production and food consumption: An empirical study. *Risk Analysis, 14*, 799–806.

Stossel, J. (1997). Science stories you should not believe. *Globe and Mail*, February 8, p. D6.

van Ravenswaay, E., Wohl, J. B., & Hoehn, J. P. (1992). *Michigan consumers' perceptions of pesticide residues in food.* East Lansing, Michigan: Michigan State University, Department of Agricultural Economics. Staff Paper No. 92-56.

West, B., Sandman, P. M., & Greenberg, M. R. (1995). *The reporter's environmental handbook.* New Jersey: Rutgers University Press.

Wohl, J. (1994). *The effect of ambiguity on consumers' willingness to pay for pesticide-residue certification on apples.* East Lansing, Michigan: Michigan State University, Department of Agricultural Economics. Unpublished Ph.D. Dissertation.

THE AUTHOR

Jennifer B. Wohl has a Ph.D in Agricultural Economics from Michigan State University. She was an Assistant Professor in the Department of Agricultural Economics at the University of British Columbia. She is currently a freelance writer, editor, and consultant. Her address is #206-3731 W. 6th Ave. Vancouver, BC V6R 1T8 Canada. E-mail: wohl@interchg.ubc.ca.

Edna F. Einsiedel

The Market for Credible Information in Biotechnology

ABSTRACT. In the past, public reactions to products have generally taken place in the post-marketing phase of development. Increasingly, however, products of controversial technologies such as biotechnology applications have undergone increasing public scrutiny in earlier stages of the development cycle. Clearly, the market for technological development includes the pool of information available to the public and the institutional arrangements in place for access to and dissemination of such information. Such arrangements have been identified as integral to creating markets for innovation. This paper addresses the following questions: What is the nature of the market for credible information for biotechnology products? What social or institutional arrangements and practices attempt to promote greater credibility of information for biotechnology products? Experiences in various European countries are examined as a context for discussing the current policy attempts in Canada to address this challenge.

In studies of the role of technology and technological choices and their contributions to social wealth, this key question is often asked: Why do some technologies take hold and others fail? What are the factors that help explain technological innovation and diffusion, or failures in the marketplace?

In the past, public reactions to products generally took place in the post-marketing phase of development. Increasingly, however, the products of innovative technologies are undergoing scrutiny by publics in earlier stages of the technological development cycle. This appears to be the case with biotechnology.[1] The coalescing of the impacts of various social changes and movements over the last several decades including consumerism, environmentalism, health activism, and the visibility of public interest campaigns relating to a variety of technologies such as toxic waste, new reproductive technologies, or nuclear technologies have pushed forward the timelines for consideration of market reactions to technological products. The visibility of social factors in the diffusion of innovations has become more palpable as consumers have increasingly challenged agricultural research agendas that are seen to be ignoring issues of nutrition, health, or environmental concerns, or as they have demanded increased government regula-

Journal of Consumer Policy **21**: 405–444, 1998.
Bartha M. Knoppers & Alan D. Mathios (eds), Biotechnology and the Consumer.

tion over health technologies to ensure public safety or individual privacy.

Not surprisingly, industries involved in potentially "controversial" technologies have shown an increased interest in public attitudes earlier on in the development cycle. There is good reason for this. In a recent study on the barriers to the diffusion of biotechnology in the food industry, an international comparison between the Netherlands, Denmark, and Germany identified the main factors as including: commercial factors and market trends, public perception, regulation, and technological options. Interestingly, the most important barrier the study identified turned out to be public perceptions (Roobeek, 1995).

It is clear that one can no longer take for granted that efficiency, markets, or profits are the only criteria for the development and success of new products or technologies. There is increasing evidence that the market for technological development includes the pool of information available to the public and the institutional arrangements in place for access to and dissemination of such information. Such networks, or organizational structures for creating a pool of social information, have been identified as integral to creating markets for innovations (Teubal, Yinnon, & Zuscovitch, 1991, p. 389).

This paper will address the following key questions: What is the nature of the market for credible information? What social or institutional arrangements and practices lead to or promote greater credibility of information? In addressing these questions, I will first describe some aspects about the nature of biotechnology products that reflect on their credibility in the marketplace. I will then briefly explore the various dimensions of credibility and what these mean to consumers. Third, I will examine what practices and institutional arrangements have been employed to address information issues and the role of the public in discussions of technology. Finally, I will discuss what the implications of these findings might be for current discussions on biotechnology in the Canadian context.

THE NATURE OF BIOTECHNOLOGY PRODUCTS

There are a number of ways in which one can describe the products of biotechnology, ways which relate to how these goods are seen in the marketplace. One approach is the economic one of describing goods in terms of the ease with which the quality of the good can be determined. These goods have been categorized as *search*, *experience*, and *credence* goods. *Search* goods are those whose quality and safety can be determined before purchasing them (children's clothing, for example). *Experience* goods are those whose attributes can be ascertained only after purchasing and consuming the product (a tropical fruit appearing for the first time in the supermarket, for example, or a can of tuna fish). *Credence* goods are goods whose qualities cannot be evaluated in normal use. The assessment of the qualities of such goods actually require additional costly information (Laband, 1991).

Many biotechnology products can be considered to be examples of credence goods. The average consumer may not have the knowledge, information, or skills to ascertain the qualities of these products. Furthermore, even if a consumer buys a product of biotechnology and uses or consumes it, the product's qualities may only be ascertained after a considerable lapse of time or may be almost impossible for the consumer to determine. For example, a can of tomato paste whose chief ingredient, tomatoes, were genetically modified, would be virtually indistinguishable from one using regular tomatoes.

In the case of biotechnology credence goods, many of the qualities that have been changed will not necessarily be visible to the consumer. Herbicide-resistance or pest-resistance are qualities which have helped at the production end but may have little impact at the consumption end in the short term. In these instances, judgments about the nature or quality of the product may rest in part on the credence qualities of the company, the regulatory system, or other factors the consumer may put into play.

Another framework within which one might describe biotechnology products is the framework of risks and benefits. Certain technologies are more controversial than others and the degree of controversy to some extent revolves around the degree of perceived risks associated with the technology and how these risks are seen to be addressed. There is a considerable volume of research on risk perception and risk communication which is discussed at greater length in another paper in this volume (Wohl, 1998). It is important, however, to draw on what is already known about processes of risk perception as these have important implications for public perceptions of a technology's credibility in the marketplace. The factors associated with risk perception and evaluation are summarized in Table I (Covello, Sandman, & Slovic, 1991).

TABLE I
Factors Important to Risk Perception and Evaluation

Factors	Conditions Associated with Greater Public Concern	Conditions Associated with Less Public Concern
Catastrophic potential	Fatalities and injuries grouped in time and space	Fatalities and injuries in time scattered and random
Familiarity	Unfamiliar	Familiar
Understanding	Mechanisms or process not understood	Mechanisms or process understood
Uncertainty	Risks scientifically unknown or uncertain	Risks known to science
Controllability	Uncontrollable	Controllable
Voluntariness of risk	Involuntary	Voluntary
Effects on children	Children specifically at risk (delayed effects)	Children not specifically at risk (immediate effects)
Manifestation of effects on future generations	Risk to future generations	No risk to future generations
Identification of victims	Identifiable victims	Statistical victims
Dread	Effects dreaded	Effects not dreaded
Trust in institutions	Lack of trust in institutions	Trust in responsible institutions
Media attention	Much media attention	Little media attention
Equity	Inequitable distribution of risks and benefits	Equitable distribution of risks and benefits
Benefits	Unclear benefits	Clear benefits
Reversibility	Irreversible effects	Reversible effects
Personal stake	Individual personally at risk	Individual not personally at risk
Origin	Caused by human actions or failures	Caused by acts of God or nature

Source: Covello, Sandman, & Slovic (1991).

In examining the factors associated with greater or less public concern, it is easier to describe why biotechnology, both as a class of products or as a particular product or process, might evoke certain public reactions. A 1993 survey of Canadians (Decima Research, 1993) categorized respondents in terms of those who see mostly risks, those who see mostly benefits, and those who see some of both. The largest category consisted of those who believe biotechnology poses an equal opportunity for *both* benefit and danger (39%). The same number associated the technology primarily with benefits as those who associated it with risks (one quarter). More recent survey and focus group data have consistently shown that Canadians are generally receptive to but cautious of products of biotechnology (see, for example, Einsiedel, 1997; Ekos Research, 1996; Optima Consultants, 1994), and their caution stems from a number of factors including unfamiliarity, perceived risks, uncertainty about benefits, as well as potential moral and ethical questions about the technology. Other elements associate biotechnology with greater public concern as well: It is an unfamiliar technology whose processes (particularly genetic modification) are not well understood. The fact that these products at the moment are not required to be labelled in Canada (despite the fact that large majorities insist they ought to be) also raises the issue of involuntary risks. The uncertainty of long-term consequences also brings up the problem of uncontrollable outcomes. In combination, these elements add up to a set of conditions that evokes greater public concern.

In general, both the economic and the risk frameworks are useful to keep in mind because both underscore the importance of moving beyond the specific characteristics of the products or processes to the social system of information surrounding the technology.

WHAT MAKES FOR CREDIBLE INFORMATION?

Dimensions of Credibility

It is worthwhile to understand the dimensions of credibility before examining information dissemination practices. On issues or products involving some element of risk, risk communication researchers have consistently stressed that *credibility* is the risk communicator's most precious asset (Covello, McCallum, & Pavlova, 1987). Early social psychological studies on persuasion have identified two dimensions as commonly associated with credibility: expertise and trust (Hovland, Janis, & Kelly, 1953). Expertise refers to how knowledgeable or competent a source is perceived to be while trustworthiness describes the degree to which audiences believe a source to be honest, sincere, or reliable. Turner and Wynne (1992) propose four dimensions of credibility which overlap with but also add to these two dimensions: impartiality, honesty, competence, and power. They suggest that the last dimension relates to issues of access to information and equity in information distribution. Recent research evidence suggests that the public does actively make distinctions in evaluating source credibility on technological risk issues (see, for example, Bord & O'Connor, 1990).

Message Sources and Credibility

That sources are being evaluated on the basis of their credibility and not just the messages they disseminate is no longer unquestioned. How knowledgeable is the source? Is he or she unbiased? That is, does the source have something to gain by taking one position versus another on an issue? Are the interests of the source congruent with my (the consumer's) own interests and needs? Is the source fair? That is, is there acknowledgement and adequate representation of the range of viewpoints available? How consistent is a given source vis-à-vis the predictability of arguments presented or the pattern of behaviors or messages? Is there recognition of the uncertainties within the knowledge base? What is the institutional context within which the information is produced and distributed?

These issues are evident when one considers the question of whom consumers trust for information. The answer seems to be "it depends." In the first national survey of Canadians on biotechnology (Decima Research, 1993), respondents were asked how much trust they would place in statements made by groups who might speak on issues surrounding the application of biotechnology. Among the most trusted were doctors and nutritionists, followed by university scientists, farmers' groups, environmental groups, and consumer associations. Least trusted were industry, the media, and federal government regulators. Trust appeared to vary based on whose interest the group was seen to represent and the perceived level of expertise on the issue (Decima Research, 1993).

More recent findings on the issue of which organizations or institutions are trusted "to tell the truth about modern biotechnology" similarly show educational institutions to be most trusted (36%), followed by consumer and environmental organizations (14% each) (Einsiedel, 1997). When specific areas of biotechnology are mentioned such as an issue around genetic modification of crop plants or around genetically modified animals for organ transplants, the most trusted sources tend to be farmers' organizations and the medical profession, respectively, again along with universities (Einsiedel, 1997). As focus group findings have elaborated (see, for example, Creative Research, 1996), the choices of whom to trust reflect broader public concerns beyond just message elements; these choices reflect broader expectations about the social context of information and the trust (or mistrust) such a context engenders.

This trust has consequences in terms of information acquisition. Because of the "costliness" of acquiring information (in terms of time, monetary costs, or motivation), consumers have been willing to trust other sources to take on the responsibility of obtaining or possessing information or ensuring that things are indeed working well enough to obviate the need for an information search. On the other hand, credibility issues may arise in connection with the sources to whom trust has been accorded.

For example, government agencies are often the "experts" to whom the public entrusts the issue of product safety. However, a number of factors are contributing to declines in public trust. First, regulatory agencies have been perceived to have been overly influenced by industry; second, agencies are seen to be inappropriately biased in favor of promoting particular technologies; third, agencies have been known to mismanage particular health or environmental activities; fourth, government officials and experts have been known to disagree amongst themselves (Covello et al., 1987). On the issue of biotechnology, Canadian surveys have shown that the most trusted sources tend to be non-government organizations (e.g., consumer organizations) or international organizations (Einsiedel, 1997; Optima Consultants, 1994). In the case of universities, while university scientists have consistently been seen to be more credible, this group's credibility might also be perceived as having been compromised in recent years by the closer ties between universities and industries (Creative Research, 1996).

These and other findings emphasize the importance of examining the social context of information. While messages are evaluated on the basis of their clarity, consistency, balance, and accessibility, the larger context within which they are received is even more important (Turner & Wynne, 1992). This seems to be especially true for more complex and abstract issues, or, one could add, for issues perceived to have high risk. In this instance, social-structural factors are more likely to come into play (Renn & Levine, 1991). A pattern of violation of trust-building efforts on issues that are more concrete and involve less risk, eventually has impacts on higher levels. By the same token, "distrust on a high level sets the conditions and determines the latitude of options for gaining or sustaining trust on a lower level." (Renn & Levine, 1991, p. 38).

PUBLIC PERCEPTIONS OF ISSUES AND THE IMPLICATIONS FOR CREDIBLE SYSTEMS OF INFORMATION

There is a sense among consumer and environmental groups and some members of the public at large that the regulatory framework for biotechnology is focused primarily on risk assessments, with insufficient attention to the socio-ethical dimensions, values, and environmental sustainability (see, for example, Canadian Institute for Environmental Law and Policy, 1995; Consumers Association of Canada, 1995; Féderation nationale des associations de consummateurs du Quebec, 1996). These issues have been summarized by the Consumers Association of Canada into five categories:

1. *Health and safety and regulatory issues*
   - Who should have input into the policy process?
   - How should long-term risks and other 'unknowns' be addressed?
   - What is the scope of the regulatory system and are there gaps in the current regulatory framework?

2. *Environmental concerns*
   - What are the environmental risks (e.g., risks from cross-breeding, of groundwater contamination, developing resistance, genetic mutations)?
   - Are testing procedures adequate?
   - What are the risks of reducing biodiversity and the gene pool?
   - What are the possibilities of out-crossing to natural species, to non-target plants, or animals? How well do we understand the science of genetic manipulation to anticipate risks and develop effective risk management procedures? Is risk management enough and how much risk is acceptable?

3. *Economic issues*
   - Is biotechnology being driven by profit motivation and not balanced by concern for public interest and environment?
   - Who will pay for failed technology, for liability, harm, or injury?
   - Who should benefit, who should pay, and how will cost-benefit ratios be determined?
   - What are the global economic and trade implications such as the effect of displacing economies of third world countries or of harmonizing needs with other trading partners?
   - How will need for new products be demonstrated? Will this be industry or consumer-driven?
   - What are the issues surrounding intellectual property and patenting of living organisms?

4. *Social issues and concerns*
   - Is biotechnology different from other technologies, and should it be treated differently? Why or why not?
   - What role should society play in determining the limitations, the boundaries of technology?
   - How will public awareness and public trust issues be treated, such as how to respect and balance the consumer right to know, the need to know, and the right to choose? How to provide public trust in industry; how to make the public aware of risks and benefits; whom will they believe?
   - How will costs and benefits be fairly distributed?
   - How will human rights be balanced with animal rights and who should decide?

5. *Ethical concerns and values*
   - Who has the right to play God and who should have control over man's destiny?
   - What are the impacts of genetic screening on privacy?
   - Who should be the ethical gatekeeper for industry?
   - How should freedom of scientists to explore the boundaries of science be balanced by values, ethics?
   - Does someone with a moral, religious stance have the right to choose a genetically engineered product, such as a Moslem or a strict vegetarian who cannot eat meat? Would they have the right to know if an animal gene has been inserted into a food product?

The Consumers' Association of Canada has distilled these concerns and questions out of its articulation of consumers' rights with regard to biotechnology: "The consumer interest in biotechnology lies in the protection of their rights: the right to information; the right to safety, quality, and choice; the right to be heard; and the right to participate in decision-making, as applications of biotechnology are developed in health care and food production" (Agriculture and Agri-Food Canada, 1993, p. 33).

These issues or questions represent a complexity of concerns that go beyond the utility of a product to the individual consumer. The considerable evidence from studies on risk perception and risk communication on controversial technologies and from the public surveys on biotechnology have demonstrated that perceptions of biotechnology issues are indeed many-sided and complex and that judgments about acceptability are not simply based on estimations of utilities and costs.

Data from the most recent survey on Canadians (Einsiedel, 1997) show that people do differentiate among different applications, that they do make differential judgments on these applications' utility, riskiness, and moral acceptability, and that these distinctions have impacts on public assessments of acceptability. Similar results have been found among European consumers as well (Biotechnology and the European Public Concerted Action Group, 1997). As data from the Canadian survey (Table II) show, six different applications were provided and for each one, respondents were asked whether they had heard of the application, whether they thought the application was useful, was risky, was morally acceptable, and finally whether it ought to be encouraged.

While medical applications were more likely to be encouraged (e.g., the introduction of human genes into bacteria for medicines and vaccines) than improving a trait in fruits and vegetables for better quality, some medical applications were less acceptable than others. For example, only 56.5% of the respondents definitely agree or agree that the introduction of human genes into animals to produce organs for human transplants ought to be encouraged.

When examining the impact of perceptions of utility, risk, and moral acceptability on the issue of whether or not the application ought to be encouraged, each of these factors significantly affected the respondent's position about encouraging the application.[2] *That is, the judgment about whether a biotechnological application ought to be encouraged was as likely to be influenced by judgments of perceived risk and moral acceptability as it was by utility.* The areas of concern expressed by respondents in focus groups as well as issues raised by public interest groups, coupled with the pattern of judgments demonstrated in the survey just described, paint a picture of a technology raising a complex range of issues and judgments.

At the same time, misperceptions and uncertainties also exist. For example, over a fifth of Canadians (22%) think that ordinary tomatoes do not contain genes; close to the same number (20%) believe that by eating a genetically modified tomato, one's genes will also be modified; a quarter (24%) think that genetically modified animals are *always* bigger than ordinary ones (Einsiedel, 1997).

Given the range of these questions, concerns, and perceptions, what communication practices are being employed to meet these challenges? We first turn to the experiences of mature industries that have faced similar kinds of issues and challenges.

LESSONS FROM OTHER INDUSTRIES

What can the biotechnology industry learn from other, more established industries? Two industries which experience similar kinds of public issues are the energy and the chemical industries.

The Energy Industry[3]

The 1970s saw the energy industry facing considerable public pressure: The oil embargo raised the spectre of scarcity and safety of supplies and the question of price dependability. The nascent but growing environmental movement raised questions about environmental pollution and energy conservation. In later years, international issues further embroiled the energy industry in controversy. For example, the acid rain issue (attributed to coal-burning plants) was highly visible in the '80s, followed by the issue of global warming. Other issues such as carcinogenicity of certain chemicals (e.g., benzene) are still under dispute. Environmental pollution issues have further dogged the industry, particularly with widespread coverage of such disasters as the Exxon Valdez oil spill.

In this environment, the Petroleum Communication Foundation (PCF) was created to deal with public questions and concerns (Bradley, 1997) and to further information dissemination activities of the industry as a whole. The foundation puts out a number of publica-

TABLE II
Perceptions of Six Biotechnology Applications

	Heard	Definitely Agree	Agree	Definitely Disagree	Disagree
Production of food, drinks	65.9				
Utility		29.1	38.1	15.0	13.5
Risk		21.2	34.0	23.6	15.3
Morally acceptable		28.5	41.3	12.7	11.8
Encouraged		26.7	32.9	17.1	18.0
Inserting genes for pest resistance	63.6				
Utility		49.3	33.0	7.6	4.8
Risk		11.0	27.1	28.5	25.3
Morally acceptable		39.2	40.3	8.4	5.9
Encouraged		41.2	35.6	10.0	7.2
Introducing human genes into bacteria for vaccines	62.3				
Utility		58.6	30.7	2.8	2.5
Risk		8.7	25.8	28.4	27.2
Morally acceptable		43.8	41.0	5.6	3.6
Encouraged		50.1	35.4	4.7	4.0
Developing gen. modified animals for medical studies	82.1				
Utility		46.1	40.1	4.9	7.3
Risk		14.7	24.1	28.2	29.9
Morally acceptable		28.5	39.1	17.3	13.0
Encouraged		35.2	35.7	13.6	13.4
Introducing human genes into animals to produce organs for transplants	66.2				
Utility		28.7	36.8	13.2	16.0
Risk		27.8	35.4	19.6	12.0
Morally acceptable		18.5	33.3	22.3	20.2
Encouraged		23.4	33.1	16.8	22.2
Using genetic testing to determine disease predisposition	63.6				
Utility		43.7	38.0	6.3	6.2
Risk		17.0	29.0	24.6	22.7
Morally acceptable		28.1	43.7	11.9	9.5
Encouraged		35.1	39.3	10.4	9.4

Note:
Respondents were asked whether they had heard of the following applications, then asked whether they definitely agreed, agreed, disagreed or definitely disagreed about the application being useful, risky, morally acceptable. Finally, they were asked whether all in all, the application ought to be encouraged. The applications: 1. Using modern biotechnology in the production of food and drinks, for example, to make them higher in protein, keep longer, or taste better. 2. Inserting genes from one plant species into a crop plant to make it more resistant to insect pests. 3. Introducing human genes into bacteria to produce medicines and vaccines, for example, the production of insulin for diabetics. 4. Developing genetically modified animals for medical studies, such as a mouse that has genes which causes it to develop cancer that can be studied in the laboratory. 5. Introducing human genes into animals to produce organs for human transplants, such as pigs for human hearts. 6. Using genetic testing to determine whether human embryos have a genetic predisposition for serious diseases such as cystic fibrosis or thalassaemia.

tions such as newsletters and more in-depth research reports and background papers. It provides story material for smaller newspaper dailies and weeklies and supplies energy writers from major media and free-lance communities with background information on the industry in general or on specific topics.

Communication considerations are also embedded in the regulatory processes at both provincial and federal levels. The regulatory process mandates the use of public consultation in a timely and open manner and applications for regulatory approval are required to be accompanied by a full description of the consultation process employed, the issues identified and resolved, and those remaining in need of resolution (Alberta Energy and Utilities Board, 1993).

Both the National Energy Board (NEB), the Alberta Energy and Utilities Board, and industry associations, particularly the Canadian Association of Petroleum Producers cooperated to develop a major public consultation handbook describing how major energy companies may effectively carry out the broad philosophical concepts of public and community consultation (Marlier, 1997). The objectives and processes for public consultation include the following attributes:

1. A recognition that the public constitutes an important stakeholder in energy development.
2. A recognition that while there are benefits from energy development, there are also sectors of the public that can and do experience the risks or costs (e.g., those in close proximity to sour gas lines; consumers who bear the costs of pricing actions).
3. A recognition that the energy industry must tell both sides of the story (the risks and benefits), in a language accessible to all.
4. A recognition that effective communication needs to run throughout the operating lifetime of the project, not just at the start or after implementation.
5. A recognition that effective public involvement begins with public understanding of the regulatory process (Alberta Energy and Utilities Board, 1989).

Understanding of the regulatory process includes an open review process so that members of the public can review an application in full, and can have an understanding of the supports available for intervention. The NEB is currently completing its on-line regulatory channel, allowing all regulatory materials including applications, interventions, and rulings to be publicly accessible on its website.

In order to ensure that these consultative activities take place effectively, both federal and provincial Boards cooperated to provide training internally for its staff to more fully understand issues on risk perception and risk communication. Externally, these workshops were also made available to industry practitioners and other stakeholders.

The Chemical Industry

If there was a wake-up call to the chemical industry, the Bhopal tragedy in 1984 sounded the alarm. By the mid-1980s, it was clear to the Canadian chemical industry that its very survival was at stake.

> Public outrage over the health and ecological impacts of incidents and chronic problems associated with the life cycle of chemicals and coalescing into calls for restrictive regulation and curtailment of the development of the industry and its products. This also translated into insurance and financing problems, community nervousness, higher incident response costs, diversion of management time to crisis control, and erosion of employee, customer and shareholder confidence in the industry (Canadian Chemical Producers Association, 1996d).

Fast-forwarding this scenario a decade later, the Canadian Chemical Producers Association (CCPA) was recognized by the United Nations with its Global 500 Environmental Award and cited by the federal government's Green Plan. What accounted for this significant change? It is worthwhile to examine the industry's Responsible Care™ program for some lessons.[4]

The Responsible Care program can be described as an ethic, backed by a code of practice and a system of self-regulatory enforcement. Member companies of the CCPA are formally bound, as a condition of membership, to the program's policies, which, in turn, have been designed to cover responsibilities spanning the total life cycle of products, from the beginnings in the laboratory to their ultimate disposal or destruction. Six codes of practice were developed in the following areas: (a) community awareness and emergency response, including a policy on the community's right to know; (b) research and development (R&D); (c) manufacturing; (d) transportation; (e) distribution; and (f) hazardous waste management. To illustrate, the first code of practice on community awareness and emergency response requires all company sites (from R&D to manufacturing, storage, distribution, and waste management sites) to have in place an active community awareness and emergency program. This program is designed to provide fora for community concerns, to have regular communication processes with the community through local outreach programs, and to provide full information about the hazards and associated risks of chemicals, chemical products, and operations to employees and members of the community. Not only are risks discussed with the community; "worst-case scenarios" are also presented, with the company explaining how it is prepared to deal with such a situation (Canadian Chemical Producers Association, 1996b).

The codes of practice in these areas reflect every step in a chemical product's "life." Also reflected in such codes and practices are values such as the community's and employee's right to know about the nature and risks of products. These codes do not just define what is expected of each member company; they spell out criteria by which the company can evaluate progress and results. Support is provided to assist companies in implementation, including provision of seminars, resources, and additional references on external sources. More importantly, a system of verification was implemented, a process involving an independent four-member team consisting of two industry experts, an activist, and a community representative. This team visits the plant, scrutinizes every aspect of a plant's operation, interviews personnel, examines company documents, and speaks to members of the community over several days. The report they write – which includes those areas in which the company falls short – is provided to the company and the public. Preparation to meet the criteria for the verification process may take a company several years and even then, some companies may fall short in a number of areas. A company cannot be "certified" without its passing this verification process. Careful records are maintained industry-wide in order to monitor industry performance. Such indicators as accidents in the shipment of goods, worker accidents in the workplace, and total industrial emissions are designed to track individual company as well as total industry performance.

As part of its performance oversight process, the Responsible Care™ program includes an independent National Advisory Panel consisting of concerned citizens, environmental and consumer representatives, academics, and other relevant experts. The Advisory Panel has been an active adviser in the design and implementation of the Responsible Care™ program. For example, it was responsible for promoting the third-party verification process of a company's implementation progress and for pushing for all CCPA members to have a community advisory process in place. It is currently promoting the idea of a zero-discharge goal on air and water emissions (Canadian Chemical Producers Association, 1996c).

Has the program paid off? In terms of a number of indicators, the answer appears to be positive: There has been a 50% reduction in emissions since 1992; frequency and severity of transportation incidents have been reduced; and the industry has also reported benefits in the area of reduced insurance costs for some companies and improved financing opportunities (Canadian Chemical Producers Association, 1995, 1996a).

On the regulatory front, the Responsible Care™ program has encouraged the federal government to allow the industry to play a primary role in setting its own targets in a number of areas and the best means for achieving them. For example, while the U.S. enacted a number of regulations in the wake of Bhopal, mandating actions by companies and communities for emergency prevention and preparedness, the Canadian government chose to create a Major Industrial Accidents Council of Canada headed by the CCPA to create and implement voluntary "best practices" approaches that were consistent with the code of practice the industry already had in place, a recognition of the high standards already promulgated by industry itself.

In essence, this self-regulatory approach incorporates notions of life-cycle stewardship of products, a precautionary approach to risk, an accommodation of all stakeholder viewpoints, and the use of mutual assistance and peer pressure among members.

The above descriptions are by no means indicative of a perfect system. One drawback, for instance, is the fact that not every chemical company is a member of the industry association. One could argue that performance standards are only imposed on and adhered to by those who see value in industry association membership and see the costs of membership as an "investment" rather than an "expense." It is generally the larger companies in both the chemical and energy sectors that actively participate in these industry associations. When voluntary codes are adopted, non-participating firms essentially gain a "free ride," benefitting from the positive image generated without an investment in meeting the standards.

In essence, both industries' approaches contain elements of frequent communication with opinion leaders and the general public, consultation processes with communities, a combination of both issues management and product life-cycle system of monitoring member performance.

There are some lessons for the biotechnology industry to be gained from the experiences of these two industries, a number of which are worth considering. These experiences suggest an obligation to consider public interests and concerns not just at the commercialization stage but from product development onwards. Public concerns are recognized as legitimate and efforts are made to address these at community and regulatory levels.

Structurally, government initiatives on public consultation efforts are jointly undertaken by federal and provincial regulatory boards and these processes are clearly spelled out and observed. There are also collaborative efforts between industry and government on public consultation initiatives. Within both industries, industry associations play major roles in the provision of market information but more importantly, these communication efforts are tied to industry-wide standards of performance.

PUBLIC COMMUNICATION APPROACHES ON BIOTECHNOLOGY – SOME INTERNATIONAL PERSPECTIVES

In general, some countries within the European community have conducted more comprehensive efforts at public consultation and participation on biotechnology issues. Some examples are presented here.

Denmark

Denmark's approach has been characterized by extensive public consultations and public involvement as early as possible in the evolutionary life of a technology. The use of consensus conferences is, symbolically and practically, the model of such a philosophy. Consensus conferences have been called an unconventional contribution to public debate and decision-making. While conventional processes of public participation have placed emphasis on professional institutions and interest groups, with public decision-making vested primarily in the work of elected representatives, the consensus conference has been designed to give voice to lay people, people whose interests are primarily those of the concerned citizen. It is a process which brings together a group of lay people and experts, allowing the former to explore the different dimensions of a given issue by exposure to background material and by further questioning of experts. It is a public participation approach which melds three concepts: (a) the judicial process with a jury; (b) the scientific meeting among peers; and (c) the town meeting with public participation (Mullan & Jacoby, 1985).

Although the term originated in the U.S. (first used in connection with organized dialogues between panels of medical experts in order to arrive at some "consensus" on the best available knowledge on a medical issue or area), it was the Danish Board of Technology in the 1980s which implemented the consensus conference as a form of technology assessment involving lay people. Such a format has been employed on such diverse subjects as food irradiation in 1989 and transgenic animals in 1992. In 1993, the Dutch followed the Danish initiative with a consensus conference on animal biotechnology, while the U.K. held its first consensus conference on plant biotechnology in 1994. The more recent consensus conference for the Dutch was in 1995 on the subject of human genetics research. Some have labelled consensus conferences in Europe as "the most popular form of technology assessment within the science and technology policy-making community" (Joss & Durant, 1995, p. 11). They have been considered successful in stimulating public debates on new technologies (Grundahl, 1995).

The procedure involves bringing together a group of lay people (usually about 15 individuals) who are provided with basic information on the topic during two preparatory week-ends. The lay panel then formulates a number of key questions which are posed to invited experts. These experts are invited to discuss the technology and its consequences by answering the key questions at the consensus conference. Different types of interdisciplinary expertise and conflicting expert opinions are juxtaposed. The lay panel then retires to write its report. These final reports are widely disseminated and the conference as well as the results receive extensive media coverage. The reports have contributed to informing politicians and decision-makers on citizens' views of, and attitudes towards, new technology and help to catalyze further public discussions on the issue at hand. It is not surprising that a large number of consensus conferences in Europe have focused on biotechnology topics – issues on human genome mapping research, plant biotechnology, genetic screening, transgenic animals, and reproductive technologies have figured prominently in these conferences.

There are other formats that have been used by the Danish Board of Technology to increase the level of information and discussion in the public arena. Among other participatory methods they have employed, scenario workshops and roundtable seminars have also played a role, in addition to consensus conferences (Kluver, 1995). Workshops have been used when a cross-disciplinary expert group requires feedback on a draft report. A mixed group – usually experts, users, and other actors – can contribute feedback that is not often available from experts.

The scenario workshop is a method in which visions and needs in a field are discussed and articulated by relevant decision-makers, stakeholders, experts, entrepreneurs, and users. It usually consists of two series of workshops, taking as its starting point two to four pre-defined scenarios for the future development of the topic. The sessions consist of a first brainstorming session, followed by a second session where plans of action are developed.

The Board also funds local initiatives, supporting about 75–100 local debate meetings a year, each attended by 25 to 100 local residents. These are usually initiated by local citizens, with expenses for transport, facilities, and expert presentations covered by the Board. A debate resource pack has also been supported by the Board which includes a booklet outlining the topic in broad terms, a list of other relevant materials and potential experts, slides, and an application form for local funding. This is distributed through *folkeoplysning* networks, or people's networks, and other relevant organizations (Kluver, 1995).

More recently, a "Dialogue Forum" was initiated by the Danish Consumers Council (*Forbrugerrådet*), with government support.[5] The purpose of this Forum is "to ensure that consumers get satisfactory information about foods and gene technology and to investigate the possibilities of a voluntary labelling system in Denmark that will supplement the coming EU rules." This was in reference to the EU labelling rules that have recently been implemented. Also participating in this process are agricultural organizations, the biotechnology industries, food workers, unions and retailers, and government ministries (Jelsøe, 1997).

The Danish approach to credible information and public participation has been described as embedded in a "consensus-seeking political culture" (Toft, 1996). It is a tradition of decision-making on technological issues which involves a broad array of interests at an early stage of the technology's development, based on an explicit recognition that assessments of technologies are, in the end, political questions (Toft, 1996).

Netherlands

The Netherlands, like Denmark, has also employed consensus conferences, sponsored by the Netherlands Office of Technology Assessment (NOTA, now the Rathenau Institute). A semi-governmental body, the Foundation for Public Information on Science and Technology, has also taken the lead in information dissemination, helping to sponsor a number of roundtables on biotechnology. These roundtable discussions have brought together representatives from consumer organizations, environmentalists, industry, and government. Finally, the government has also supported a hotline run by *Consumentenbond*, the largest Dutch consumer organization, to deal with questions on genetically modified products.[6]

Most notable in the Dutch context is the use of a consultation group – in existence for the last three years – consisting of representatives from the consumer and environmental organizations as well as industries involved in biotechnology products. This group provided important input into the drafting of the Consumer Products Regulation on Novel Foods of July 17, 1993 which regulates the safety testing of such products and into two recommendations by the Consumer Products Advisory Committee on the labelling of novel foods.

This Informal Consultation Group agreed that the use of uniform language can in itself promote clarity for consumers. The phrase they recommended and which was adopted by the Dutch government was "verkregen met behulp van moderne biotechnologie," or "produced with the aid of modern biotechnology." As part of this process, an agreed-upon list was created for (a) products which must be labelled, (b) products which need not be labelled, and (c) points for further consultation.[7]

Businesses involved in the informal consultation process have also agreed to undertake the following where possible:

- to inform and hold consultations with social organizations, including specifically the Consumer and Biotechnology Foundation (*Stichting Consument en Biotechnologie*) as soon as possible prior to the market introduction and prior to the start of the formal clearance procedure;

- to draw up where necessary an information plan for consumers as part of the consultation. In addition to product-related information, this plan may consist of public information and communication activities.

The Informal Consultation Group (1996b) has also been instrumental in creating a checklist for drawing up evaluation reports for foodstuffs, ingredients, and additives produced via genetic modification processes. The checklist includes a range of items including the use of the product, safety issues (e.g., toxicology, allergenicity), environmental aspects, consumer concerns (e.g., nutritional consequences), and socio-economic issues (e.g., effects on employment, on international trade relationships, consequences for developing countries).

This checklist, though not legally binding, has been adopted by an Advisory Committee to the Dutch Ministry of Health. Applicants who seek permission to market GMO-derived foodstuffs are strongly urged to make use of this checklist.

These agreements and guidelines are regarded by the Informal Consultation Group as "working hypotheses," with the intention of gaining experience with them over a period of about two years and to further adapt or fine-tune these guidelines as circumstances dictate. This stage has been described as the social embedding of a technology (Von Schomberg, 1996). It is essentially the third phase in a policy development process of technology assessment which started in the mid 1980s, beginning with the stimulation of public debate on all aspects – socio-economic, political, ethical, and scientific – of biotechnology.

While there might be disagreement on the end-policies adopted by the Dutch (e.g., both Canada and the U.S. have so far avoided a labelling policy), what is important in terms of the Dutch experience is the active participation of relevant social groups in the policy formation process, a process that has been called "*a deliberations-based approach*" (Von Schomberg, 1996). The result of this transparency in the market and participatory process is a broad consensus in the Netherlands over the policy adopted, accompanied by a commitment to share responsibilities for outcomes (Von Schomberg, 1996).

Germany

The social context surrounding the issue of biotechnology in Germany was a highly polarized one in the 1980s (Jasanoff, 1995). The Green Party was elected to the Bundestag in 1984. Its alliance with the Social Democrats led to the creation of a working group on gene technology. Subsequent public debates and Green opposition to biotechnology led to litigation against a plan by the Hoechst company to open a facility for genetically engineered insulin. An ensuing gene law (since rescinded) led to further public debates and it was in this context that a notable experiment making use of a participatory technology assessment process was carried out between 1991 and 1993 on the subject of crop plants with genetically engineered herbicide resistance. This focus was chosen because it seemed to be sufficiently relevant and controversial and because by this period, half of all release experiments involved transgenic, herbicide-resistant plants in Germany. As an emerging technological development, the assessment process was intended to focus on the possible consequences of this development.

Organized by *Wissenschaftszentrum Berlin* (Science Center Berlin) and funded by the German Ministry for Research and Technology, participants included representatives from industry, environmental organizations, consumer groups, alternative agriculture associations, regulatory authorities, media representatives, and experts from relevant disciplines.

The process revolved around 20 expert reports as a starting point. Participants met during a total of 10 days over the year-and-a-half time period and discussed and debated various aspects of the technology, including its risks and benefits, and the alternatives to this technology. The process is essentially a discursive and investigatory one, arriving at some description on the state of knowledge on a given controversial subject. In the words of the report, "a participatory technology assessment commits the parties of the political conflict over new technology to cooperation in a procedure with an uncertain outcome" (Van den Daele, 1995, p. 30). It may have been this uncertainty that prompted the environmental groups to withdraw from the process close to the end. Although no conclusive arguments against the use of transgenic herbicide resistant plants were formulated, there was consensus in the group that a significant issue arising from the process is the quest for democratic control of the process of technological innovation.

In the last few years, Germany has tried to regain its competitive position in biotechnology and has "re-established a working state-industry partnership that formally bowed to citizen concerns but closed down the kind of open-ended political debate that had preceded the enactment of the genetic engineering law" (Jasanoff, 1995, p. 128). With state and federal funding, centers for technology assessment were created to conduct research on relevant technologies, facilitate processes of political mediation on technological issues, and to conduct public education and participation activities. Representation in these centers are broadly based. For example, the Center for Technology Assessment in Baden-Wuerttemberg has two oversight committees representing such social organizations as trade unions, churches, industry associations, political parties, and scientists.[8]

United Kingdom

There are a number of advisory committees in the U.K. dealing with various aspects of biotechnology and genetics. At least one body has been providing counsel since 1976: this was the Genetic Manipulation Advisory Group (GMAG), which earlier on provided advice on health and safety of workers working in genetic labs. This committee's name was changed in 1984 to the Advisory Committee on Genetic Manipulation (ACGM), with its focus on worker safety essentially remaining the same. With contentious issues associated with release of genetically modified organisms, another advisory body was created called the Advisory Committee on Releases to the Environment (ACRE). This advisory body also exercised regulatory control in terms of providing approval to GMO release field trials. ACRE currently has statutory powers under the U.K.'s 1992 Environmental Protection Act, providing advice to the environment and agriculture ministers and the Health and Safety Commission. In the area of GMO foods, a different advisory body is operational, called the Advisory Committee on Novel Foods and Processes (ACNFP).

On the medical biotechnology side, two committees, both non-statutory, were created to provide advice to the Department of Health: the Gene Therapy Advisory Committee (GTAC) and the Advisory Committee on Genetic Testing (ACGT). The former was established in response to proposed gene therapy trials for cystic fibrosis. A Human Genetics Advisory Committee was created in 1996 "to consider broad social, ethical, and/or economic consequences of developments of human genetics, for example, in relation to public health, insurance, patents, and employment. They will also advise on ways to build public confidence in the new science."[9]

In terms of larger social concerns and public representation, these Advisory bodies are broadly representative. Their membership consists of experts *and* non-experts, with the latter including a range of stakeholders (e.g., consumer and environmental groups, or a medically vulnerable group) and the media. The mandates of these advisory bodies also include public understanding and public consultation.

Two notable activities involving public discussions and debate on biotechnology include the first U.K. consensus conference on plant biotechnology in 1994 and a National Biotechnology Conference in March, 1997. In general, one might characterize the institutional arrangements on socio-ethical issues in the U.K. as primarily expert-driven, located within the format of broadly-based advisory committees, with representation and input from other stakeholder groups.

European Union

At the European level, the European Federation of Biotechnology has had in place since 1991 a Task Group on Public Perceptions of Biotechnology, with members from industry, the media, communications and survey research organizations, and environmental and consumer organizations from almost all European countries. The mandate of this task force includes: promotion of greater public awareness and understanding of biotechnology; review of current knowledge of the relationship between biotechnology and society in order to identify deficiencies and recommend appropriate research; identification of target audiences, issues and messages; and development and implementation of strategies to meet these objectives. The Task Force has produced a set of briefing papers, a body of research, data bases on public opinion, and other information resources for the general public.

The EU also has an advisory body, a Group of Advisors on the Ethical Implications of Biotechnology, established in 1991. Consisting of nine members, this body includes geneticists, philosophers, theologians, physicians, scientists, and lawyers.

What can be drawn from this overview of approaches in Europe? While the information environments have varied considerably, especially according to political culture differences and the specific regulatory context within which biotechnology has developed, several attributes of the European approaches can be identified.

First, there has been a concerted examination of a range of technologies – including biotechnology – in their larger socio-political contexts. Technological innovation and development have thus tended to be more explicitly and systematically linked to, rather than decoupled from, the spectrum of social, political, and ethical human values.

Second, technology assessment processes have also tended to be broadly based and consensual, resulting in greater public involvement much earlier on in the process. Technological development has thus tended to be more democratized. While there have been variations in procedures in the countries briefly reviewed here, this observation generally holds. For example, while the U.K. was characterized as being expert-driven and while mechanisms for broader public involvement have been more limited there than in Denmark or the Netherlands, advisory committees have also been broadly based to include a range of stakeholder interests beyond government, industry, or science. In Germany, the social interests engaged in the process of technology assessment can also be characterized as highly diverse.

Finally, the process of technology assessment employs two complementary processes: that of enhancing public awareness and understanding and that of facilitating public involvement and participation.

United States

Activities in the U.S. designed to enhance or shape the market for credible information can be described as "multi-channel, multi-level, multi-actor." Various interests have used a range of fora including the media and the courts to influence biotechnological directions. The actors involved in actively shaping the information environment are equally diverse, including federal, some state, and some local governments, professional associations (mainly scientific), industry groups, some universities, as well as a range of public interest groups.

Among the initiatives that have provided a broad base of information on biotechnology including discussions on social, ethical, and environmental issues, a number have come from the university sector. The National Agricultural Biotechnology Council (NABC) was formed in 1988 by four agricultural research, extension, and educational institutions as a consortium to provide a neutral forum for open dialogue on agricultural biotechnology issues. It now consists of some twenty member universities in the U.S. and Canada. The forum holds workshops whose proceedings are subsequently published. The workshops are notable for their exploration of diverse viewpoints. For example, one workshop explored biotechnology's potential for improving the world's food supply but at the same time, also examined the contradictions in the positive claims regarding the increased world food production and productivity of grains. These contradictions include the significant structural changes in food and fiber production, international trade and cooperation, changes which historical contexts have shown have not always been beneficial.

Individual universities have also established successful public education programs on biotechnology. Most notable among these are the University of Wisconsin's Biotechnology Education Program for the Public (BCEPP) and Iowa State University's Public Education Program on Biotechnology. Both programs act as information resource centers, provide educational resources for schools and the general public, and run active websites.

Food industry and nutrition groups have played an active role in promoting biotechnology. The Grocery Manufacturers of America (GMA) and the Produce Marketing Association (PMA) have issued position papers expressing confidence in the safety of foods developed through biotechnology (Grocery Manufacturers of America, 1992). The American Dietetic Association and the Institute of Food Technologists have equally been supportive (American Dietetic Association, 1993). These professional organizations have been active on both the public education and regulatory fronts, supported by the size and resources of their organizations. The Institute of Food Technologists, for example, coordinated a scientific review of a policy proposed by the Environmental Protection Agency to regulate "plant pesticides" (or plants with genetically engineered traits for pest resistance). The review was conducted by representatives from 11 professional scientific societies including such groups as the American Institute of Biological Sciences, the American Society for Microbiology, and the Crop Science Society, and recommended against the use of statutory oversight mechanisms. In a similar vein, the Institute made successful interventions at the local level to stop the Chicago City Council from instituting labelling requirements.

There has, of course, been visible opposition in the U.S., most vocal being Jeremy Rifkin's Foundation for Economic Trends. Other interest groups have expressed their positions as a number of issues have come and gone (see, for example, cases described in Krimsky & Wrubel, 1996). For a variety of reasons, many of the controversial cases on biotechnology products such as ice-minus bacteria or rBST have so far been debated in the media fora or fought in judicial arenas.

Another U.S. initiative at the government level has been the creation of a high-profile Bioethics Advisory Committee. This Committee's purview relates to "bioethical issues arising from research on human biology and behavior, and applications, including the clinical applications, of that research" (National Bioethics Advisory Commission Charter, July 26, 1996). These issues include the protection of the rights and welfare of human subjects as well as issues relating to the management and use of genetic information including human gene patenting. It is not clear at this point how effective this Advisory Committee is going to be in the long run. In the short term, however, the presence of such advisory bodies has been highly useful, particularly as high-profile issues break out, such as the first mammalian cloning effort (Dolly). President Clinton turned to his Advisory Committee with a request for guidance on the issue, as did the British government to their own Advisory Committee.

Finally, the U.S. has, for more than two decades, employed the multi-stakeholder process, particularly in attempts to resolve environmental disputes. Government agencies have taken the responsibility to formally convene multi-stakeholder groups, organize the proceedings, and facilitate a consensus-building process that can be and has been the basis for legally binding rules under administrative law. Called "regulatory negotiation" or "negotiated rule-making," the process has been used on environmental policy issues at the regional or federal level, where representatives of interest groups are brought together "to identify areas of agreement and disagreement and develop mutually agreeable proposals for policy makers" (Pritzker & Dalton, 1990, pp. 36–37). Negotiated rulemaking has been in use in the U.S. for close to 20 years on such diverse issues as penalties for vehicle emissions (EPA) and benzene exposure at the Occupational Safety and Health Administration.

The highly pluralist, activist nature of U.S. politics is reflected in the information environment for biotechnology. There is a high level of stakeholder activity covering a full range of interests and viewpoints. Industry associations are strong but so are organizations representing consumer and environmental interests. What differentiates the U.S. context from the European one is that the former is essentially reliant on forces in the marketplace to play important roles in information dissemination. Many organizations with interests in biotechnology are able to mobilize a range of resources to try to influence opinions of various publics or to influence regulatory directions. Other than individual federal agencies exercising consultative or educational roles, the state (i.e., the federal government) does not play a primary role in establishing mechanisms for public participation and debate.

INDUSTRY INITIATIVES ON BIOTECHNOLOGY

There has been a considerable number of initiatives undertaken by the biotechnology industry. Two generic approaches have been utilized by the biotechnology industry to target the public: the product-specific approach and the general-technology approach. The product-specific approach describes the promotional approaches around a particular product. Despite regulatory approval of bovine somatotropine by the U.S. FDA, there was considerable public controversy over hormone-treated milk from diverse consumer, small farmer, and animal welfare groups as well as large segments of the general public. This encouraged Monsanto to invest heavily in marketing BST directly to dairy farmers and to advertise to consumers in the mass media (Krimsky & Wrubel, 1996). Promotional campaigns were also conducted in conjunction with the introduction of the early consumer products such as the Flavr-Savr tomato.

There are, of course, other "signalling" devices available to industry such as advertising and labelling. In some instances, labelling has occurred voluntarily, working as a tool for information as well as for promotion. The first genetically engineered product on the market in the U.K., canned tomatoes, have been labelled thus by the grocery store chain, Sainsbury's: "Made with genetically modified tomatoes." A separate boxed message on the can says: "The benefits of using genetically modified tomatoes for this product are less waste and reduced energy in processing."

Novo Nordisk of Denmark provides the following information on its genetically engineered human insulin product: "Biosynthetic human insulin is produced by recombinant DNA methods using S. cerevisiae (baker's yeast) and followed by unique purification processes. Human insulin (biosynthetic) is structurally identical to natural human insulin."

In Canada, Monsanto Canada is market-testing transgenic potatoes in selected supermarkets with a voluntary label about being genetically modified. Not surprisingly, many segments of the industry have indicated that labelling could be detrimental to sales because of potential consumer rejection.

The more general approach can be characterized as promotional investments directed at opinion leaders, the schools, and the general public. The idea behind the first approach is to "educate" opinion leaders about the technology in general, with information about specific applications as part of this general education approach. Monsanto Canada has used this approach with its food dietitians network. Dietitians from various provinces attend a fully-funded workshop designed to address health and nutrition aspects of food biotechnology. The company has also provided financial support to the Agriculture in Education program of the Key Foundation for production of educational resources on biotechnology. Special-event sponsorships are commonly used as is the production of informational materials for schools.

An indirect approach is for industry to participate via membership in an industry association. BIO in the U.S. and the Canadian Institute of Biotechnology (now called BioteCanada, based in Ottawa) are industry associations and are actively engaged in programs designed to promote the technology.

OTHER STAKEHOLDERS AS PUBLIC/CONSUMER REPRESENTATIVES

Overall, public and consumer representation in biotechnology remains inadequate. The primary consumer organization, the Consumers Association of Canada, is a nongovernment association with a membership base of 10,000, and a staff of five full and part-time employees, and it has to rely primarily on a volunteer network across the country. It produces a newsletter three times a year, in addition to distributing the U.S. Consumers Reports magazine. Its primary activities are policy interventions on health, food, communications, and financial services, areas of current concern to its membership. It does not have sufficient resources for these types of activities and is highly dependent on federal government support, primarily from Industry Canada. The same can be said of environmental organizations in this country.

Other professional organizations also contribute to the pool of information on biotechnology. The National Institute of Nutrition, for example, is an organization that sees itself at the interface between nutrition science, consumers, and industry. NIN is part of the National Biotechnology Network Secretariat, a "network of networks" which ensures coordinated action in health and environmental areas. The organization aims to educate the public on the science of nutrition (implicated in this instance would be food biotechnology) and to educate industry and government about consumer concerns (National Institute of Nutrition, 1996).

In general, the representation of consumer interests specifically and the public interests generally through organizations such as consumer and environmental groups occurs on a playing field that is highly uneven. These groups are poorly resourced and, while often invited to provide input into regulatory issues, have little influence in shaping regulatory policy when compared to industry.

INITIATIVES FROM THE CANADIAN PUBLIC SECTOR

In general, approaches in the domain of creating a "market for credible information" can be described as ad hoc. A variety of approaches have been used by different ministries that range from tapping the public pulse to consultative mechanisms that include multi-stakeholder consultation meetings, technical workshops, or townhall meetings.

Environment Canada has used multi-stakeholder negotiations as part of its efforts to draft the Canadian Environmental Protection Act and of more recent efforts to revisit the Act and to include a section on biotechnology. Other stakeholder negotiations have been conducted over the major overhaul of the pest management regulatory framework.

Agriculture and Agri-Food Canada sponsored a number of workshops on biotechnology regulation, labelling, and novel foods that also included a variety of stakeholders. Efforts have been made to ensure that consumer and environmental interests as well as those from other sectors – organic farming associations, animal welfare, labour – were heard and considered. In some instances, some media representatives were also invited to attend the working group sessions. Other workshops were co-sponsored with other ministries such as Health, Industry, and Fisheries and Oceans (see, for example, Agriculture and Agri-Food Canada, 1993, 1994).

At the provincial level, a few provincial governments have begun to establish mechanisms for public education and consultation. Saskatchewan has established a successful body, Ag-West Biotech, which is primarily publicly funded and has responsibilities in research, commercialization, and public education efforts. A similar effort in British Columbia is the BC Biotechnology Alliance, an industry association supported almost entirely by the provincial government, and with similar mandates to speak for provincial biotechnology interests and conduct public education activities.

Recent efforts to look more critically at issues related to biotechnology have emanated from House of Commons parliamentary committees. It is these efforts that have raised more critical questions about the technology while attempting to engage the public and to account for public questions and concerns. For example, the House of Commons Standing Committee on Human Rights and the Status of Persons with Disabilities recently completed a consultation process on Privacy Rights and New Technologies, with genetic screening as one of the technologies in question. This process involved town hall meetings in a number of sites around the country, with the meetings structured around scenarios involving video surveillance, smart cards, and genetic testing technologies. The Committee's report raised questions about the adequacy of protections afforded against invasion of privacy or against discrimination when these technologies are employed (House of Commons Standing Committee on Human Rights and the Status of Persons with Disabilities, 1997). Other House parliamentary committees have similarly conducted serious inquiries into and broad consultations on the applications and regulation of biotechnology (Standing Committee on Agriculture, April, 1994; Standing Committee on Environment and Sustainable Development, November, 1996).

ENHANCING THE MARKET FOR CREDIBLE INFORMATION

In this paper, I have focused on the issue of credible information in the marketplace. I first explored the nature of "credibility" for consumers. I then discussed biotechnology products as a category of goods requiring higher levels of credible information in the marketplace. Different approaches to raising or maintaining levels of credible information were then examined in terms of the experience of comparable but mature industries, the experiences of other countries, and the Canadian context.

In this final section, I will discuss some of the challenges of the Canadian context that need to be considered when examining the social system for credible information. I will then describe the attributes of and the challenges posed by a credible social information system.

Recognizing the Challenges to New Strategies

Before making recommendations on strategies for enhancing the information environment on biotechnology, it is important to understand the context for these recommendations, including the challenges to public information, consultation, and participation. These are challenges that exist in connection with public discussions on "value-laden scientific disputes" in Canada (Science Council, 1982):[10]

1. The difficulty of communication and diversity of interests due to the size and regional differences of this country.
2. The powerful economic incentive for industrialized nations to seek and exploit technological innovation, even in the face of recognized scientific uncertainties and ethical dilemmas.
3. Overlapping responsibilities of various levels of governments, resulting in duplication in some areas, neglect of others.
4. The lack of interest or the skepticism that scientific issues generate amongst the general public.
5. Inadequate research or experimentation into ways to resolve value-laden scientific disputes. The Canadian government has not done enough to support or encourage consensus-finding mechanisms.
6. The lack of an agreed-upon theory and methodology even to determine what the value choices are, and to establish norms, and to do so in a fair and consistent manner.
7. The low priority given to early warning and preventive mechanisms in the decision-making process.
8. The short-term nature of political attitudes which hinders our ability to deal with long-term scientific and ethical problems.

9. The frequent absence of formal procedures specifying required steps, including public participation, in the preparation of regulations.
10. Insufficient encouragement or funding of public interest groups to allow for fair representation of diverse interests.

This set of factors echoes much of the earlier discussions. The size of the country and the significant regional differences that exist pose more challenges to public participation mechanisms. Hearings held around the country will be more likely to take account of these regional interests but will also tend to be very expensive. On the other hand, one can rarely do justice to public input and participation at a national conference.

There are different degrees of activity at the provincial level at the moment but for the most part, provincial initiatives on biotechnology are just starting to take shape. Coordination among federal and provincial levels in terms of information provision and public consultation is only just beginning but could be pursued more systematically.

The first National Biotechnology Strategy has focused solely on developing the biotechnology industry in terms of its financial and research and development capacities and only recently has more attention been focused on socio-ethical and public participation dimensions. This has resulted in responses to public concerns that have been stop-gap and reactive. The active resistance to rBST in Canada is an example where public opinion has been vocal and effective but also illustrates the inadequacies in the current system in terms of mechanisms for public participation.

Attributes of Best Practices

In considering the question of what might be considered excellent approaches to enhancing the environment for credible information, I will first lay out my conceptions about technology and society. Already in earlier sections, I have discussed the idea of the significance of the social, political, and cultural context of technology and the importance of analyzing technology within this larger web.

New technologies which have been identified as "strategic" such as information technologies and biotechnology provide avenues for the production and distribution of wealth and have been promoted vigorously by many national governments. At the same time, control or regulatory functions are being exercised to ensure that goals of safety and equity are met and to allay public concerns about risk. These functions of promotion and regulation can be seen to be competing but they may also be viewed as complementary. That is, technology in society is essentially "the co-production of technology and its effects" (Rip, Misa, & Schot, 1995, p. 3). If the goal is to develop technologies with desirable positive impacts and fewer negative effects, an ideal strategy would be to "integrate the anticipation of technological impacts with the articulation (and promotion) of technology development itself" (Rip et al., 1995, pp. 3–4), a process called "constructive technology assessment." What attributes might characterize this process?

First, public information and public involvement processes need to take place as early in the evolutionary cycle as possible.

Second, the effort requires nothing less than the mobilization of networks of social actors in the process. These networks exist along the entire opinion spectrum and as many as possible need to participate or be represented "at the table." When a dominant voice is consistently being heard, measures need to be taken to ensure that other voices have the capacity to be heard as well. On the government side, better integration of activities between various levels of government ought to be instituted. Scientist networks as well as citizen networks need to be active participants as well.

Third, support needs to be available both for the mechanisms of public input and participation and for potential intervenors or participants in order to ensure equity, fairness and access. While Canada has extensive experience with multi-stakeholder consultations, we need to explore the use of other participatory mechanisms. It is easy to dismiss approaches in countries such as Denmark and the Netherlands as being "irrelevant" to the Canadian context because these countries are small and homogeneous, but these approaches can be modified and at least experimented with so as to make them more suitable to the Canadian context. For example, instead of national consensus conferences, a regional model could be examined.

This effort to explore mechanisms of public input and participation ought to be part of an institutionalized process, as has been employed by other regulatory agencies such as those overseeing energy and communications.

Fourth, the regulatory process can only benefit from greater transparency. Part of the public education process needs to focus on the mechanisms for assuring safety and for dealing with current uncertainties. While firms will naturally want to protect their competitive positions by keeping as much information classified, some have found that only small portions of their submissions (e.g., risk information) are truly sensitive and that their credibility is more seriously damaged by secrecy.

Fifth, the regulatory and political processes can benefit from a high-level advisory system or body which can be called on to provide expertise on issues or questions as they arise. Such an advisory body should be broadly constituted to represent the diversity of interests involved. It can further be an instrument for building public confidence. In many countries like the U.S., the U.K., or Norway, such advisory committees have contributed to fostering public discussion and dialogue, in addition to providing expert advice when needed (Moreno, 1995).

Sixth, an education process needs to be implemented that makes full recognition of public concerns and is applied not just to members of the public but to regulators and other stakeholders. The latter need to understand processes of risk perception and communication so that they do not succumb to the old myth that members of the public are empty vessels, waiting to be filled with wisdom. There are legitimate concerns, moral issues, ethical questions, and competing social values invoked in public discussions of technology.

Unresolved Issues

A number of issues loom on the horizon or are now with us but remain unaddressed. The adoption of a Regulatory Framework for Biotechnology in 1993 included stipulation of the principle that biotechnology products must have *environmental and human health and safety assessments before commercialization.* However, the Framework does not provide any mechanisms for taking account of socio-ethical issues. This is a gap that clearly needs to be addressed.

The recent international attention to the issue of cloning is focusing public attention on processes and outcomes of biotechnology as never before. Media attention, not only on domestic issues but also on events occurring outside Canada, will continue to keep biotechnology on the public agenda. Because there are other issues which also are widely recognized and discussed internationally such as biodiversity and environmental sustainability, the linkages of these issues to biotechnology will become more prominent, posing challenges as well as opportunities for public debate.

Finally, with a number of products gearing up for market entry (e.g., transgenic potatoes), issues of labelling and regulatory assurance are going to become even more prominent. The perception of the inadequacy of current regulatory mechanisms is widespread. Although still in its infancy, a Citizens' Biotechnology Network of Canada, organized by the Federation Nationale des Associations de Consummateurs du Quebec (FNAQ), has been created "to pursue a common interest in mandatory labelling of all products derived from biotechnology," another indication of the perceived insufficiency of current measures.

It is clear that if the market for credible information is to be enhanced, the efforts require nothing less than the involvement of all the relevant social actors. Institutional actors within the regulatory systems as well as market structures (such as industry mechanisms for addressing credibility issues) should be mobilized in this effort. Federal regulators and industry will continue to be active in these processes but the public needs to have a more prominent place at the table. To leave the management of credible information solely to forces in the marketplace would not do justice to the nature of the technology, the requirements of industry for successful commercialization, or to what we know about Canadian public attitudes and expectations. Fostering public confidence in biotechnology will demand nothing less.

NOTES

[1] The Canadian Environmental Protection Act (CEPA) definition of biotechnology is: "The application of science and engineering in the direct or indirect use of living organisms or parts or products of living organisms in their natural or modified form."

[2] Stepwise multiple regression analysis was used to examine the impact of perceived utility, risk, moral acceptability on the respondent's position about whether the application ought to be encouraged. In the case of all six applications, these three factors were significant predictors. Interestingly, whether or not the respondent had heard of the application had no impact on the same dependent variable. This survey was conducted by telephone between February 5 and 15, 1997. One thousand randomly selected respondents participated in the survey, with fieldwork carried out by Ekos Research Associates. The study was supported by a grant from the Social Sciences and Humanities Research Council and the Canadian Institute of Biotechnology.

[3] Information was provided by the following for this section on the energy industry: Simone Marlier, Alberta Energy and Utilities Board; Dennis Tremblay, National Energy Board; Len Bradley, Executive Director, Petroleum Communication Foundation.

[4] This section on the Canadian chemical industry's Responsible Care program was compiled from documents on the program and interviews with the following individuals: Vice-President of Responsible Care, Brian Wastle, Frank Marston of the Alberta division of the Canadian Chemical Producers Association, and Dr. Steve Hrudey, University of Alberta and member of the National Advisory Panel of the CCPA.
[5] Information on the Dialogue Forum was provided by Prof. Erling Jelsøe of Roskilde University, Denmark.
[6] Much of the initial information on the Dutch context was provided by Anneke Hamstra of SWOKA, a consumer research organization in Amsterdam. Further information was provided by the consumer organization *Consument & Biotechnologie*.
[7] The guidelines developed show agreement on both procedures (the needs for timely information and for consultation are specified and spelled out procedurally) and content. The labelling agreement suggests that products requiring labelling include transgenic animals; transgenic plants (genetically modified plants which either contain proteins which do not occur naturally or whose consumption properties have been changed); genetically modified micro-organisms; ingredients obtained from genetically modified plants which provide new consumption properties to the end product; and additives originating from genetically modified organisms. No labelling is required for: processing aids which do not occur in the finished product or which occur in inactivated form and does not give the finished product new properties; additives derived from gmo's but identical to existing additives; and ingredients which, based on existing labelling regulations are not subject to compulsory declaration; and products originating from animals fed with animal feeds which contain gmo's, except where this relates to growth-promoting substances such as BST. A third category includes products requiring further consultation such as chemically identical ingredients (such as oil from modified oilseeds) (Informal Consultation Group on Biotechnology, 1996a).

The author thanks Arie van Genderen of Consument and Biotechnologie for her assistance on this section.
[8] The assistance on this section of Dr. Jürgen Hampel of the Center for Technology Assessment, Baden-Württemberg and Dr. W. Van den Daele of Science Center Berlin is gratefully acknowledged.
[9] Press Release, Ministry for Science and Technology. December 2, 1996.
[10] I have adapted the Science Council's original list of 17 items to include the 10 most relevant to the focus of this paper.

REFERENCES

Agriculture and Agri-Food Canada (1993). *Workshop on regulating agricultural products of biotechnology*. November. Ottawa: Agriculture and Agri-Food Canada.
Agriculture and Agri-Food Canada (1994). *Technical workshop on the labelling of novel foods derived through genetic engineering. Proceedings*. November. Ottawa: Agriculture and Agri-Food Canada.
Alberta Energy and Utilities Board (1989). *Public consultation and participation*. Calgary, Alberta: AEUB.
Alberta Energy and Utilities Board (1993). *Oil and Gas Developments, Eastern Slopes – Information Letter*. Calgary, Alberta: AEUB.
American Dietetic Association (1993). *Press Release*, November 8, Chicago, IL.
Biotechnology and the European Public Concerted Action Group (1997). Europe ambivalent on biotechnology. *Nature*, *387*, 845–847.

Bord, R., & O'Connor, R. (1990). Risk communication, knowledge and attitudes: Explaining reactions to a technology perceived as risky. *Risk Analysis*, *10*, 499–506.
Bradley, L. (1997). Personal comunication. January 23.
Canadian Chemical Producers' Association (1995). *Reducing emissions*. Ottawa: CCPA.
Canadian Chemical Producers' Association (1996a). *Does Responsible Care™ pay?* Ottawa: CCPA.
Canadian Chemical Producers' Association (1996b). *Responsible Care™ Way of Life: expectations of member and partner companies*. Ottawa: CCPA.
Canadian Chemical Producers' Association (1996c). *Are we there yet?* Ottawa: CCPA.
Canadian Chemical Producers' Association (1996d). *Responsible Care™: A total commitment*. Ottawa: CCPA.
Canadian Institute for Environmental Law and Policy (1995). *Citizen's guide to biotechnology*. Toronto: CIELAP.
Consumers Association of Canada (1995). *Background paper on food biotechnology in Canada*. Prepared for the National Roundtable on the Environment and the Economy, Ottawa, April 29.
Covello, V. T., McCallum, D. B., & Pavlova, M.T. (1987). *Effective risk communication: The role and responsibility of government and nongovernment organizations*. New York: Plenum.
Covello, V. T., Sandman, P., & Slovic, P. (1991), Guidelines for communicating information about chemical risks effectively and responsibly. In: E. Mayo & R. Hollander (Eds.), *Acceptable evidence: Science and values in risk management*, pp. 95–118. New York: Oxford University Press.
Creative Research International (1996). *Environmental applications of biotechnology: Focus groups*. Report to Environment Canada. Ottawa: Environment Canada.
Decima Research (1993). *Report to the Canadian Institute of Biotechnology on public attitudes toward genetic engineering*. Ottawa: CIB.
Einsiedel, E. F. (1997). *Canadian attitudes to and perceptions of biotechnology*. Calgary: University of Calgary. Unpublished report.
Ekos Research Associates Inc. (1996). *Focus groups on agri-food: Applications of biotechnology. Report to Consumer Affairs, Industry Canada*. Ottawa: Industry Canada.
Féderation nationale des associations de consummateurs du Quebec (1996). *To protect our quality of life and that of our children: Recommendations for the development and marketing of biotechnology*. Ottawa: Office of Consumer Affairs.
Grocery Manufacturers of America (1992). *Press Release*, July 18, Washington, D.C.
Grundahl, J. (1995). The Danish consensus conference model. In: S. Joss & J. Durant (Eds.), *Public participation in science*, pp. 31–40. London: Science Museum and EC DG XII.
House of Commons Standing Committee on Human Rights and the Status of Persons with Disabilities (1997). *Privacy rights and new technologies consultation package*. Ottawa: Library of Parliament Research Branch.
Hovland, C. I., Janis, I. L., & Kelley, H. N. (1953). *Communication and persuasion*. New Haven: Yale University Press.
Informal Consultation Group on Biotechnology (1996a). *Market introduction and labelling of foods produced with the aid of modern biotechnology*. Amsterdam: Informal Consultation Group on Biotechnology. Unpublished report.
Informal Consultation Group on Biotechnology (1996b). *Checklist Novel Foods I*. Amsterdam: Informal Consultation Group on Biotechnology.
Jasanoff, S. (1995). Product, process, and programme: Three cultures and the regulation of biotechnology. In: M. Bauer (Ed.), *Resistance to new technology: Nuclear*

power, information technology, and biotechnology, pp. 185–210. London: Cambridge University Press.

Jelsøe, E. (1997). Roskilde University. Personal communication. March 30.

Joss, S., & Durant, J. (1995). *Public participation in science: The role of consensus conferences in Europe*. London: Science Museum and EC DG XII.

Kluver, L. (1995). Consensus conferences at the Danish Board of Technology. In: S. Joss & J. Durant (Eds.), *Public participation in science: The role of consensus conferences in Europe*, pp. 41–49. London: Science Museum and EC DG XII.

Krimsky, S., & Wrubel, R. (1996). *Agricultural biotechnology and the environment: Science, policy and social issues*. Urbana, IL.: University of Illinois Press.

Laband, D. (1991). Search vs. experience goods. *Economic Inquiry, 34*, 497–509.

Marlier, S. (1997). Alberta Energy Utilities Board. Personal communications. January 24.

Moreno, J. D. (1995). *Deciding together: Bioethics and moral consensus*. New York: Oxford University Press.

Mullan, F., & Jacoby, I. (1985). The town meeting for technology: The maturation of consensus conferences. *Journal of the American Medical Association, 254*, 1068–1072.

National Institute of Nutrition (1996). *Annual report*. Ottawa: NIN.

Optima Consultants (1994). *Understanding the consumer interest in the new biotechnology*. Ottawa: Office of Consumer Affairs.

Pritzker, D., & Dalton, D. (1990). *Negotiated rulemaking sourcebook*. Washington, DC: Administrative Conference of the United States.

Renn, O., & Levine, L. (1991). Credibility and trust in risk communications. In: R. Kasperson & P. Stallen (Eds.), *Communicating risks to the public*, pp. 175–218. Dordrecht: Kluwer Academic Publishers.

Rip, A., Misa, T., & Schot, J. (1995). Constructive technology assessment: A new paradigm for managing technology in society. In A. Rip, T. Misa, & J. Schot (Eds.), *Managing technology in society: The approach of constructive technology assessment*, pp. 54–80. London: Pinter.

Roobeek, A. J. M. (1995). Biotechnology: A core technology in a new techno-economic paradigm. In: M. Fransman, G. Junne, & A. Roobeek (Eds.), *The biotechnology revolution?*, pp. 86–102. Oxford: Blackwell.

Science Council (1982). *Value-laden scientific disputes and policy*. Ottawa: Supply and Services.

Teubal, M., Yinnon, T., & Zuscovitch, E. (1991). Networks and market creation. *Research Policy, 20*, 381–392.

Toft, J. (1996). Denmark: Seeking a broad-based consensus on gene technology. *Science and Public Policy, 23*(3), 171–174.

Turner, J., & Wynne, B. (1992). Risk communication: A literature review and some implications for biotechnology. In: J. Durant (Ed.), *Biotechnology in public*, pp. 109–141. London: Science Museum.

Van den Daele, W. (1995). Strategies of dealing with with the risks of genetic engineering: Lessons from a participatory technology assessment in Germany. In: J. Landsmann & R. Casper (Eds.), *Key biosafety aspects of genetically modified organisms*, pp. 121–140. Berlin: Blackwell Wissenschafts-Verlag.

Von Schomberg, R. (1996). Netherlands: Deliberating biotechnology regulation. *Science and Public Policy, 23*(3), 158–163.

Wohl, J. (1998). Consumer decision-making and risk perception regarding foods produced with biotechnology. *Journal of Consumer Policy, 21*. Also included in: B. M. Knoppers & A. D. Mathios (Eds.), *Biotechnology and the consumer*.

THE AUTHOR

Edna F. Einsiedel is Professor of Communication Studies in the Graduate Program in Communication Studies at the University of Calgary. Address: 2500 University Dr. N.W., Calgary, Alberta T2N 1N4. E-mail: einsiede@ucalgary.ca.

Thérèse Leroux, Marie Hirtle, and
Louis-Nicolas Fortin

An Overview of Public Consultation Mechanisms Developed to Address the Ethical and Social Issues Raised by Biotechnology

ABSTRACT. Biotechnology, whether applied to plants, animals, or humans is a rapidly expanding industry. Citizens, often organized in groups or organizations, wish to express their views and even to participate in the decision-making process that authorized to the marketing of new biotechnology derived products. In order to allow participation of the public in the debate on the ethical and social issues raised by biotechnology, it is important to select appropriate consultation mechanisms. Much may be learned from the experiences and mechanisms used in various countries.

Part I of this report surveys various institutional models for public consultation activities. Part II then reviews various mechanisms used to address ethical and social issues with public participation and Part III highlights the strengths and weaknesses of these mechanisms in a Canadian context.

As the National Biotechnology Network (1997) notes, "[s]ince 1994, the number of Canadian companies whose core business is biotechnology has nearly doubled, from 121 to 224 in 1997." These numbers illustrate the rapid growth of this sector of activity. Nevertheless, the expansion of biotechnology ultimately depends on its acceptance by the public; hence the importance of taking into consideration the reactions of the public to these scientific developments. More and more citizens, often in organized groups, wish to express their viewpoints and take part in the decision-making process that authorizes the marketing of products emerging from the biotechnology industry. Already the US-EC Task Force on Biotechnology Research (1992) recognized the importance of the role of citizens, and it recommended that mechanisms be implemented to ensure the participation of all who wished to express views on the development of biotechnology.

The goal of this study is to discuss various mechanisms, established in certain countries, that promote interactions between companies,

Journal of Consumer Policy **21**: 445–481, 1998.
Bartha M. Knoppers & Alan D. Mathios (eds), Biotechnology and the Consumer.

government, public interest groups, and individual citizens. We focus on mechanisms used to examine the social and ethical concerns raised by biotechnology applications with regard to plants and animals as well as human beings.

In the first part of this paper, we describe various institutional models found in other countries for consulting the public. In the second part, we survey different methods that have been used to address the ethical and social issues raised by biotechnology. Drawing on these experiences, the third part highlights strengths and weaknesses of these tools when transposed onto the Canadian situation. This overview should facilitate the choice of appropriate public consultation approaches.

PART I: DESCRIPTION OF VARIOUS INSTITUTIONAL MODELS

Over the past 15 years, many countries have established either permanent or ad hoc structures whose role is to address the ethical and social issues raised by the development of biomedical sciences (Leroux, Le Bris, & Knoppers, 1998) and modern biology. Within the wide array of issues, some bodies have focused on concerns that are more specific to biotechnology. We shall examine examples of permanent or ad hoc national bodies that, in the course of their work, consult, are concerned with, or take into account public opinion in their work. We have excluded the far more numerous institutional bodies that educate and inform the public without however consulting it, for example, France's Comité Consultatif National d'Éthique pour les Sciences de la Vie et de la Santé (CCNE), described and explained by Byk (1993, p. 265).

Permanent Councils

Among the mechanisms developed to examine the ethical and social issues in biotechnology, the permanent national structures of two countries offer interesting public consultation models. Accordingly, we shall focus on the Australian and Danish models, although we have consulted the national structures established in other countries, for example, France's CCNE, Sweden's National Council on Medical Ethics, and the Health Council of the Netherlands. Their mandates do not include public consultation, but this does not exclude these bodies from undertaking activities in the fields of public education and information.

The permanent ethics councils of Australia – the Australian Health Ethics Committee (AHEC) – and Denmark – the Danish Council of Ethics and the Danish Council of Technology Assessment – were established by legislation that also stipulates their operating procedures. The mandates of these national authorities are of particular interest since they create a legal obligation to consult the public and to provide citizens with the opportunity to express their views on the ethical and social issues raised by life sciences in general, including biotechnology.

Australia. The Australian Health Ethics Committee (AHEC) was established in 1992 as a principal, permanent, and independent committee of the National Health and Medical Research Council (NHMRC). The NHMRC is an independent consultative council established by legislation.[1] At the federal level, any health-related ethical question must now be studied by this central committee.[2] The AHEC's mandate notably includes consultation with the public[3] during the drafting of guidelines for research on human subjects, to be adopted by the NHMRC.[4]

The act establishing the AHEC also stipulates that 1 of the 13 members of the permanent committee must be familiar with consumer concerns.[5] In order to carry out its tasks, the AHEC can set up ad hoc task forces and sub-committees as it sees fit.[6]

Denmark. Denmark has two bodies which are specifically mandated to stimulate public debate on biotechnology. One, the Danish Council of Ethics, focuses on problems emerging from biomedicine; the other, the Danish Council of Technology Assessment, is concerned with plant and animal related biotechnology.

The Danish Council of Ethics, constituted by legislation in 1987, is a permanent council mandated to provide the Danish Parliament, official authorities, and the public with ongoing information and notices about ethical issues raised by the national health program as well as by biomedicine. The enacting legislation provides that the Council must inform the public about its work and about developments in the field of biomedical science. When ethical issues arise, the Council must ensure that they undergo public debate.[7] To this end, the Danish Council of Ethics has developed several public debate mechanisms. Some of these mechanisms will be examined in the second part of this paper.

In order to fulfil its legal obligations to promote public debate, the Danish Council of Ethics has developed an information policy that confirms this role. The Council considers that its task is not so much to make known its own positions but rather to undertake initiatives to assess public opinion. Accordingly, the Council's information policy objectives encourage local initiatives for spreading information and provide that public debates be structured so as to promote useful dialogue between participants, while preventing the jelling of antagonistic positions. The Council does not seek to establish consensus but instead promotes the expression of various views. Moreover, its information policy specifies that public debate activities on ethical questions must take other major social debates into account. In practice, debates focus on one or several defined topics and are followed by a series of activities. This approach of linking information activities with debate seeks to provide the public a way to exert an influence on Parliamentary decisions (Danish Council of Ethics, 1992b, pp. 4–5).

Except for the requirement that 8 of the Council's 17 members be chosen for their knowledge of ethical, cultural, and social aspects of work performed by the Council,[8] there are no specific provisions as regards membership representation, be it of the general public or other interest groups. Despite this silence, it is interesting to note that Council members have diverse backgrounds, such as journalists, writers, nursery school teachers, and archivists, as evidenced by the list of members published in Danish Council of Ethics (1995). To complete its expertise, the Council can create working groups to address specific issues such as the 1994 working group on genetic engineering and gene therapy.

In May 1995, the Danish Parliament mandated a separate body, the Danish Council of Technology Assessment, to stimulate public debate on plant and animal biotechnology. This Council originates from a restructuring of the Danish Board of Technology, an independent institution that was created in 1986 and worked in close collaboration with the Ministry of Education. Under the new legislation, the Council of Technology Assessment has become a permanent body; it remains independent but now is under the aegis of the Ministry of Research. Its mandate includes supervision of technological developments, assessing the potential of technologies and their impacts on society and individual citizens, and communicating the results of its work to Parliament, decision makers, and the public. In addition, it must support and stimulate public debate on technology.[9]

The Council of Technology Assessment has a board of directors consisting of 11 members, with the chair and 3 members appointed by the Minister of Research. It also has a board of representatives of 50 people, which meets twice a year.[10]

One should note that the original Board of Technology promoted consensus conferences (one of the mechanisms studied in the second part of this paper) as a means of better understanding public opinion and promoting interchanges between experts, decision makers, and the public. The new Council of Technology Assessment can thus draw on very useful experiences.

Ad Hoc or Provisional Commissions

Some countries have set up ad hoc or provisional commissions instead of permanent bodies. There are a few models that allow public participation in discussion or policy development. For example, such participation is often limited to the opportunity of making interventions rather than a true consultation and taking into account of public opinion. At the national level, ad hoc or provisional commissions are found in the United States and the United Kingdom, to be described below.[11] At the regional level, we shall examine the Nordic Committee on Bioethics, formed by the Scandinavian countries.

The United States. At the U.S. federal level, the ethical and social issues raised by biotechnology for use with humans is generally examined by ad hoc commissions.[12] At present, there are two types of commissions dealing with these matters. The first type falls under the aegis of the joint National Institutes of Health (NIH) – Department of Energy (DOE) Program on the Ethical, Legal, and Social Implications of Human Genome Research. The aim of the program is to foresee and examine the individual and societal issues raised by human genome sequencing, as well as to stimulate public discussions and to draft policies. To this end the NIH-DOE Working Group on Ethical, Legal, and Social Implications of Human Genome Research (ELSI Working Group) was created to address and make recommendations on various issues.[13] Within the ELSI Working Group, the Task Force on Genetic Testing was created in order to prepare recommendations to ensure the safe and effective development of genetic tests, the quality of laboratory analyses, and their appropriate use by health experts and consumers.[14]

The second type of ad hoc commission in the United States is the National Bioethics Advisory Commission (NBAC), established by the President of the United States in October 1995 for a period of two years.[15] Its mandate is to provide advice and make recommendations to the federal government and the public, notably on ethical questions raised by research on human subjects and the application of this research, including clinical applications. Although the NBAC was not specifically mandated to consult the public, it nevertheless has various forms of public involvement. The NBAC has at most 15 members, of which 3 represent the general public and have a type of expertise differing from that of the other members.[16] Furthermore, since this is a federal commission, all NBAC advice and recommendations must be accessible to the public,[17] which also has the right to attend meetings unless otherwise stated.[18] However, we have no information as to how effective participation by the general public is ensured at these meetings.

United Kingdom. In the United Kingdom, there are many bodies that examine bioethical issues. According to a report of the House of Commons Committee on Science and Technology (1996), there are eight bodies that are studying genetics-related issues.[19] Despite the numerous bodies, however, the report notes the need:

> . . . for a forum in which geneticists could engage in constructive discussion with the public. At present this does not exist. There is no one body to which a concerned member of the public can refer for impartial advice or express misgivings even though public concern is rising and public confidence in the regulatory system is essential. Nor is there a body intended to act as source of knowledge with a consistent role in broadening public confidence in genetics.

This observation must not undermine the work of the various expert groups, but it does highlight that the work is often performed by experts, behind closed doors, without the participation of the public.[20] After tabling the report of the Science and Technology Committee,[21] the U.K. government responded in December 1996 by creating the Human Genetics Advisory Commission. The mandate of this consultative body is to advise government ministers on mechanisms for increasing public confidence in, and understanding of, new genetic technologies.[22]

Such ad hoc or provisional bodies also exist at the supranational level. An example is the Nordic Committee on Bioethics.

The Nordic Committee on Bioethics. The Nordic Committee on Bioethics was established by the Nordic Council of Ministers[23] for a period of two years, from 1996 to 1998. The government of each Nordic country (Denmark, Finland, Iceland, Norway, and Sweden) appointed two experts to the Committee. Notably, its objective is to organize conferences and publish reports in order to stimulate and encourage discussion on ethical issues. The Committee has already published reports on the ethical issues raised by the assessment of risks associated with biotechnology, the use of genetically modified organisms, the use of predictive genetic tests and gene therapy (Josefsson, 1996, p. 31).

Sponsored Bodies

Under this heading we shall examine bodies that, without being governmental, are partly financed by public funds and may thus be mandated by government to consult the public. We shall look at the models developed in the Netherlands and the United Kingdom.

The Rathenau Institute. The Rathenau Institute of the Netherlands is an independent body whose mandate is to organize social debates on the issues raised by scientific and technological advances in general, regardless of whether they apply to plants, animals or human beings. In 1994, the Rathenau Institute was entrusted with new responsibilities, and since then it has dedicated less time to its research activities in order to concentrate on its tasks of stimulating social debate and developing public opinion.[24] Hence, its traditional role of technology assessment was extended quite easily to include that of addressing ethical and social issues raised by novel technology. The result has been a broadening of the Rathenau Institute's activities, for reasons and objectives that go far beyond the scope of this paper.[25]

The Rathenau Institute is a small body with approximately 14 employees, which provides support for organizing debate activities. It works in co-operation with other bodies, institutions, and universities identified according to the subject selected. The questions addressed may give rise to diverse debate activities that may run for as long as a year.

While its work agenda is determined by the Dutch Parliament every two years,[26] the Institute decides which questions will be examined under each major theme. Furthermore, it can take the initiative of raising questions that it considers to be neglected by the public or parliament. For example, under the major theme of "Intervention in man, animal, and ecosystem," the question of genetic tests on human beings was studied, as well as that of technology assessment and industry. Next, under the theme "Platform on Science and Ethics," the consensus conference mechanism (examined in the second part of this paper) was used in order to study the question "Predictive genetic research, where are we going?" (Rathenau Institute, 1995b). The consensus conference, called "public debate" in the Netherlands, seeks to offer a forum for individual citizens and organizations wishing to contribute to public debates on science and technology. The role of the Rathenau Institute is to organize and manage activities rather than undertake basic research, which has become the task of other bodies.

The Nuffield Council on Bioethics. The Nuffield Council on Bioethics, established in the United Kingdom in 1991, is an independent body receiving most of its funding from private agencies, with the remainder provided by the Medical Research Council (Nuffield Council on Bioethics, 1992, p. 4). Its mandate is to foresee and study ethical concerns raised by recent advances in medicine and biology. The Nuffield Council seeks to promote understanding and discussion by the public on certain questions (p. 10). It publishes reports and makes representations as it sees necessary in order to influence the decision-making process. The government may ask the Council to study particular issues. In its work, the Nuffield Council is not required to consult the public but it nevertheless undertakes some information strategies.

PART II: OVERVIEW OF TOOLS USED TO EXAMINE ETHICAL AND SOCIAL ISSUES IN BIOTECHNOLOGY

Whatever the institutional model selected to examine the ethical and social issues associated with biotechnology, the body has various means to properly fulfil its mandate. This part of the paper describes some of these tools, which are grouped according to whether they consult the public or rather inform and educate it.

Consulting the Public

Consulting the public can take various forms, some being more structured and complex than others. Mechanisms generally used are notification, distribution, and solicitation of comments, or surveys. A more elaborate model, what we call a sequential consultation, is used in Australia. Furthermore, a fairly recent consultation method, the consensus conference, has been tried in Denmark, has spread to other European countries, and has recently re-surfaced in the United States. Last, we shall highlight the use of new information technologies, such as the Internet, as tools to appreciate/estimate public opinion.

Notification, distribution, and solicitation of comments. The most common public consultation mechanism remains the circulation of preliminary texts to interested groups, persons, or agencies for comment and critique. In Sweden, the government can decide to circulate reports it receives from its commissions to interested groups and organizations. The question of using genetic technology on human beings has been assigned to this type of study.[27] In the United Kingdom, two of the consultative agencies under the U.K. Minister of Health – the Gene Therapy Advisory Committee and the Advisory Committee on Genetic Testing – are not required to consult the public in preparing their recommendations.[28]

In the United States, the documents of certain public agencies must be published in the *Federal Register*.[29] The notification is sometimes followed by round table meetings to discuss comments and criticism. For example, in early 1997, the Task Force on Genetic Testing published a notice in the *Federal Register* to the effect that the second version of its recommendations was available for written comments by the public within a 38-day period. The preliminary report in the form of *Interim Principles* of the U.S. Task Force on Genetic Testing was distributed to interested groups for their comments and made available to the general public via the Internet. Interested groups could present their reactions either orally (within an allotted period of 26 days) or in writing within two months; these comments were required to be considered in the preparation of a second version of the document. The public was invited to participate in the final meeting but could speak only if time permitted; no time was specifically allotted for comments by the public. Given such restrictions, one could conclude that this meeting was held publicly but was not a public hearing. This situation perhaps reflects the distinction made by a U.S. government official:

[t]here is a vast difference on a spectrum that is anchored on the one end by informing the public and allowing for their spontaneous comments and at the other end by seeking specific input from "the public."[30]

Opinion surveys and focus groups. Opinion surveys and focus groups are often used to learn the views of the public on various questions (Priest Hornig, 1994; Zimmerman, Kendall, Stone, & Hoban, 1994). Despite its popularity, the opinion survey has major weaknesses. Of course it gives a snapshot of the situation; it helps to identify the range of opinions at a given moment, but it does not allow interaction with the public. Most countries hold surveys which, if conducted on a regular basis, provide an indication of how public opinion develops.

For example, with the Eurobarometer series of studies, conducted on behalf of the European Commission, it is possible to have a better grasp of Europeans' attitudes and concerns about biotechnology developments. The public opinion survey unit of Eurobarometer organizes opinion surveys of the general public and specific focus groups in all European Union member countries. In 1991, for example, it issued a report entitled *Opinion of Europeans on Biotechnology in 1991*,[31] and in 1993 it published a second report entitled *Opinion of Europeans on Biotechnology/Genetic Engineering in 1993*;[32] these documents show that perceptions vary from country to country as well as with time. The survey findings have been commented on (Algoud & Joly, 1995; Barbot, 1994; Jank, 1995) and combined with similar studies undertaken in Japan (Macer, 1992, 1994) and the United States (Hallman, 1996; Hoban, 1997) in order to better identify public trends.

Sequential consultation. A particular consultation model on health-related ethical questions, developed in Australia, is noteworthy. As mentioned earlier, the Australian Health Ethics Committee (AHEC) is required by law to develop and issue guidelines for research on human subjects, for consideration by the NHMRC. Furthermore, the AHEC must ensure ongoing revision of these guidelines. The law provides that, when possible, the Council should adopt a policy of consulting the public about issues under consideration;[33] accordingly, the NHMRC and its principal committees have designed a public consultation strategy that aims at developing the most appropriate consultation model for each subject or item studied, while taking into account the time and resources available. The consultation strategy can be adapted to take into consideration the particular needs of special groups; for example, the time allotted for participating in the consultation can be adjusted. In any consultation process, the public must be advised of the reasons for the consultation, in addition to receiving all the information documents available as well as the name and address of a contact person at the AHEC.

The consultation process seeks to ensure participation by the general public, including professionals and consumers. This process aims to promote the collection of high-quality, fact-based information and opinions. More particularly, when the NHMRC and its committees intend to propose regulatory changes, or to draft/revise guidelines, the law specifies several steps for consulting the public. This sequential approach involves, first, notifying the general public of the Council's intention to draft/revise notices, guidelines, or recommendations on a particular topic, and next encouraging the public to make comments on the issue. The comments must then be taken into account while drafting the preliminary version of a notice, guideline, or recommendation, and this version will be submitted to the public to allow it to react and make further comments.[34]

In 1996, when work began on revising the guidelines on somatic gene therapy, as well as several other guidelines,[35] the sequential consultation process was being followed.[36] During the first stage, the NHMRC sets up a task force charged with studying somatic gene therapy. Once the task force is formed, its intention of studying the question is made public,[37] explanatory documents are distributed,[38] and comments of the public are solicited.[39] At this preliminary stage, the comments will relate either to already existing guidelines, as in the case of somatic gene therapy, or the general issue and the timeliness of drafting guidelines, when no text exists. Once the first consultation stage is completed, the task force prepares the preliminary version of the guidelines or recommendations, taking into account the comments received and the expertise of the panel. The document prepared must include a list of the groups and individuals who submitted comments, as well as a summary of the points raised.

The next consultation stage focuses on this preliminary version, following a process similar to the first stage. The task force must then prepare the final version of the document, taking into account the comments received. This is then submitted to the NHMRC for approval.

In certain cases, the public consultation process is conducted on an ongoing basis. For example, with regard to gene therapy, the *Supplementary Note on Somatic Cell Gene Therapy* was adopted in 1987 after three years of discussion. As early as 1988, however, it was the subject of a round table conference that had three objectives: (a) to broaden the debate on the subject; (b) to inform governments about developments on the matter; and (c) to test the position that gene therapy carried out in compliance with the guidelines is not ethically objectionable and that research on animals may be undertaken. The guidelines were revised in 1992 and should be revised in 1997.

The consensus conference. The consensus conference model has the special feature that it puts in contact a group of citizens having different and varied activities and interests, and gathers their opinions on specific scientific and technical issues. The holding of consensus conferences is based on the principle that any average citizen having the necessary time and resources to learn about a subject is capable of understanding complex concerns and making sound, well-grounded decisions on that subject. In this way, public debate is carefully sustained and public authorities can direct their work and determine what the needs of the public are in terms of information on these questions and the methods of responding to them.

The operating principles of consensus conferences, as they are known in Europe, were established by the Danish Board of Technology toward the end of the 1980s (de Chenay & Therre, 1995). The conferences must allow certain members of the public who form a panel, to learn about and discuss issues that involve complex scientific and technical aspects

This consultation mechanism differs from more common methods seeking to obtain the viewpoint of the public. It does not rely on interest groups or professional bodies making representations on behalf of their members, nor on elected officials exercising their political power. It involves only individuals drawn from the general public.

After a consensus conference is announced, interested persons may apply to participate. Panel members are selected by the institutional body responsible for the consensus conference. From that point onward, the panel members are in control of the process from beginning to end: They choose the key issues to be examined, the questions to be considered, the experts to be consulted, and the conclusions to be drawn (Joss & Durant, 1994).

In concrete terms, the conferences involve a preparatory phase, which allows panel members to interact and during which they familiarize themselves with the issues under review. Once they formulate the questions that will be submitted to the experts they choose, the panel members may question the experts. They evaluate the experts' responses and reach a consensus on particular topics, which they then communicate during a press conference.

This type of conference is being used increasingly in Europe. In Denmark, for example, consensus conferences organized by the Danish Board of Technology were held on food irradiation in 1989 and on transgenic animals in 1992. Following the Danish example, in 1993 the Netherlands organized a consensus conference on biotechnology applied to animals (Joss & Durant, 1994).[40] Drawing on the consensus conference model, in 1995 the Rathenau Institute organized a public debate on predictive genetic research (Rathenau Institute, 1995b).

In November 1994, a consensus conference on biotechnology applied to plants was held in London.[41] The Science Museum[42] appointed a committee to organize the conference. Participation of the public was solicited through 11 regional newspapers and via local, independent public radio stations. This process was reported in *Nature* magazine that emphasized the call for public participation.[43] Of the 350 candidates who offered to participate, 16 were selected on the basis of socio-demographic criteria such as age, sex, employment, and education. There were two preparatory meetings lasting two days in order to introduce the subject, determine the issues to be studied, and choose the questions to be raised as well as the experts charged with answering them. The panel fixed a schedule stretching over three days, conducted the investigation, and issued a written report giving its opinion on the use of biotechnology on plants. In addition to the foreseeable conclusion that food labeling should be improved in order to help the public in its consumption choices, the panel also agreed on the need for more effective international regulation (Verrall, 1994).

In the United States, the Loka Institute organized a consensus conference pilot project, a National Citizens' Panel largely inspired from European experiences.[44] This pilot project is similar to the Citizens Jury Process that has been used by U.S. organizations since 1974 but, to our knowledge, has never been applied as such to examine biotechnology.

Finally, the Rathenau Institute has developed a different discussion mechanism called a "propositions debate" (Rathenau Institute, 1995a). Although this consultation mechanism resembles the consensus conference, the organizers take more responsibility for the event, decisions are made by voting rather than consensus, and interest groups can contribute between the time when discussions are held and the final conference. The "propositions debate" first of all requires the holding of two preparatory sessions bringing together 20 participants selected from the general public, who are then informed about the chosen topic. After discussions between the participants, the organizers draft propositions on which the participants vote. The organizers also consult interest groups to obtain their opinions on the same propositions. A final public conference is organized, where politicians are invited to participate. Until now, the Rathenau Institute has not used the propositions debate model in the field of biotechnology.

The Internet. A new mode of public consultation warrants review, although with certain reservations. Some bodies make use of the Internet in order to gather opinions from the public. For example, while preparing its report, the U.S. Task Force on Genetic Testing made an appeal over the Internet for first-hand accounts of genetic testing from health professionals, patients, or any other interested person.[45] Similarly, the Nuffield Council on Bioethics posted on the Internet the question currently under study: the ethical, social, and legal implications of research on the genetic aspects of mental illness. The site includes a popularized account of various aspects of the question; the public is invited to submit comments, reports of experiences, and criticisms.[46] An international forum, If gene, has been established to encourage the exchange of ideas.[47] France, the Netherlands, Germany, the United Kingdom, and Switzerland are taking part in this initiative. However, until a study has been conducted on the effectiveness of the Internet and on the representativeness of respondents, it seems premature to describe this mechanism of interaction with the public as genuine consultation.

Informing and Educating the Public

The vast majority of organizations studied are working toward the objectives of educating the public and distributing information. Several tools are used to this end.

The most common remains the holding of conferences on particular themes, to which the public is invited. For example, the Australian Health Ethics Committee organises workshops approximately every two years, in all the capitals of Australia's states and territories. The workshops are open to members of the public, who may receive information kits; one of the aims of this workshop is to provide a forum for discussion of ethical aspects of research on human subjects. The 1993 and 1995 workshops focussed on certain aspects of how research ethics committees operate. According to the report on the 1995 workshop, participation in the workshops was quite varied and the general public made up around one fifth of the audience. Participants were generally very much in favour of this type of activity.[48] A workshop focusing on ethical concerns raised by genetics was planned.

The annual ethics days ("Journées annuelles d'éthique") organized by France's Comité Consultatif National d'Éthique pour les Sciences de la Vie et de la Santé (CCNE) is another very well-known example. The CCNE also organizes regional days ("Journées régionales") to reach the public outside the Paris area (Byk, 1993).

Another favoured education and information tool is the publication of reports and studies produced by various committees and commissions. Increasingly, these documents are available on the Internet.[49]

The most innovative model for distributing information and educating the public undoubtedly is found in Denmark. As mentioned, the Danish Council of Ethics is legally required to ensure public debates on major ethical questions, including those raised by biotechnology. To this end, it has adopted the information policy described above. To implement the policy, the Council uses a strategy that involves a series of events and activities dealing with the ethical issue in such a way as to reach the widest possible audience.

For example, the Council relies heavily on publication of explanatory brochures, books, and newspaper articles, and on debates carried on television, radio, or in the newspapers. Council members give from 200 to 400 oral presentations per year at the local and national level, and also publish newspaper articles (Danish Council of Ethics, 1992b,

p. 6). The Council not only organizes public debates and national conferences, but also promotes the holding of such events at the local level, provided that the community undertakes the organizational work. Its support includes financial assistance. In addition to more traditional tools, the Council is trying new methods: It insists that artists be involved in the information process, for example, through painting competitions or production of short or feature-length films.[50] It also seeks the help of journalists in producing booklets on ethical and legal aspects in a more accessible style. Public libraries are favoured distribution points for this material.[51]

It is interesting to note that various tools are adapted to the target audience. For example, a series of activities aimed at the general public sought to stimulate debate on determining clinical death, while young people were the target audience for debates on protecting human reproductive material and on the use of information obtained from human genome sequencing. Activities in the debate on protecting human reproductive material targeted mainly young people in secondary schools and academic institutions; the Danish Council of Ethics (1992b, p. 10) reasoned that this generation would face decisions involving such issues. Activities aimed specifically at young people included poetry and music competitions broadcast on Denmark's national television network, writing competitions on various themes,[52] multimedia shows, and production of material for use in schools.[53] Furthermore, two long-term projects were launched to reach specific audiences: "Communications on the ward" sought to improve communication between patients and health professionals, while "Secondary school ethics" sought to provide information to tomorrow's decision makers (Danish Council of Ethics, 1992a, pp. 18–20).

Another example in a Scandinavian country, but by a different type of organization, is Norway's National Committee for Research Ethics in Science and Technology (NENT). This is a body comparable to an ethics committee, which has a mandate to inform the public about ethical issues raised by research in fields such as biotechnology (NENT, 1995, pp. 30–38). In 1994, the Committee organized a national technology assessment conference; one of the topics discussed was the participation of the general public in assessment processes (p. 30). At the same time, the Committee was considering the possibility of using the Danish model of consensus conference (p. 34).

Moreover, there are also organizations that promote public education about developments in biotechnology. For example, the European Initiative for Biotechnology Education seeks to improve education at the high school and college levels.[54] Sharing this objective is Britain's National Centre for Biotechnology Education.[55]

If it is better informed, the public will no doubt participate more actively in the debate on the ethical and social issues raised by biotechnology. Its contribution will be all the more reasoned because it will be able to understand the nature of scientific developments and their impacts. Hamon (1979) maintains that information is a means of training and education, thereby encouraging citizens to participate and improving their capacity for judging and reasoning.

After having described the various institutional models and a wide array of methods for examining the ethical and social issues in biotechnology, in the next part of this paper we refer to experiments abroad in discussing the approaches to be chosen here in Canada.

PART III: CONSULTATION TOOLS AND THE SITUATION IN CANADA

We continue our study by systematically looking at each of the various tools for public consultation described earlier. Each will be analysed from three angles: public representativeness, the quality of the information collected, and feasibility. We shall also note the strengths and weaknesses of each, taking into account the Canadian situation. Tables I and II summarize our conclusions.

Notification, Distribution, and Solicitation of Comments

The most widespread public consultation mechanism in the countries studied is the notification, distribution, and solicitation of comments. The relevance of this mechanism to Canada is obvious. Moreover, the Canadian notification system is cited as a model by the Organisation for Economic Co-operation and Development (OECD).[56] Regarded as one of the simplest forms of consultation, notification followed by solicitation of comments nevertheless has limitations that ought to be noted.

Public representativeness. In theory, the general public can be reached through notification in official publications, announcing the government's intention to examine an issue, or presenting the text of proposed legislation. There is no preselection of interested groups or persons, as is generally the case with direct solicitation of comments from certain groups or individuals with recognized interests.

Although notification is public, the weak incentive to participate, "perhaps aggravated by a degree of cynicism about government motives" (Jacobs & Korhonen, 1995, p. 30), can result in effective participation that is uneven, since certain groups are difficult to reach. There is a risk that this type of consultation will be dominated by the best organized interest groups, particularly since they are likely to have access to official publications. Other viewpoints will be overshadowed. An OECD expert explains the situation as follows:

> Even where the consultation procedures are, in principle, open to the wider public, they can be dominated by groups who are familiar with the regulatory structure in question, who are well organized and specialized, and who are experienced in making themselves heard (Jacobs & Korhonen, 1995, p. 30).

Solicitation of comments on increasingly technical and complex problems or texts is likely to introduce an additional constraint influencing the level of public response to the call for consultation. The risk is that the public contribution will come mainly from groups of experts and scientists, opening the door to more technocratic decisions.

Hence, notification, distribution, and solicitation of comments can, in theory, reach the general public, but in practice those who express themselves are groups and people familiar with the issue. This aspect must be considered by those responsible for conducting a consultation.

Quality of information collected. To be effective, the mechanism of consultation by notification, distribution, and solicitation of comments requires that the question asked or the text submitted be very clear and well defined. The comments of the public should also be targeted, thereby offering valuable information to the consulting organization. Comments transmitted in writing encourage groups and individuals to deepen and elaborate their thinking. In contrast, however, allowing comments to be submitted on audio tape can widen the circle of participants to include those for whom writing comments is an obstacle to expressing their concerns. If the consultation mechanism that is used seeks to gather high-quality information representing diverse viewpoints, it tends to promote discussion of the facts and assessment of the documents submitted by the authorities, rather than the political acceptability of a proposed action (Jacobs & Korhonen, 1995, p. 29).

Also affecting the quality of information collected is the time chosen to notify the public and solicit comments. For example, if the consultation process is launched at the time a bill is tabled, as is often the case, the regulatory process can already be too engaged for a genuine discussion on basic aspects such as defining the problem itself and alternative approaches.[57] This is one of the reasons why the system of notification and solicitation of comments has been criticized: Despite broad participation, the interaction between the concerned public and the authorities is often very limited, with no real possibility for dialogue or negotiation (Jacobs & Korhonen, 1995, p. 32).

Feasibility. Currently, the mechanism of notification, distribution, and solicitation of comments is well integrated into Canada's regulatory process, at least at the advanced stage of the tabling of proposed legislation.

When notification is via publication in an official source, such as the *Canada Gazette*, the transparency[58] of the notification mechanism is guaranteed. The information is accessible to all citizens without political, geographic, or linguistic restriction. The process is less transparent when the solicitation is directed toward targeted groups and individuals.

It should be recalled that, for public consultation through this mechanism, sufficient time must be allowed for responses; otherwise the process may be criticized as being a sham consultation.

Finally, if the consultation is effective and if citizens and groups take the opportunity to express their concerns and opinions, the body responsible for the consultation must analyse a prodigious quantity of information; this crucial stage will thus require considerable resources.

Opinion Surveys and Focus Groups

The public opinion survey is a tool commonly used to identify different trends about a particular issue. The focus group involves exchanges between a small number of selected participants to reveal the opinions of a sampling of a defined population. Despite their popularity, these methods are still not ideal models for consulting the Canadian public about biotechnology issues.

Public representativeness. Because it does not require extensive resources, the survey makes it possible to collect the opinions of a large number of individuals; it thus offers broad representation if a diversified, statistically representative sampling of the population is used. This feature overcomes the obstacle presented by Canada's vast geographic area.

The aim of focus groups is to consult a group of diverse citizens that is statistically representative. However, the selection criteria can themselves create a bias in eliciting opinions on a particular subject. Furthermore, because the number of participants must be restricted to facilitate interaction between them, the opinions that emerge are limited in the degree to which they are representative (Aubry & Saint-Arnaud, 1975, pp. 15–16; Zimmerman et al., 1994, p. 76).

Quality of information collected. The particular methodology of the public opinion survey affects the quality of the information collected (e.g., Barbot, 1994). The process of formulating specific questions that will make up the opinion survey can give rise to biases with respect to the target topic. Added to these inaccuracies are the limits placed by the type of responses to be analysed – responses that are short, circumscribed, and unidirectional. Last, the survey provides a snapshot of opinions frozen in time. These factors make the opinion survey a less than an ideal mechanism. Accordingly, it is important to remember that its use is limited to eliciting the general opinion of the public on certain issues of a chosen topic.

In the case of focus groups, the dynamics of the group are intended to trigger discussions, from which the participants' perceptions will emerge. The exchanges between the people present should promote more detailed and articulated expressions of opinion. However, the interaction between the participants also means that the ideas expressed by each are likely to be modified by the ideas of the others. The role of influence can censor opinions that happen to be original. Moreover, organizers must respect the freedom of participants to express their opinions, and this requirement can interfere with obtaining responses that are sufficiently complete or specific for the purposes of the consultation.

Feasibility. The resources needed to administer an opinion survey are relatively limited, and this technique is already very familiar to expert firms. At the statistical level, to achieve a representative sampling of the Canadian population, the survey must cover a very large number of respondents, but modern communications technology makes this possible, as can be seen from the surveys conducted by Eurobarometer.[59]

The use of focus groups can consume considerable resources. Since the Canadian population is spread over a vast area, it may be inevitable to create a number of these groups. Furthermore, group members must be brought together under circumstances of time and place that favour their interaction, and doing this can be costly. Since the discussion process is semi-guided and does not necessarily end with a consensus of the persons present, organizers may face difficulties analysing the copious remarks and opinions collected.

Sequential Consultation

The sequential consultation method seeks to enable the public to participate by sending in comments at various stages in the preparation of framework texts on a given subject. Soliciting comments at various moments presents advantages and disadvantages when seeking to obtain public opinion.

Public representativeness. Comments about public representativeness within the sequential consultation process are similar to those made about the process of notification, distribution, and solicitation of comments. One has to be aware of the limited size of a population that can be effectively consulted through this mechanism. The process of expressing opinions in several stages requires that participants monitor the issue. Monitoring generally implies an interest in the issues under discussion as well as the necessary resources, which are found more often with organized groups than with individuals.

Quality of information collected. This mechanism promotes more effective circulation of information since the comments of participants are integrated into the process of consultation and drafting of a framework document. Furthermore, consultation in several stages allows solicited members of the public to reiterate their views and make corrections or new comments after reading preliminary texts. In this way certain gaps can be avoided which occur in a one-time solicitation of comments.

Feasibility. The sequential consultation mechanism involves multiple stages for collecting comments. The authorities responsible for the consultation program must process the comments. This methodology makes it necessary to use a complex system for sending and managing collected information in order to draw the maximum benefit from the process.

The Consensus Conference

In recent years, experiments with consensus conferences have been conducted in Europe and North America, as described above. Since this is a relatively new mechanism, some writers and organizations have questioned their usefulness and effectiveness (Joss & Durant, 1995; Mayer, de Vries, & Geurts, 1996, Rathenau Institute, 1995a).

Public representativeness. The main objective of the consensus conference is to enhance public awareness and information – or at least several members of the public – about a given topic in order to ascertain their opinion (Joss & Durant, 1995, p. 9). The consensus conference has two major stages and the targeted public varies for each. The first stage consists of meetings with experts, discussions, and works towards consensus; it involves a small group of lay individuals (for example, between 10 and 20 people). The second stage is a conference during which the main observations and conclusions are presented to the media and the general public.

Public participation in the first stage of the consensus conference is usually solicited through advertisements published in broadly distributed daily newspapers (Lee, 1995) or scientific publications. This approach can encourage participation of members of the general public who do not have any ties to the agency responsible for the consensus conference. The recruitment method, however, may not ensure representative participation (Grundahl, 1995, pp. 33, 39; Joss, 1995, pp. 101–102) since, in general, people already interested in the topic will offer to take part in the conference. In addition, to facilitate exchanges between participants, the size of the group taking part in the consensus conference should be close to 10 people. Consequently, the organizers must select candidates according to preestablished criteria with the aim of obtaining a more representative group.

The holding of a general information conference at the end of the process is aimed at a wider audience. It must be stressed that the purpose of this stage is to communicate findings and not discussion. At this stage, it is similar to the simple conference model used in France. The media are often invited to publicize the major findings emerging from the group consultation process. Media coverage varies depending on the topics addressed and coverage may fall short of expectations (Glasmeier, 1995; Grundahl, 1995, Mayer et al., 1996).

This mechanism thus has a major constraint in its first stage, since it involves a small, preselected group (Rathenau Institute, 1995a, p. 29). In the final stages of publicizing the information via a conference, the public is not consulted but simply informed.

Quality of information collected. The consensus conference uses two main strategies: communicating information and debating ideas on a given topic. The information phase comprises a strong educational component as participants meet with experts to acquire the knowledge needed to develop an informed opinion on the topic under review. Then, during the discussion session, participants debate their opinions striving, however, to reach a consensus. This will be reported in a document drafted by the participants. Finally, conclusions and the consensus position are communicated to the general public and widely publicized through media coverage (Joss, 1995, p. 100; Joss & Durant, 1995, p. 25). The entire process makes it possible to obtain the opinions of lay persons who have been educated about the issue, as well as to better inform the general public about the topic (Rathenau Institute, 1995a, p. 31).

In principle, to ensure greater objectivity of the process, consensus conference organizers must not be involved with the substantive contents during the phases of information, discussion, and debate by the participants (Grundahl, 1995, p. 32). All decisions taken in order to reach consensus must come from the lay members so as to ensure the respect of their independence. According to one author, "Ideally, a conference should be funded and organized by an independent and publicly accountable national institution that is recognized as an instigator of public debate on science and technology" (Joss, 1995, p. 104).

A National Institutes of Health study of the process conducted concluded "Consensus statements were generally written in a clear style, discussed significant issues, and provided adequate and useful information on the efficacy, safety, and the use of technology" (Joss & Durant, 1995, p. 25).

This attractive consultation process has certain drawbacks, however. Notably, it may be extremely difficult to reach a consensus on issues as complex as biotechnology. In reality, the consensus conference may instead serve to identify emerging opinions on the topic addressed.

Feasibility. The consensus conference may be described as a flexible process. The topics addressed may be diversified, and the public consulted quite broad.

Although it offers an interesting method for identifying the opinion of lay persons on a given subject (Frewer, Howard, & Shepherd, 1995), the consensus conference is an elaborate mechanism. In the Canadian context, ensuring that opinions collected are representative of its entire population may require holding multiple conferences in order to arrive at conclusions expressing a nationwide consensus on a given topic. Moreover, since the consensus conference takes place in several stages and requires successive meetings over a period of days, it requires logistical planning and involves travel and accommodation expenses for participants and experts. This public consultation mechanism thus requires substantial resources.

The Internet

The information highway, notably the World Wide Web, offers a new approach for consulting the public. Web sites now have interactive tools such as questionnaires, simulations, and surveys that can focus on specific subjects.

Public representativeness. It is generally recognized that there is no way of ensuring a specific audience to a website. Any Web surfer can visit the site, unless there is a password. Checking users' identity is not yet possible. In these circumstances, there could even be the unsought participation of respondents outside the targeted population. It must also be recognized that the degree to which the audience is representative is limited by the fact that not all people in Canada have access to a computer connected to the Internet, and that some groups are more exposed to this technology than others.

Quality of information collected. It is important to consider the type of information that can be collected by an Internet site. The information collected will vary according to the consultation tools made available to users in Canada and abroad. Some tools are simply questionnaires offering a static picture of the respondents' opinions. If we do not know the characteristics of site users, such tools are less useful than conventional surveys. Other tools may allow more interactive exchanges; examples are simulations and development questionnaires, or real-time exchanges. This possibility is discussed in Rathenau Institute (1995a, p. 31) and in Joss and Durant (1995, p. 12). These last forms facilitate the collection of more detailed information on a punctual basis, but lead to the problem of confidentiality when answers and comments come from individuals. The quality of the information collected can be affected by controlling user identity, and verifying validity and relevance of the information provided. Moreover, since access to such consultation tools is unrestricted, large quantities of information inputted by a possibly very large public will need to be processed.

Feasibility. Internet surfing makes use of an increasingly widespread technology. Cyberspace-based tools are thus easily distributed by the agency wishing to conduct a consultation. They require financial and technical resources that can be limited or substantial, depending on the means of the distributing agency.

However, there is no assurance that such tools will be used by the general public. Web site consultation is limited to those with access to a computer, which is a technology so far not within the reach of all. Furthermore, the number of visitors to a site depends mainly on the marketing strategies. Advertisements may be placed on the Web, for current users, or it may be through more traditional media such as television, radio, and billboards.

Web sites have limitations in terms of distribution and accessibility, and also in terms of the information collected. We must therefore conclude that they are not yet a reliable resource for consulting the public on a large scale.

Summing Up

To summarize this overview of different public consultation tools, we have prepared Tables I and II. Table I compares tools in terms

TABLE I
Application of Public Consultation Tools to the Canadian Situation

	Representativeness	Quality of information	Feasibility
Notification, distribution, and solicitation of comments	+++	++	+++
Opinion survey	++++	+	+++
Focus groups	+	++	++
Sequential consultation	+++	+++	++
Consensus conference	++	++	++
Internet	?	?	+

(+ = weak/++++ = excellent/? = unknown).

TABLE II
Advantages and Disadvantages of Public Consultation Tools

	Advantages	Disadvantages
Notification, distribution, and solicitation of comments	Simple methodology Detailed written or oral comments	Participation usually limited to a knowledgeable public Fragmentary, partial information
Opinion survey	Accessible Wide range of respondents	Static snapshot of public opinion Limited, unidirectional information
Focus groups	Process favouring discussions and exchanges Emergence of new opinions	Limited representation Factors of time and place
Sequential consultation	Circulation of information Opportunity to verify if positions considered	Participation limited to a knowledgeable public Management of information collected
Consensus conference	Dynamic participation by interested persons Consensus on a given subject	Limited number of participants in a one-time event Logistics
Internet	Technology offering a variety of tools Investments in accordance with the means of the agency	Limited access for the general public Lack of control over identity of participants

of the three features considered: public representativeness, quality of information collected, and feasibility. Table II summarizes the main advantages and disadvantages of each tool.

CONCLUSION

This study has enabled us to observe that several countries have bodies specialized in public consultation. Several of them have opted for a single structure responsible for examining the social and ethical concerns raised by biotechnology applications on plants, animals, and human beings; others have created different structures for each of the different possible impacts of these scientific developments. Furthermore, the bodies that have been created are in some cases permanent, while in others they are temporary.

Regardless of their status and of their diverse nature, these institutional structures all have an array of methods for consulting the public on ethical and social issues in biotechnology. The most commonly used methods are notification, distribution, and solicitation of comments and opinion surveys. But other forms of consultation are also used. One of these, sequential consultation, is similar to solicitation of comments; it seeks to encourage members of the public to contribute by providing them with the opportunity for reacting once a decision is taken to draft or revise notices or recommendations on a given topic. Individuals can express their positions both before the first draft is prepared, and then following its release. Individuals have the opportunity of making comments and reiterating them on more than one occasion. The consensus conference is distinctive – among other things – because of its participants. It does not systematically make use of interest groups or professional bodies; instead it seeks contributions from the lay public. As discussed above, it involves a more limited number of individuals engaging in very intensive interaction. Thus, the way in which the public participates differs depending on the consultation mechanism that is used.

In deciding which consultation mechanism to use, consideration must always be given to a country's social, economic, political, and technological circumstances. A key factor to be taken into account in deciding which consultation mechanism to choose is the targeted public. The very concepts of "public" and "public opinion" are highly complex (Marsh & Fraser, 1989; Tremblay, 1991). Is the aim to learn the opinion of people already aware of the issue and indeed concerned about it, as individuals or organized groups? Or is it to survey a public that is completely unaware of biotechnology and its uses? As the third part of this paper shows, the degree of public representativeness varies considerably from one consultation method to another. For example, using the Internet opens the door to an immense but faceless public, while the consensus conference brings together a small number of chosen people. Both these methods will certainly provide us with information on the concerns of the public about biotechnology developments, but it is highly likely that each will provide different information.

The process of selecting a consultation mechanism must also take into account its accessibility. For equitable access, operating procedures must be as simple and inexpensive as possible for participants. If they are effective, public participation will be facilitated. A very elaborate, highly complex tool may in fact paradoxically limit contributions.

Finally, we must not overlook the major role of educating and informing the public. If better informed, the public can participate more actively in the debate raised by the ethical and social issues in biotechnology. At the time of consultation (using a carefully chosen method), the public will be better able to express its positions on trends in Canada's biotechnology industry.

All the efforts undertaken in research and development are aimed at the same objective: to develop a new product that will be accepted positively by the public. Investments in biotechnology will be profitable only if consumers opt for the new technology. At the moment, consumers have not yet made up their minds; hence the importance of consulting them and encouraging them to make their opinions known. The success of these technological breakthroughs depends on the dynamic participation of the public in discussions about their development.

ANNEX: LIST OF INTERNET SITES RELEVANT TO BIOTECHNOLOGY

- *Sites of International Organizations*

World Health Organization (WHO). Home page: search for information carried on the site with the help of key words, including "biotechnology."
http://www.who.ch/

OECD. Regulatory developments dealing with biotechnology in member countries of the Organization for Economic Co-operation and Development.
http://www.oecd.org/ehs/country.htm
BINAS Online. The Biosafety Information Network and Advisory Service (BINAS) is a service offered by the United Nations Industrial Development Organization (UNIDO). BINAS follows developments relating to biotechnology regulation issues.
http://binas.unido.org/binas/index.html

- *Governmental Sites*

U.S. Environmental Protection Agency. The mission of the U.S. Environmental Protection Agency (EPA) is "to protect human health and to safeguard the natural environment – air, water, and land – on which life depends."
http://www.epa.gov/

- *Sites of Organizations Concerned with Biotechnology*

Center for Biotechnology Policy and Ethics. This centre, located at Texas A&M University, was established in 1990 with the aim of promoting research on a range of subjects related to science and technology policy, particularly to do with agricultural biotechnology.
http://www.tamu.edu/cbpe/
Eubios Ethics Institute. The Eubios Ethics Institute is a non-profit organization that seeks to stimulate international debate on ethical questions, as well as on "how we may use technology in ways consistent with 'good life' (eu-bios)." It promotes a general, intercultural bioethics approach and the creation of an international network.
http://www.biol.tsukuba.ac.jp/~macer/index.html
National Centre for Biotechnology Education. The NCBE offers information and training for the academic and industrial sectors, professional associations, and the general public. The centre (the first of its kind in Europe) was established in 1985 as a component of the Department of Microbiology at the University of Reading in Britain.
http://www.NCBE.reading.ac.uk/
European Federation of Biotechnology. The objectives of the EFB are: to develop biotechnology for the benefit of the public, to promote awareness raising, communication and co-operation in all sectors of biotechnology, and to offer to governmental and supra-national agencies information and informed opinions on biotechnology.
http://sci.mond.org/efb/home.html
Canadian Biotechnology. The Canadian Institute of Biotechnology (CIB), the Industrial Biotechnology Association of Canada (IBAC) and the Biotechnology Human Resources Council (BHRC) have merged to create a single national organization focussing on biotechnology.
http://www.biotech.ca/cib/index.htm
Canadian Institute for Environmental Law and Policy. During the drafting of legislation and environmental policies, this organization promotes the public interest and sustainable development principles, including protection of the health and well-being of present and future generations and of the natural environment.
http://www.web.net/cielap/

- *Sites Containing Interpretative Documents*

Nature Biotechnology. Research engine.
http://biotech.nature.com/

• *Sites Containing Lists of References (Other URL Addresses or Bibliographies)*

InfoBiotech.Canada. InfoBiotech Canada (IBC) is the outcome of a partnership between the public, private and university sectors. Its objective is to offer access to information on biotechnology in Canada and abroad.
http://www.nrc.ca/ibc/
Bioline Publications. Bioline Publications is an electronic publication service for scientists. In collaboration with science publishers, editors of newsletters, and authors of reports, it provides easier, cheaper access to scientific material.
http://www.bdt.org.br/bioline/
Biotechnology Information Center. The Biotechnology Information Center (BIC) is one of 10 information centres within the National Agricultural Library of the U.S. Department of Agriculture's Agricultural Research Service. The BIC offers access to a variety of publications covering several aspects of biotechnology as applied to agriculture.
http://www.nal.usda.gov/bic/

NOTES

[1] National Health and Medical Research Council Act No. 225 of 1992. The NHMRC's mandate includes the development of guidelines, giving opinions, and making recommendations on health-related ethical questions. See Sections 7, 8(1) and 8(2) of the Act.
[2] There previously existed the Medical Research Ethics Committee and the National Bioethics Consultative Committee (Report of the review of the role and functioning of Institutional Ethics Committees, 1996, p. 8).
[3] See Section 12 of the National Health and Medical Research Council Act No. 225 of 1992:
12.(1) Subject to sections 13 and 14, before the Council:
(a) makes a regulatory recommendation; or
(b) issues guidelines; or
(c) engages in any other prescribed activity;
the Council must consult persons or bodies in accordance with the steps set out in this section.
12.(2) As soon as practicable after deciding that, subject to consultation processes, it intends to make a regulatory recommendation, issue guidelines or engage in a prescribed activity, the Council must publish a notice, in the manner and form specified in the regulations:
(a) informing of its intention to make the recommendation, issue the guidelines or engage in the activity; and
(b) inviting persons or bodies to make submissions relating to the proposed recommendations, guidelines or activity in accordance with the procedures, and within the period, specified in the notice.
12.(3) As soon as practicable after the end of the period specified under paragraph (2)(b), the Council must, having regard to any submissions received:
(a) prepare a draft of the regulatory recommendation or guidelines that the Council proposes to make or issue or an outline of the prescribed activity in which the Council proposes to engage and publish a notice, in the manner and form specified in the regulations:
(i) containing the draft or outline; and

(ii) inviting persons or bodies to make submissions relating to the draft or outline in accordance with the procedures, and within the period, specified in the notice; or

(b) publish, in the manner and form specified in the regulations, a notice informing that it no longer intends to make the recommendation, issue the guidelines or engage in the prescribed activity.

[4] Section 8, paragraphs (1) and (2) of the National Health and Medical Research Council Act No. 225 of 1992 stipulates: "(1) Without limiting any of the matters on which the Council may issue guidelines – the Council must issue guidelines – for the conduct of medical research involving humans. (2) The guidelines for the conduct of medical research involving humans must be issued precisely as developed by the Principal Committee known as the Australian Health Ethics Committee and provided to the Council for that purpose."

[5] National Health and Medical Research Council Act No. 225 of 1992, Section 39(1).

[6] National Health and Medical Research Council Act No. 225 of 1992, Section 36(1).

[7] Section 9, Law No. 353 of 3 June 1987 on the Establishment of an Ethical Council and the Regulation of Certain Forms of Biomedical Research (1988), 39 No. 1 *Int. Dig Hlth. Leg*. 95. (See also Danish Council of Ethics, 1995.)

[8] Section 2, Law No. 353 of 3 June 1987 on the Establishment of an Ethical Council and the Regulation of Certain Forms of Biomedical Research (1988), 39 No. 1 *Int. Dig Hlth. Leg*. 95.

[9] Von Ingrid, Von Berg (Institute for Technology Assessment and Systems Analysis), "The Danish Board of Technology forever!" from the Web site <www.tab.fzk.de/deu/TADN/TADN895/inst2.html>, consulted in March 1997.

[10] Hansen, L., "The Danish Board of Technology established on a permanent basis," from the Web site <www.tab.fzk.de/eng/PAN/TADN1295/inst.html#inst2>, consulted in March 1997.

[11] Sweden also has an ad hoc institutional body to examine policy aspects of ethical questions. Before tabling a bill, the government can appoint a commission with a specific mandate to examine an issue and make recommendations. Once the government receives the commission's recommendations, it can circulate them to elicit comments and criticism from a wider number of organisations or interested authorities. Communication from Lena Jonsson, Deputy Director, Ministry of Health and Social Affairs, March 4, 1997.

[12] Previously there was also the Office of Technology Assessment, which had a very broad mandate including health and biotechnology. One of its roles was to facilitate communication between experts, the public, and decision makers. This federal agency used mechanisms such as surveys to discover the opinions of various groups on a given topic. The Office of Technology Assessment published many reports (e.g., Congress of the United States, Office of Technology Assessment, 1984, 1987, 1992).

[13] NIH-DOE Working Group on Ethical, Legal and Social Implications of Human Genome Research, *Report of the Task Force on Genetic Information and Insurance, Genetic Information and Health Insurance*, May 10, 1993, Appendix B.

[14] The National Human Genome Research Institute, Promoting Safe and Effective Genetic Testing in the United States, Final Report of the Task Force on Genetic Testing, September 1997, on the Web site <www.nhgri.nih.gov/ELSI/TFGT_final>.

[15] "Executive order 12975 of 3 October 1995, Protection of Human Research Subjects and Creation of a National Bioethics Advisory Commission" (1995) 160 No. 193 *Federal Register* 52063.

[16] More particularly, expertise in philosophy or theology, social or behavioural

science, law, health, and biology. See National Bioethics Advisory Commission Charter.

[17] Notably, by publication in the *Federal Register* of the United States.

[18] In the United States, under the Federal Advisory Commission Act, any federal commission must hold its meetings in public. Personal communication from Dr. Eric Meslin, Director, ELSI Research Program, NIH, February 21, 1997.

[19] Examples of non-statutory committees are the Gene Therapy Advisory Committee, established in 1993 following the report of the Committee on the Ethics of Gene Therapy (1992); and the Advisory Committee on Genetic Testing, established in January 1996 in response to the House of Commons Science and Technology Committee (1995).

[20] Some commissions do not hold public meetings such as the Gene Therapy Advisory Committee, Committee on the Ethics of Gene Therapy.

[21] The first report dates from 1995: House of Commons Committee on Science and Technology (1995).

[22] Human Genetic Advisory Commission, Terms of Reference, document sent by Mr. Mark Noterman, Department of Health, Genetics Secretariat, Advisory Commission on Genetic Testing, personal communication, February 21, 1997.

[23] The Nordic Council of Ministers, established in 1971, is an intergovernmental co-operation agency. The five Nordic countries are members, as are the autonomous regions of Greenland, Åland, and the Faroe Islands. Communication with Dr. Ylva Tilander, Nordic Council of Ministers, April 20, 1997. See also the Web site of the Nordic Council and Nordic Council of Ministers: <http://192.38.240.206/f_uk/fukindx.htm/>, consulted August 4, 1997.

[24] The Rathenau Institute replaced the Netherlands Organization for Technology Assessment (NOTA), a body established in 1986 by the Minister of Education and Science. In 1994, NOTA's objectives and operating method were changed by the Dutch Parliament through the *Institutional Order of 19 April 1994*. The mandate of the Rathenau Institute must be revised every two years by Parliament. See Web site <www.tab.fzk.de/eng/PAN/TADN1295/inst html#inst2>, consulted in March 1997.

[25] See the internal document of the Rathenau Institute "Plurality and Interrelation, Proposal for the Establishment of a Platform for a Public Discussion of the Ethical Aspects of Science and Technology"; and Joss & Durant (1994).

[26] For 1995–96, the four major themes on the agenda were: (1) Technology and sustainable development; (2) Intervention on man, animal, and ecosystems; (3) Information and communication; and (4) Platform on Science and Ethics. See the Web site <www.tab.fzk.de/eng/PAN/TADN1295/inst.html#inst2>, consulted in March 1997.

[27] Personal communication from Lena Jonsson, Deputy Director, Ministry of Health and Social Affairs, March 4, 1997.

[28] However, the Gene Therapy Advisory Committee organized a round table conference, most of the publicity focusing on interested groups. As for the Advisory Committee on Genetic Testing, it is using Internet communication methods to distribute its preliminary report and collect comments. See Advisory Committee On Genetic Testing, Draft Code of Practice for Human Genetic Testing Offered Commercially Direct to the Public, November 1996, Web site <www.scicomm.org.uk/biosis/acgt/ACGT1.html>, consulted August 4, 1997.

[29] National Institutes of Health, Task Force on Genetic Testing, Proposed recommendations of the Task Force on Genetic Testing, Notice of Meeting and Request for Comment (1997) 62 No. 20 *Federal Register* 4539.

[30] Personal communication from Dr. Eric Meslin, Director, ELSI Research Program, NIH, February 21, 1997.

[31] Eurobarometer 35.1, *Biotechnology*, for the Commission of the European Community, DGXII, INRA Europe, 1991.

[32] Eurobarometer 39.1, *Biotechnology and Genetic Engineering: What Europeans think about it in 1993*, Eric Marlier, European Commission, and INRA Europe, 1993.

[33] NHMRC, Public consultation procedures for making submissions. This booklet states, "Where practicable, the Council should adopt a policy of public consultation on matters it is considering."

[34] As summarised in a personal communication from Dr. Cindy Wong, Secretary of the Australian Health Ethics Committee, March 7, 1997. See Section 12 of the National Health and Medical Research Council Act No. 225 of 1992.

[35] Since guidelines consist of several documents called Supplementary Notes, on specific subjects (for example, somatic therapy or new reproductive technologies), each document can be reviewed independently. This was the case for *Supplementary Note 4: In vitro fertilization and embryo transfer*.

[36] The NHMRC has also specified two variations on the consultation strategy. When the question raised is either of minor importance or urgent, the process can be shortened. For example, consultation stages can be eliminated completely, or can take place following release of temporary guidelines on an urgent basis. In both cases, the reasons for making such decisions must be published. Sections 13 and 14 of the National Health and Medical Research Council Act of 1992.

[37] Section 12(2) of the National Health and Medical Research Council Act of 1992. Notices must be published in the *Commonwealth Gazette*, and can also be placed in newspapers and on radio or television. Furthermore, the notice must be sent directly to professional associations and consumer/community groups, as well as to any individual likely to be interested in the question. See NHMRC brochure entitled *Public consultation procedures for making submissions*. Personal communication from Dr. Cindy Wong, Secretary of the Australian Health Ethics Committee, March 7, 1997.

[38] For example, on gene therapy, the NHMRC has published a document entitled, *Human Gene Therapy and Related Procedures –An information paper to assist in the consideration of ethical aspects of human gene therapy, 1994*. Use of technical language has been minimised and the vocabulary simplified in order to reach a wider readership.

[39] The consultation notice must contain the following information: that comments can be submitted either in writing or on audio cassette; the deadlines for submitting comments (usually between four and eight weeks); the address to which to send comments; and whether the consultation process will take place through channels other than from submissions, notably through hearings. NHMRC, *Public consultation procedures for making submissions*.

[40] See also "Results of U.K.'s first National Consensus Conference (on Plant Biotechnology)," from the Science Museum Web site <www.open.gov.uk/scmus/finalrep.htm>, consulted August 8, 1997.

[41] The information that follows comes from "Results of U.K.'s first National Consensus Conference (on Plant Biotechnology)," from the Science Museum Web site <www.open.gov.uk/scmus/finalrep.htm>, consulted August 8, 1997, and from "U.K. National Consensus Conference on Plant Biotechnology Final Report (November 1994)" available at <ftp://ftp.open.gov.uk/pub/docs/sci_museum/consensus.txt>, consulted August 8, 1997.

[42] The mission of the Science Museum is "to promote the public's understanding of the history and contemporary practice of science, medicine, technology and industry." For further information consult the Science Museum Web site: <www.open.gov.uk/ scmus/scmhome.htm>, consulted August 8, 1997.

[43] *Nature*, 369, as reported in the NBBBP Web site <www.biol.tsukuba.ac.jp/~macer/NBBBP.html>, consulted August 8, 1997.
[44] Sclove (1996); "Pilot Citizens' Panel," document taken from the Loka Institute Web site: <www.amherst.edu/%7Eloka/panel/panel.htm>, consulted August 8, 1997.
[45] See the Web site of the Task Force for Genetic Testing: <ww2.med.jhu.edu/tfgtelsi/request.html>, consulted August 8, 1997.
[46] See the BIOSIS Web site: <www.scicomm.org.uk/biosis/nuffield.html>, consulted August 8, 1997.
[47] See the Web site of the *If*Gene International Forum: <www.peak.org/~armstroj/>, consulted August 8, 1997.
[48] Personal communication from Dr. Cindy Wong, Secretary of the Australian Health Ethics Committee, March 7, 1997. See, for example, NHMRC, Report of 1995 Ethics Workshops, prepared by the Australian Health Ethics Committee, a principal committee of the NHMRC, 1995.
[49] See, for example, the texts published by the U.K. Advisory Committee on Genetic Testing, which has distributed on the Internet the Draft Code of Practice for Human Genetic Testing Offered Commercially Direct to the Public, on the BIOSIS Web site <www.scicomm.org.uk/biosis/acgt/ACGT2.html>, consulted August 8, 1997; and the text of the NIH-DOE ELSI Task Force on Genetic Testing, on the Web site of the Task Force: <ww2.med.jhu.edu/tfgtelsi/fedregister>, consulted August 8, 1997.
In addition, certain research centres, such as the Center for Biotechnology Policy and Ethics, publish reports or newsletters in order to provide information to the public and encourage it to consider the issues. See, for example, the Web site of the Center for Biotechnology Policy and Ethics: <www.tamu.edu/cbpe/cbpe.html>, consulted August 8, 1997.
[50] "As a rule, the Council tries to persuade artists of all kinds to contribute to the debate material. Once well informed, they are able to voice the ethical dilemmas involved more articulately than any expert" (Danish Council of Ethics, 1992b, pp. 4–5, 8). The Council has notably organized poster competitions for artists and writing competitions for philosophers and authors; the winners have their works reproduced in the Council's annual report and also receive a cash prize.
[51] For example, the Council commissioned journalists to prepare a version for the general public of the preparatory discussions for the Council of Europe's *Bioethics Convention* (then in preparation) (Danish Council of Ethics, 1992a, pp. 22–23).
[52] For example, following the writing competition on the consequences of human genome sequencing, the Danish Council carried out a qualitative study of the 520 submissions received; this revealed the opinion of some young people (Danish Council of Ethics, 1992b, p. 11).
[53] For example, the publication *Ethics and Human Genes*, 1994, for students in some 240 schools, and other material such as videos sent to teachers. The teaching material was designed to be used in 17 different subjects studied at all levels (Danish Council of Ethics, 1995, p. 21).
[54] For further information, consult the Web site of the European Initiative for Biotechnology Education: <http://134.225.167.114:8001/>, consulted August 8, 1997.
[55] For an overview of its many activities, consult the Web site of the European Initiative for Biotechnology Education: <http://134.225.167.114/>, consulted August 8, 1997.
[56] "In Canada and the United States, periodic publications describe proposed federal regulations planned for the coming year, and provide the names and telephone numbers of the regulators" (Jacobs & Korhonen, 1995, p. 31).
[57] OECD, Public Management Committee, Public Consultation, URL address.
[58] See Note 52, paragraph 6.

[59] For example, Eurobarometer 35.1, *Biotechnology*, for the Commission of the European Community, DGXII, INRA Europe, 1991.

REFERENCES

Algoud, D., & Joly, P. B. (1995). SOS plantes transgéniques, Une nouvelle filière en péril. *Biofutur*, pp. 23–26.

Aubry, J.-M., & Saint-Arnaud, Y. (1975). *Dynamique des groupes*. Ottawa: Editions de l'Homme.

Barbot, J. (1994). Que pensent les Européens des biotechnologies. *Biofutur*, pp. 30–33.

Byk, C. (1993). Comité consultatif national d'éthique. In: *Dictionnaire Permanent Bioéthique et Biotechnologie*, pp. 265–290B. Montrouge, France: Éditions législatives.

Committee on the Ethics of Gene Therapy (U.K.) (1992). *Clothier Report on the ethics of gene therapy*. London: HMSO.

Congress of the United States, Office of Technology Assessment (1984). *Human gene therapy*. OTA-BP-BA-32. Washington, DC.: U.S. Government Printing Office.

Congress of the United States, Office of Technology Assessment (1987). *New developments in biotechnology 2: Background paper – Public perception of biotechnology*. OTA-BP-BA-45. Washington, DC.: U.S. Government Printing Office.

Congress of the United States, Office of Technology Assessment (1992). *Cystic fibrosis and DNA tests: Implications of carrier screening*. OTA-BA-532. Washington, DC.: U.S. Government Printing Office.

Danish Council of Ethics (1992a). *4th Annual Report 1991*. Copenhagen: Danish Medical Research Council.

Danish Council of Ethics (1992b). *Public discussion about bioethics activities*. Copenhagen: Danish Medical Research Council.

Danish Council of Ethics (1995). *7th Annual Report 1994*. Copenhagen: Danish Medical Research Council.

de Chenay, A., & Therre, H. (1995). L'information du public: Un droit et une nécessité. *Biofutur*, 38–42.

Frewer, L. J., Howard, C., & Shepherd, R. (1995). Genetic engineering and food: What determines consumer acceptance? *British Food Journal, 97*(8), 31–36.

Glasmeier, A. (1995). Consensus conferences, the media, and public information in the Netherlands. In: S. Joss & J. Durant (Eds.), *Public participation in science: The role of consensus conferences in Europe*, pp. 67–73. London: Science Museum.

Grundahl, J. (1995). The Danish consensus conference model. In: S. Joss & J. Durant (Eds.), *Public participation in science: The role of consensus conferences in Europe*, pp. 31–40. London: Science Museum.

Hallman, W. K. (1996). Public perceptions of biotechnology: Another look. *Bio/Technology, 14*, 35.

Hamon, L. (1979). Information et gouvernement. In: J. Antoine et al. (Eds.), *L'information*, pp. 71–96. Paris: Desclée de Brouwer.

Hoban, T. J. (1997). Consumer acceptance of biotechnology: An international perpective. *Nature Biotechnology, 15*, 232–234.

Hornig, P. S. (1994). Structuring public debate on biotechnology. *Science Communication, 16*, 166–179.

House of Commons Committee on Science and Technology (1995). *Third Report, Human genetics: The science and its consequences*. London: House of Commons.
House of Commons Committee on Science and Technology (1996). *Third Report, Human Genetics: The Government's response*. London: House of Commons.
Jacobs, S., & Korhonen, J. (1995). Opening up regulation. *OECD Observer, 195*, 29.
Jank, B. (1995). Biotechnology in European society. *Trends in Biotechnology, 13*, 42–44.
Josefsson, J. (1996). The Nordic Committee on Bioethics. In: *World Congress of Bioethics, San Francisco, USA, November 22–24*, p. 31. San Francisco.
Joss, S. (1995). Evaluating consensus conferences: Necessity or luxury? In: S. Joss & J. Durant (Eds.), *Public participation in science: The role of consensus conferences in Europe*, pp. 89–108. London: Science Museum.
Joss, S., & Durant, J. (1994). *Consensus conferences: A review of the Danish, Dutch, and U.K. approaches to this special form of technology assessment, and an assessment of the options for a proposed Swiss consensus conference*. London: Science Museum.
Joss, S., & Durant, J. (Eds.) (1995). *Public participation in science: The role of consensus conferences in Europe*. London: Science Museum.
Lee, G. (1995). A consensus conference from the point of view of a lay-panel member. In: S. Joss & J. Durant (Eds.), *Public participation in science: The role of consensus conferences in Europe*, pp. 81–86. London: Science Museum.
Leroux, T., Le Bris, S., & Knoppers, B. M. (1998). The feasability of a National Canadian Ethics Advisory Committee: Points to consider. In: *Canada Health Action: Building on the legacy*, Vol. 5. pp. 439–644. Ottawa: National Forum on Health.
Macer, D. (1992). *Attitudes to genetic engineering: Japanese and international comparisons*. Christchurch, New Zealand: Eubios Ethics Institute.
Macer, D. (1994). Perception of risks and benefits of in vitro fertilization, genetic engineering and biotechnology. *Social Science and Medicine, 38*, 23–33.
Marsh, C., & Fraser, C. (1989). *Public opinion and nuclear weapons*. London: Macmillan.
Mayer, I. S., de Vries, J., & Geurts, J. (1996). *Effects of participation: A Quasi-experimental evaluation of a consensus conference on human genetics research.* Tilburg, The Netherlands: Tilburg University, Work and Organization Research Centre.
National Biotechnology Network (1997). Canadian biotechnology: Diverse and dynamic. *Biotech Liaison, 2*(1), 1–2.
NENT (1995). *Årsmelding 1994*. Oslo: National Committee for Research Ethics in Science and Technology (Den nasjonale forskningsetiske komité for naturvitenskap og teknologi).
Nuffield Council on Bioethics (1992). *Annual Report 1991–92*. London: Nuffield Council on Bioethics.
Rathenau Institute (1995a). *The Rathenau Institute and the debate, Annual Report 1994*. The Hague: Rathenau Institute.
Rathenau Institute (1995b). *Report to Parliament, Predictive research, where are we going?* The Hague: Rathenau Institute.
Report of the review of the role and functioning of Institutional Ethics Committees (1996). Report to the Ministry for Health and Family Services, March 1996, Commonwealth of Australia. Australian Government Publishing Service.
Sclove, R. E. (1996). Town meetings on technology. *Technology Review, 99*(5), 24.

Tremblay, G. (1991), L'opinion publique. In: M. Beauchamp (Ed.), *Communication publique et société repères pour la réflexion et l'action*, pp. 149–181. Boucherville, Québec: Gaëtan Morin Éditeur.

U.S.-EC Task Force on Biotechnology Research (1992). *Methods of communicating biotechnology with the public*, Final Report. Washington, DC.: United States Department of Agriculture, Office of Agricultural Biotechnology.

Verrall, M. (1994). Lay panel back gene-modified plants but urge stricter monitoring. *Nature, 372*, 122.

Zimmerman, L, Kendall, P., Stone, M., & Hoban, T. (1994). Consumer knowledge and concern about biotechnology and food safety. *Food Technology, 48*(11), 71–77.

THE AUTHORS

Thérèse Leroux is a Professor and Researcher (E-mail. leroux@crdp.umontreal.ca), Marie Hirtle a Research Associate (E-mail: hirtle@droit.umontreal.ca), and Louis-Nicolas Fortin a Research Assistant (E-mail: fortinni@crdp.umontreal.ca) at the Centre de recherche en droit public, Faculté de droit, Université de Montréal, P.O. Box 6128, Succursale Centre-ville, Montréal, Canada H3C 3J7.

The text was originally written in French. Research for this study was completed in August 1997.

The authors wish to thank Ms. Guila Delouya for her contribution to the preparation of this study.

Furthermore, the authors wish to express warm thanks to the following persons and organizations for their kind help:

Dr. David Bennett, Secretary, Task Group on Public Perceptions of Biotechnology, Netherlands.

Prof. Ruth Chadwick, Euroscreen, Centre for Professional Ethics, University of Central Lancashire, United Kingdom.

Dr. Wybo-Jan Dondorp, Secretary, Standing Committee on Medical Ethics and Health Law, Health Council of the Netherlands.

Prof. John Durant, Assistant Director (Science Communication Division), Science Museum, London, United Kingdom.

Mr. Jan Ejlsted, Teknologi-Radet, Denmark.

Mrs. Linda Gowing, Executive Secretary, Australian Health Ethics Committee, National Health and Medical Research Council, Australia.

Dr. Lena Jonsson, Deputy Director, Ministry of Health and Social Affairs, Sweden.

Dr. Jonas Josefsson, Nordic Committee on Bioethics, Kungshuset, Sweden.

Dr. Eric Meslin, Director, ELSI Research Program, National Institutes of Health, United States.

Mr. Mark Noterman, Department of Health, Genetics Secretariat, Advisory Commission on Genetic Testing, Department of Health, United Kingdom.

Dr. David Shapiro, Director, Nuffield Council on Bioethics, United Kingdom.

Mme Natalie St-Pierre, Fédération nationale des associations de consommateurs du Québec Inc., Canada.

Dr. Ylva Tilander, Nordic Council of Ministers, Copenhagen, Denmark.

Dr. Ne van de Poll, Chair, Platform Wetenschap en Techniek, Rathenau Institute, Netherlands.

Dr. Cindy Wong, Secretary, Australian Health Ethics Committee, Australia.

Timothy Caulfield

The Commercialization of Human Genetics: A Discussion of Issues Relevant to the Canadian Consumer

ABSTRACT. Private sector interest in the practical applications of genetic research continues to grow. Indeed, the increasing availability of clinical genetic services is, to a large degree, the result of the involvement of commercial entities. However, not all are optimistic about the impact of market forces in this context. This paper analyses many of the concerns which have been articulated in relation to the commercialization of human genetics. In particular, it reviews the literature, commentaries, position papers, and relevant law which touch on issues of concern relevant to the Canadian consumer.

> If you can make a condition a stigma, you can make money on it.[1]

Since the start of the Human Genome Project (HGP) in 1990, Canadians have been inundated with reports of the wonders and possibilities of genetics.[2] It is, indeed, an exciting time. Research has progressed at an incredible pace with the discovery of new genes being announced on almost a daily basis. Many believe that more than 90% of the human genome will be sequenced by the end of this century – supplying researchers around the world with an unprecedented tool for the understanding of human diseases. Of course, one of the overarching goals of this loosely coordinated international initiative is to develop practical interventions – be they diagnostic aids, predictive tests, genetic therapies, or genetically derived pharmaceuticals – that will ease human suffering.

We now seem to be on the cusp of an era when this goal of practical application may become a pronounced reality – a time when genetic services leave the laboratory and become available to Canadian health care consumers. This availability will, to a large degree, be the result of the involvement of commercial entities. Driven by the same understanding of the immense possibilities of genetic research which started the HGP (with the important twist of the desire for profit), the private sector's interest in the applications of genetic

Journal of Consumer Policy **21**: 483–526, 1998.
Bartha M. Knoppers & Alan D. Mathios (eds), Biotechnology and the Consumer.

research continues to grow. However, not all are optimistic about the impact of market forces in this context.[3] This paper will review many of the concerns which have been articulated in relation to the commercialization of human genetics. In particular, it will examine these concerns from the perspective of the Canadian consumer.

When one undertakes an overview of this nature there is a risk of sounding overly pessimistic or coming off as a "genetic Luddite." As such, it is essential to recognize the importance and irreplaceable role of industry support to genetic research. It has been noted that most genetic products will inevitably be refined and delivered by industry. Private funding can flow more rapidly and is often more responsive and flexible than other funding mechanisms (Science and Technology Committee, 1995). Indeed, it is difficult to imagine successfully completing and disseminating the results of genetic research without this support. The purpose of this paper, however, is not to praise the commercialization process (a process which has generated enough momentum on its own), but to review the literature, commentaries, position papers, and relevant law which touch on issues of concern relevant to the Canadian consumer. Nevertheless, this discussion must be set against the reality of commercial involvement.

At the outset, I would also like to highlight the limits and scope of this paper. For example, there is little on the topic of consumers as research subjects. Likewise, although I have endeavoured to use a broad range of material, much of the information is from the U.S. However, given our proximity – geographically, economically, and culturally – the information is nevertheless highly relevant. In fact, the overwhelming force of the U.S. market may make it difficult to develop a social policy that insulates Canada from U.S. commercialization activities. I have also taken a broad view of the concept of "genetic services." That is, I rarely explicitly differentiate between the many different types of genetic services (e.g., diagnostic testing, carrier screening, etc.). To some degree this can be misleading. The commercialization of susceptibility testing services will obviously have different implications than the commercialization of DNA paternity testing. There are, however, common themes, concerns, and trends that can be associated with commercialization generally. In addition, many of the areas are inextricably linked.

As we shall see, there are numerous conflicting forces, values, and interests at play. And, as such, it will be a challenging endeavour to develop policy which both meets the needs of industry and addresses the concerns relevant to the consumers of "genetic products." One of the central goals of this paper is to provide a broad picture of the various issues relevant to the development of such policy. To this end, the paper begins by reviewing a number of the factors which have set the tone of the current commercialization environment. This is followed by an examination of some of the most frequently raised concerns. Finally, a few tentative suggestions for policy development are provided.

SETTING THE SCENE

Before delving into a discussion of the associated concerns, this section will briefly describe a number of the (often conflicting) factors which are creating the complex environment in which genetic technologies are emerging. As we will see, many of these factors are highly relevant to the success of both the commercialization process and regulatory policy. More importantly, however, they both reflect and will shape the attitudes of the Canadian consumer.

The State of the Biotechnology Industry

> All biotechnology companies, by dint of the power of genomics, will have to tailor their research programs toward genomics.[4]

Biotechnology has become a vibrant and an increasingly important component of the current Canadian economy. The number of biotechnology companies has grown substantially;[5] the value of stocks in biotech companies continue to soar;[6] large cooperative deals have been struck between emerging genome research companies and established pharmaceutical corporations;[7] there are substantial connections between for–profit clinical genetic diagnostic companies and academic institutions;[8] over 1,175 patents were granted on human genetic sequences between 1981 and 1995;[9] and universities and government research centres are increasingly turning to the private sector for collaborative financial support.[10]

While much of this activity is centred in the United States, the Canadian biotech industry has also experienced substantial growth.

Some example facts concerning recent activity in the Canadian biotech sector include:[11]

- Market capitalization of publicly traded companies increased from \$2.7 billion in 1994 to \$8 billion in 1996.
- From 1989 to 1993 biotech's share of the Canadian GDP grew from 0.08% to 0.17%.
- In 1996 the biotech industry raised over \$1 billion in capital (the same total amount raised between 1991 and 1995).
- The total number of "core" biotech companies has risen from 121 in 1994 to 224 in 1997.
- Employment growth in this area was 17% – well above the industry norm.
- In 1993, genetics activity accounted for \$465 million in sales. This represents 22% of the total of Canadian biotechnology sales in 1993 (approx. 2 billion in 1993 and 3 billion in 1994). Most of these sales were in health care – a sector which has grown at a pace of 27% per year between 1989 and 1993.[12]

There seems little doubt that the Canadian biotech sector is doing very well – and the excitement which surrounds the emerging genetic technology is largely responsible for its growth. Whether it is the development and direct marketing of diagnostic services, the exploration of therapeutics, the sale of information to large biotech companies,[13] or, perhaps most importantly, the development of genetically derived drugs, human genetics has become big business.

However, while it could be said that "genetech" is currently booming, it is a market that some feel has been over-hyped and is thus extremely fragile. The Science and Technology Committee (1995, para. 150) notes that most of the early enthusiasm has been replaced by a degree of caution (cf. also Erramouspe, 1996). For their part, Malinowski and O'Rourke (1996, p. 247) feel that U.S. federal policy – which has encouraged commercialization – has created a false market built upon "unobtainable expectations," and note that most companies do not have a product and therefore must survive on capital. In fact, investing in the biotech sector can hardly be described as a sure thing – to date, only a very small percentage of biotech companies have made a profit. Morton (1995) thus reports that in the U.S., "[o]nly 1% of the biotech companies make a profit" (cf. also Lanthier, 1996a). And while the sale of genetic information has proven to be extremely lucrative (e.g., selling access to databases of expressed sequence tags), there are as yet no real examples of a successful direct marketing and sale of a pure genetic service (e.g., a diagnostic test) (Malinowski & O'Rourke, 1996). Likewise, the experience with cystic fibrosis carrier testing provides some indication that the uptake of genetic services may not be as robust as some have foretold (Bekker et al., 1993; Loader et al., 1996).[14] Will this fragile market force companies, particularly small ones, to become increasingly desperate to produce and market services which will allow them to maintain investor confidence and interest? What impact will this have on the information Canadian consumers receive about genetic services?

There are a number of other factors which should be considered in this context.

First, genetic technology is emerging in an era when governments around the world are cutting health care budgets (see, e.g., Angus, 1995; Donaldson & Gerard, 1993). As a result, it will likely be difficult to find public funding to support the integration of genetic testing, drugs, etc. – unless those services are marketed as cost saving mechanisms. Science and Technology Committee (1995, para. 135) notes that in Britain there have been problems finding public funding for some genetic diagnostic services. These services may have been available as part of a research project but are lost in the move to the public system funding. Meanwhile, the public funding of basic science is also being restricted. These cost containment strategies create fertile ground for the injection of private funds into both the research and clinical setting.

Second, the commercialization of genetics will largely be taking place within a part of the economy – the health care sector – where market forces are generally very ineffective (Banks, 1996; Donaldson & Gerard, 1993, p. 26). This economic reality both contemplates a potentially chaotic commercialization process and heightens the need for some form of external regulation. Malinowski (1996, p. 344) argues that the first wave of genetic technologies will increase health care costs. He also notes that many genetic technologies have no market substitute and are thus "price prohibitive." In addition, the very reasons why market forces fail (e.g., the asymmetry in knowledge between health care provider and patient), create unique concerns for the Canadian consumer of genetic services.

The Pro-Commercialization (or Anti-Regulation) Environment

> Prohibitive government legislation would be highly detrimental to the genetics market. To prohibit any use of genetic information by legislation would stifle investment and test development.[15]

At the current time, encouraging the growth of the biotech sector is viewed as good social policy. Indeed, in many respects, and rightfully so, the expansion of a region's biotech industry is considered an important part of economic development.[16] As a result, any policy which slows this process, even if implemented with the purpose of protecting the consumers of genetic services, will likely meet a degree of resistance. The importance of the regulatory environment was recently noted by the Science and Technology Committee of the British House of Commons:

> All those witnesses who addressed the subject felt that regulation was a key factor in determining a biotechnology company's success. However good the capital, management and intellectual expertise available to a company, it may still fail if it is subject to over regulation which prevents it operating effectively or if it operates from a country where regulation is perceived to be too lax to protect the public (Science and Technology Committee, 1995, para. 174).

Naturally, this is the position which is also forwarded by commercial entities. In addition, it is argued that the mechanisms of commercialization are essential to successful research (e.g., the patent regime) and any regulations which impede the commercial process will erode the entire research infrastructure (see, e.g., Burk, 1994, p. 837). In fact, given the cost of this research, both government and the private sector seem more concerned about exploring what regulations are necessary to *help* development. In its arguments against what it perceived to be the over-regulation of the European biotech industry, Pfizer Ltd. noted:

> A perceived unwillingness to lead the world in exploiting new genetic knowledge will have a major impact on Europe's future competitivity in the health-care market. This in turn will lead to increasing out-sourcing of biotech manufacturing and research capabilities, which in the long term will undoubtedly affect both the employment prospects and the quality of care available to our citizens (Science and Technology Committee, 1995, para. 175).

We have seen similar positions emerge in Canada. For example, the Alberta Science and Research Authority (1996) has recommended that we "reduce, simplify and standardize regulations in the Province for biotechnology" arguing that the "current delays and regulatory uncertainties are discouraging new research and investment, driving up the cost of innovation and undermining public confidence" (p. 2, p. 5). This is also the perception of Canada which is held by U.S. biotech companies. See, for example, Chaloner-Larsson, Haynes, and Merritt (1993), where it is reported that: "The U.S. companies' general impression of Canada's diagnostic market was that it was highly regulated under the medical system, was fragmented, and had tight reimbursement policies" (pp. 259–260). Likewise, a recent report by Goudey and Nath (1997) on Canada's biotech industry concluded: "While Canada has a strong scientific base, the complexity of getting a product through the Canadian regulatory process is an ongoing challenge, adding significantly to the time and capital required for development," noting that while Canada has improved its turnaround time for approvals (pp. 15–16), it still lags the U.S. and Europe (in 1995 it took an average of 682 days, an improvement over 1,142 in 1994).

Such perceptions, particularly those held by government, will obviously have a profound effect on the potential development and implementation of commercialization regulations. However, they are, at least, indicative of an emerging consensus that the regulatory environment needs to be re-examined. There is no reason to assume that a stream-lined and more efficient regulatory process must necessarily equal a scheme which does not also protect the interests of the Canadian consumer.

Concern Regarding the Commodification of the Human Body, Human Genetics, and Reproduction

> [T]he Parliament of Canada acknowledges the health and ethical dangers inherent in the commercialization of human reproduction . . .[17]

Although there is clear support for commercial initiatives, there also exists an underlying, somewhat amorphous, fear of the impact of the commercialization process. In fact, numerous national and international documents have made statements regarding this concern – be it in the context of genetic patent policy, reproductive technologies, or otherwise. For examples of international documents that set a cautious tone in the context of the commercialization of human tissue, reproduction, and genetics, see Committee of the Health Council of the Netherlands (1994); Council of Europe (1994); Danish Council of Ethics (1994); Nuffield Council on Bioethics (1995). And we have seen similar statements emanate from the Canadian federal government. For example, on June 13, 1996, the federal government introduced Bill C-47, the *Reproductive and Genetic Technologies Act.*[18] While the sections of this Bill that deal directly with commercialization are focused more on new reproductive technologies than on genetic research, the Bill nevertheless serves as an illustration of a basic resistance to market forces in this context (a somewhat paradoxical resistance given the foregoing discussion on the extent of commercialization).

At some level there seems to be an acceptance that commercialization in the realm of health care (and, to some degree, science) has particular qualities that are inherently bad. While this more general concern for the commodification of human tissue has yet to significantly slow the growth of the biotech sector (except, perhaps, with respect to patent policy), it is probable that future policy makers will need to address this paradox in biotech policy.

The Current View of Patient Autonomy

> Genuflections were made to the virtues of free speech and tolerance, but to a great degree, freedom came to be understood as access to consumer choice: the right to acquire purchasing power and to dispose of it as one liked.[19]

Few principles hold the sway of individual autonomy. In the realm of health care it lies at the heart of consumer rights, as manifested in case law and legislation.[20] A concept which is consistently relied upon by our courts, the Supreme Court of Canada recently noted: "In a free and democratic society, the individual must be left room for personal autonomy to live his or her own life and to make decisions that are of fundamental personal importance."[21] And, as highlighted by the research of individuals such as Dorothy Wertz (e.g., Wertz, 1995), autonomy is also considered a paramount value by both health care providers and patients. However, to some extent, our culture's emphasis on patient choice has arguably metamorphosed the traditional conceptions of autonomy in the health care context –

a right to determine what is done to your body – into a perceived right of access (Veatch, 1991). Schneiderman and Jecker (1995), who have characterized this phenomena as "the glare of autonomy," note as follows: "Some patients have interpreted their right to control what is done to them by physicians as a right to demand anything they want from physician" (p. 47). Emanuel and Dubler (1995, p. 324) observe that "the tremendous emphasis Americans place on choice can sometimes appear as a fetish . . ."; cf. also Emson (1995).

Such statements are supported by Wertz's pilot studies on patients. Sixty eight percent of patients surveyed by Wertz thought that withholding any requested genetic service is paternalistic and 61% thought they were entitled to any service they could pay for out of pocket. While this patient survey was done on a relatively small sample size (99), its results are supported by the other pilot research on American patients, for example Fletcher and Wertz (1994):

> Most [patients] thought withholding any service was a denial of patient rights (80%), providers unwilling to do some procedures for moral reasons should offer referrals (79%), prenatal tests should be available on request (75%), and consumers were entitled to whatever services they can pay for out of pocket (61%) [Abstract #1712].

While such statements undoubtedly have more application in the U.S., they nonetheless remain relevant to Canada (cf., for example, Dr. Arnie Aberman, Dean of the University of Toronto Medical School, who argued: "I don't understand why we don't allow patients to spend their income on what they want to spend it on, including health care," see Lowry, 1996). Wertz's survey of Canadian genetic professionals found that 94% thought that they had an obligation to refer a patient to another physician if they were unwilling to perform the procedure for moral reasons (1995, p. 82). Fifty-three percent of the surveyed professionals thought they should refer a patient outside of Canada if domestic law forbids a requested genetic service – a particularly potent statistic given the prohibitions proposed in Bill C-47. She also found a surprising number of professionals who were willing to either refer or perform prenatal diagnosis for the purposes of sex selection (50–67% depending on the scenario) (p. 83). These findings demonstrate *not* an acceptance of such activities; rather, I believe they highlight the tension between a perceived right of access and the professional's own moral views.

As a result of this intense focus on patient choice, any regulation that can be seen as restricting the principle of autonomy may face a degree of resistance. As one biotech executive surmised, referring to the marketing of cancer and ovarian predisposition tests in Canada: "If people decide they want to be tested, who has the right to tell them not to be?"[22] Indeed, the public reaction to Bill C-47 serves as an example of the possible response to overt prohibitions; see, e.g., McIlroy (1996); Moysa (1996); Reproductive technology bill (1996); Selick (1996).[23]

THE PERCEIVED PROBLEMS

While far from a comprehensive list,[24] the following section reviews a number of the more salient issues frequently raised in association to the commercialization process.

Non-Disclosure Agreements and Other Patenting Issues

> Just as a physician has a moral responsibility to do no harm, so does a scientist engaged in medical research. Deliberately withholding useful information or reagents is a violation of this principle. If secrecy slows progress, then human suffering may be prolonged and unnecessary deaths may occur. Although these harms are not the intention of scientists who withhold information, they are a logical consequence of such secrecy.[25]

Although we do not immediately envision the patenting process creating direct consumer issues, as we will see below, the incentives which are built into the intellectual property system have the potential to create some interesting and highly relevant consumer dilemmas.[26]

Few would argue with the notion that intellectual property protection is essential for the encouragement of increasingly expensive genetic research. As was noted in a review of the Canadian biotech sector, "patent status of a biotechnology product is one of the first issues discussed with financial analysts, investment bankers, venture capitalists, or potential pharmaceutical partners" (Goudey & Nath, 1997, p. 17). However, in order to reap the benefits of patent protection, an investor or sponsoring company must ensure that vital information is not disclosed to the general public (or to other scientists) prior to the filing of a patent application. As a result, important scientific, and, perhaps clinically relevant, information may be withheld longer than would be usual but for the patent process; for a review of this issue, see, e.g., Blumenthal (1996); Blumenthal, Causino, Campbell, and Louis (1996); Healy (1992). Researchers who have received industry support are often required by contract to maintain a degree of confidentiality regarding their research in order to facilitate the filing of a patent. In fact, Blumenthal et al. (1996) found that 56% of the U.S. companies surveyed required academic researchers to keep information "confidential to protect its propriety value beyond the time required to file a patent" (p. 371).

More recently, this same group of researchers asked U.S. life science academics about their "data-withholding behaviour" (Blumenthal, Campbell, Anderson, Causino, & Louis, 1997). They found that involvement in commercial endeavours was strongly associated with "delays in publication and refusals to share research":

> Our findings confirm the widespread impression that involvement with commercialization and participation in AIRRs [academic-industry research relationships] are significantly associated with the tendency to withhold the results of their research (p. 1227).

The most commonly cited reason for the delay in publication was the "need to allow time for filing a patent application." The need to negotiate licensing agreements was also cited as a reason. It is important to note that "commercialization is also associated with denying research results and materials to other faculty." From the perspective of this paper it is interesting to notice "that investigators in the field of genetics are more likely to engage in data-withholding behaviour" (Blumenthal et al., 1997, p. 1227)

Although this environment of secrecy undoubtedly has an adverse affect on collaborative research and the teaching of graduate students (Blumenthal, 1996; Blumenthal et al., 1996), more worrisome is the possibility that confidentiality agreements will impact the reporting of adverse events or less than positive findings associated with clinical trials. One example is a case described by Percy (1997). In this case, a legal battle has developed over a clinical researcher's decision to disclose adverse results associated with industry sponsored research; see also Frankel (1996). As a result, consumers of genetic services and products, for example, may have delayed access to information relevant to their current treatment (this would be particularly relevant to those involved in clinical trials).

Given the importance of patent protection to continued genetic research (Galas & Eisenberg, 1996), the issue becomes: How can we ensure that Canadian consumers receive (perhaps through their physicians) important information while still allowing the patent process to effectively reward the investment of the researchers and industry?[27]

Premature Implementation

From the perspective of the Canadian consumer, probably one of the most directly relevant concerns is that commercial pressures will result in the premature entrance of genetic services into the market place. There are many statements to this effect. Malinowski and O'Rourke point out that "[task forces] have acknowledged that commercial and academic institutions are making genetic testing capabilities easily accessible to the public too quickly and that many genetic tests are subject to misinterpretation" (1996, p. 239), and Motulsky fears that "the promise of a large and lucrative markets for predictive testing of common diseases makes for strong pressures by commercial testing companies to offer such tests before they are thoroughly tested in field studies" (1994, p. 605); cf. also Kolata (1995) and Moran (1996, p. 139).

The fear is thus that not only will genetic services be introduced before their efficacy is etablished, but that they will become part of standard practice, and part of patient expectations, before the ethical, social, and legal ramifications can be assessed. While currently the most controversial genetic tests remain in the domain of research centres, genetic testing is slowly making its way into the clinical setting (see Cole et al., 1996, p. 153).[28]

Given the degree of investment in the biotech industry (as noted above) this concern seems well founded. Corporations such as Myriad Genetic Laboratories, Inc. – the first company to market tests for BRCA1 and BRCA2 – have already constructed facilities to handle the anticipated and, no doubt, hoped for demand for genetic tests (Skolnick, 1996). One can hardly expect this "if we build it they will come" approach to be entirely passive. The building of laboratory infrastructure arguably intensifies the need to create a market. Indeed, from the perspective of industry, any process which delays getting a service to market will likely be viewed as an unnecessary handicap.[29] Similarly, we should be cognizant of the potential effect of industry on the developing norms relating to the breadth of application for a given genetic test. Decisions about who should be considered an "at risk" individual, and therefore within the criteria to be offered a genetic test, should be based on sound scientific, psychological, and social information and not on a desire to increase the potential market. For instance, there is still some debate regarding who should be provided with the option of testing for the BRCA1/2 mutation test. Generally, it is considered appropriate to discuss the test with those individuals at "high risk," e.g., a strong family history of breast or ovarian cancer (see, for example, Working Group of the Stanford Program, 1996, p. 8). The definition of "high risk," however, is malleable. In addition there are, of course, numerous potential harms connected with increasing the market. For example, creating a false sense of security in those women who test negatively, the cost of clinically unnecessary testing, etc.

Recent research around the actual benefits associated with BRCA1/2 testing underscores the need for caution when implementing testing programs prior to appropriate outcomes studies (including a consideration of ethical and legal implications). For example, Couch et al. (1997) found that if too broad a selection criterion is used (e.g., those individuals with any family history of breast cancer), the odds that one of the women screened would have the BRCA1 mutation was 7 in 100. As a result, routine screening of this population would be wasteful since "9 times out of 10 the test will be negative" (Healy, 1997, p. 1449). See also Krainer et al. (1997) and Schrag, Kuntz, Garber, and Weeks (1997). The latter study argues that the life years gained by prophylactic mastectomy and oophoretectomy, one of the only preventative steps available for individuals who test positive for one of the BRCA genes, may be minimal, thus, casting further doubt on the usefulness of the genetic tests.

In addition, there is the worry that those individuals who test negative will be lulled into a false sense of security concerning their risk of developing breast cancer – which would remain at least the same as for the general population – and will therefore be less careful in their future preventive measures (such as self examinations). Such concerns have prompted many to conclude that we should proceed cautiously. For example, an advisory committee to the National Institutes of Health has "recommended that genetic testing for breast cancer be conducted only within strictly defined research protocols. This reverses an earlier position encouraging wider use of testing"

(Gene tests, 1996, p. 202). And Healy provides the following recommendation:

> It is too early to use BRCA gene testing in everyday clinical practice, because it violates a common-sense rule of medicine: Don't order a test if you lack the facts to know how to interpret the result. When it comes to the care of an individual woman facing many health risks, including the risk of breast cancer, genetic information should become a crucial factor only when woven into the tapestry of her overall health (Healy, 1997, p. 1449).

The earlier commercial introduction of a genetic test for APOE4, an allele identified as a risk factor for Alzheimer disease (AD), serves as another example of the potential problems associated with premature implementation. As Relkin, Kwon, Tsai, and Gandy have observed, almost as soon as the gene was identified in 1993, "commercial interests in the United States and Europe began marketing APOE genotyping as a means of predicting the future development of AD in asymptomatic individuals" (1996, p. 154). Because of the mass media coverage, patients began asking for the "Alzheimer gene test." However, after further scientific investigation and a consideration of the ethical, legal, and social implications of predictive testing, it has since been recommended that the test should still be viewed as an ineffective and inappropriate predictive tool (American College of Medical Genetics, 1995; National Institute on Aging, 1996). As a result, it is currently only available as a diagnostic aid. The experience with commercial AD testing led one group of researchers (Post et al., 1997) to the following conclusion:

> Technology transfer from basic genetic research to commercial use with little outcomes-based research in between should be avoided. The availability of an AD genetic test before there is scientific consensus about its clinical usefulness raises questions about the appropriateness and consequences of its use, particularly the potential for discrimination, adverse psychological effects, and inappropriate effect on life decisions (p. 835).

On the positive side, it is encouraging to note that the further scientific research on AD, together with a consideration of ethical and legal implications, did have an impact on the commercial marketing of APOE4 testing. At the current time, Athena Neurosciences, the company marketing the test, will only provide the test to assist in the differential diagnosis of AD. In fact, "before the company will run the test, the ordering physician must sign a statement attesting that the patient is demented" (Post et al., 1997, p. 835).

Given the media excitement which often surrounds the announcement of predisposition genes for such serious and common diseases as Alzheimer and breast cancer, it is understandable that consumers would want access to information about their personal risk, thus making the initial marketing of these tests a relatively easy task. In one study, the researchers found that 87% of the general population surveyed was interested in personal "genetic predisposition for cancer" (Andrykowski, Munn, & Studts, 1996). However, consumer demand may also be tied to reports of the actual effectiveness of the test. Post et al. found that "consumers from our focus groups were not interested in testing that could provide neither predictive data nor a reasonably precise answer about their individual risk of developing AD at a particular age" (1996, p. 835).[30] In fact, it has been suggested (Marshall, 1997) that the unresolved nature of the legal and ethical issues – fear of genetic discrimination, for example – may also affect the uptake of genetic services. So, there may be an ironic situation brewing: The push to market genetic tests, such as BRCA1/2, has created policy concerns which could affect the marketability of the product – despite the initial excitement associated with the discovery of a gene. Although to date there is little evidence to confirm the existence of this phenomenon, the U.S. biotech industry has joined the lobby for stronger privacy laws (Marshall, 1997, p. 782).

Another concern associated with premature implementation is that commercialization may affect the availability of appropriate counselling services – a process which is considered essential to the delivery of any genetic procedure (Knoppers & Robertson, 1995). Natowicz and Ard (forthcoming) have noted that "most of the clinical genetic corporations have had limited or no genetic counselling services, especially physician provided genetic counselling." Such a development in Canada would be truly unfortunate, as the complexity of the information which may impact uptake of a given test (e.g., effectiveness, predictability, psycho-social implications, etc.) heightens the need for comprehensive genetic counselling (and/or education) both before and after the test is performed. In Canada, it is likely that the vast majority of genetic services will, at least initially, be provided through the public health care system. As such, it should be easier to ensure that adequate counselling programs are in place.

"Eugenic" Pressures and the Skewing of Consumers' Views of Disease and Disability

> While contemporary eugenics may not be dictated from above by the political programmes of dictatorial regimes . . . the pressures of the market are as forceful and destructive, if not more forceful and destructive, than what we have had in the past.[31]

The word "eugenics" carries a large amount of historical baggage – and rightfully so. As such, it is a term that should be used with a degree of hesitancy as it can easily invoke emotional and less than reasoned dialogue. Nevertheless, with this caution expressed, I will use the term throughout this section because it has been used, as we shall see, by so many authors in relation to the process of commercialization.

Eugenics as a "science" was described by its founder, Sir Francis Galton, as: "The study of aegis under social control that may improve or impair the racial qualities of future generations either physically or mentally."[32] However, Canadians are more likely to be familiar with eugenics being used as part of as coercive social policy with the goal of improving the genetic make-up of the human race. While the adoption of eugenic philosophy by Nazi Germany is undoubtedly its most notorious manifestation, eugenics also had a large role in Canada's horrendous sterilization laws (e.g., Alberta's *Sexual Sterilization Act*).[33]

Although few fear the implementation of such explicit government designed eugenic initiatives, there are those who believe the recent rapid increase in our genetic knowledge coupled with subtle social pressure, including commercial forces, could result in a *consumer led* "new eugenics" movement.[34] Indeed, this has emerged as the most encompassing (and controversial) fear associated with commercialization. The concern is as follows: In order for there to be a genetics market there must be a perceived need. Will biotech companies, through their efforts to create a robust market for the emerging genetic services, affect our perception of normalcy?[35] As argued by Testart (1995): "For reasons associated with the economy and the market, normality of human beings will be more and more defined according to the needs of industry and insurance contracts" (p. 310).

Few in our society need to be convinced of the power of capitalist forces to create markets. From cigarettes to running shoes, our society seems to have an unending capacity for selling and consuming. At the very least, if left completely unregulated, there is the *potential* that the same forces will have an impact on how genetic services are utilized and thus how we perceive those with disabilities and define normalcy.[36]

There are several ways in which commercial forces could play a significant role in the controversy surrounding the emergence of a "new" eugenics movement. First, as mentioned, markets are needed to sell, for example, genetic tests – be they for the purpose of screening, individual testing, or prenatal diagnosis. The bigger the market (i.e., the wider the definition of disease), the easier it will be to justify and recoup research costs and to maximize profits. Genentech's approach to its recombinant growth hormone, Protropin, serves as an illustration of this marketing phenomenon. While the traditional clinical U.S. market for this drug includes 20,000 individuals with pituitary dwarfism, the company effectively increased its market to include all those individuals in the third percentile in height (Leopold, 1993, p. 219): "By creating the perception that normal shortness is a disease, Genentech has tapped into a potentially major market of 90,000 children born in the United States annually who will fall under the third percentile for height." Cuttler et al. (1996) conclude that "future GH use will likely be determined not only by the results of controlled trials, but also by family preferences, producer pricing, and physician perceptions of the value of height and GH therapy" (p. 531). Using the same growth hormone example, Daniels (1997) analyses this problem through the lens of treatment versus enhancement. He notes:

> Information from the genome project might make the distinction between disease (including genetic disease) and the normal distribution of capabilities seem more arbitrary. . . . Identifying [shortness] genes may then tempt us to think of them as "bad" ones: they lead to Johnny's unhappiness in a "heightist" world. We will then be sorely tempted to think of them very much on the model of genetic defects or disease, especially if they work through mechanisms that have some analogy to pathological defects (p. 191).

If market forces could similarly buttress the demand for genetic tests by redefining what is considered normal or healthy, then is this not, perhaps, a "eugenic" force? By causing individuals to re-evaluate the notion of disease and normalcy in the context of genetics services, we are, at some level, implying that certain genes are "bad" and, if possible, should not be passed on to future generations. Moreover, we are providing the tools which allow individuals to act on these evolving social perceptions. What is potentially worrisome is not that individuals will be making increasingly informed reproductive decisions (a fundamentally important patient right) or that certain disease genes may not be passed on (the potential eradication of a serious disease such as Tay Sachs can only be viewed as a positive achievement). Rather, the concern is that the commercial market could play a significant, profit oriented, unregulated role in this crucial decision-making process. However, whether this *laissez-faire* eugenics should be viewed in the same negative light as the social eugenic programs of the past is an issue worthy of further debate (Kitcher, 1996, p. 196).

By simply emphasizing the significance of genetics to our conception of "who we are," commercial forces can also inadvertently encourage social perceptions essential for eugenic ideology. For example, Lewontin (1992) has argued that the mere existence of the Human Genome Project "has validated the determinism on which eugenics was based" (pp. 72–73). Arguably, commercialization will compound this effect by further increasing the visibility of the "genetic revolution" and by strengthening the social sway of biological information. Of necessity, marketing, be it to physicians or otherwise, will send the message that: *You need this information because it is important*. This, in itself, may help generate a simplistic and overly deterministic view of genetic information. As argued by Crossley (1996): "By encouraging us to conceptually break down persons into traits, prenatal genetic testing threatens to reinforce our existing and destructive reductivist tendencies" (p. 1234).[37]

Admittedly, whether the individual consumer of genetic services will actually become the engine of a new eugenics movement fuelled by commercial pressures is far from certain. Likewise, it is unclear what the social trade-offs of such a process would be (e.g., the funding of scientific research versus the skewing of social perceptions). And although this author is sceptical of the potential for market forces to play a constructive role in this context, one does not want to lose potential scientific benefits by invoking the spectre of past atrocities – unless, of course, such invocations are justified.

Marketing and Advertising

> **If the practice of medicine is an art, pharmaceutical advertising is its predominant public expression.[38]**

The marketing of genetic technologies can happen on a variety of levels. A scientist may promote a discovery, or the potential for a discovery, for the benefit of venture capitalists. A biotech company may hype emerging genetic services in order to ensure a successful public offering of shares or to attract investors.[39] Physicians and geneticists may receive glossy mail advertisements for a new genetic technology. When reporting on a publicly funded project, researchers may feel the need to simplify and give a practical spin to the potential outcomes of their research in the hope of justifying the work in these times of private sector dominance. While most of these marketing strategies are not aimed directly at the consumer (in fact, some are not traditionally thought of as "marketing"), they nevertheless form part of the climate in which genetic services are emerging and, as such, many Canadians will be exposed to their pressures. Indeed, the inter-industry hyping of genetic technologies often ends up in the popular media (e.g., newspapers, the Internet, etc.)[40] and therefore becomes part of the public's "genetic" education. But how accurate and fitting is this education?

In a critique of modern marketing strategies, Neil Postman has said that "what the advertiser needs to know is not what is right about the product but what is wrong about the buyer" (1993, p. 170). This comment is especially fitting in the health care context. In fact, because of the great potential for abuse, the promotion of medical technologies, such as prescription drugs and diagnostics, has been restricted by the Health Protection Branch of Health and Welfare Canada.[41] In general, the biotech/pharmaceutical industry may only directly advertise to the those in the medical profession (Chaloner-Larsson et al., 1993, p. 263). As noted by the Working Group of the Stanford Program in Genomics, Ethics, and Society (1996) with respect to testing for breast cancer susceptibility, the need for the regulation of advertising seems particularly appropriate in the field of genetics:

> **Many women are terrified of breast cancer. Marketing campaigns might easily prey on this fear, by first increasing women's anxiety and then offering testing as a solution to this heightened concern. Genetic testing for BRCA1/2 mutations holds too many dangers to allow unrestricted advertising (p. 12).**

Apprehension about the appropriateness of advertising in the realm of health care is obviously not new. In a U.S. study of the promotional material published in medical journals, summarized by Shapiro (1997), it was concluded that "a large portion of the advertisements minimized concerns about adverse reactions, were misleading with respect to efficacy, used statistics inappropriately, promoted the use of products in inappropriate populations and used headlines and subheads that were not supported by the remainder of the advertisement" (p. 360). Shapiro also notes that in another study, "42% of materials distributed by pharmaceutical manufacturers directly to physicians were found not to comply with relevant U.S. federal regulations."

Although traditional media such as medical magazines remain the primary venue for biotech advertising, this is an era of expanding communication technologies. It has become increasingly common for the biotechnology industry to use 1–800 numbers, the Internet, and even fax machines.[42] However, these conduits of information also have their own unique benefits and risks (Pines, 1997):

> The Internet's international scope and immediate consumer access to such information sources as medical data and informal chat boxes offer unprecedented opportunities for obtaining highly reliable or, on the other hand, distinctly unreliable information. The possibilities in using advanced technologies to reach consumer and physician audiences even more broadly are enormous (p. 65).

Despite the consistent presence of concerns regarding the impact of advertising in the realm of health care, there is pressure to change our present restrictive marketing approach. In fact, in the United States, the pharmaceutical industry has successfully lobbied the FDA to reduce restrictions on how prescription drugs may be advertised on television (Ingersoll, 1997). Drug companies will now be allowed to extol a drug's potential benefits while only needing to warn consumers of the "product's main risks." In other words, TV ads will not need to list all the possible side effects associated with a prescription drug. This new policy has been initiated at a time when there has already been "exponential growth of DTC [direct-to-consumer] programs for prescription products, particularly those using newer technologies" (Pines, 1997, p. 64). As Pines notes, in 1996, for the first time ever, the amount spent [in the U.S.] for DTC advertising exceeded the amount spent on direct-to-physician advertising. We will have to wait to see what impact this American decision has on Canadian policy.

Currently, in Canada, pharmaceutical companies are not allowed to advertise prescription drugs directly to consumers (e.g., magazines, TV ads, etc.). There is, however, similar pressure to allow more "direct-to-consumer" advertising in Canada, a policy which is currently being pushed by the Pharmaceutical Manufacturers Association of Canada (PMAC).[43] Evidence has shown that direct-to-consumer advertising is an effective means of increasing the utilization of a given service (Pines, 1997, p. 65). Such marketing schemes are often justified as mechanisms of patient empowerment, however, they are also criticized as an inappropriate means of conveying medical information:

> As various investigations into the drug industry have demonstrated, such advertising seldom provides clear, concise information that fully informs the reader. It is designed mainly to sell goods and services, not educate and warn of risks (Armstrong & Armstrong, 1996, p. 207).

For similar criticisms, see Kessler, Rosa, Temple, Schapiro, and Griffin (1994) and Shapiro (1997, p. 361).

Likewise, although research done in Canada and the United States seems to indicate that patients support this type of advertising, health care providers believe that it undermines the physician/patient relationship by encouraging unreasonable patient demands for services. In one study, 61% of those surveyed supported direct advertising (Whyte, 1993, p. 150); however, 63% said they would not be able to tell if they were being misled by such ads. See also Allentuch (1996), where it is reported that a study done on behalf of the PMAC "showed that 72% of Canadians think people should be able to get information on drugs from manufacturers. As well, more than two-thirds of the 1,509 persons surveyed said they oppose laws that prohibit drug makers from telling the public what drugs do." It should be emphasized that this Canadian research was done for PMAC, a group which obviously has a vested interest in the outcome. See also Salloum and Franssen (1993, p. 1058), where it is reported that 89% of the physicians surveyed thought that "patient expectations and demands influenced their test ordering behaviour"; Langley, Machellen, Sutherland, and Till (1992, p. 665) where it was concluded that "patient wishes emerged as the most important [nonmedical] factor" in referral rates; and Williams and Beresford (1991, p. 308) where it is noted that physicians are reluctant to deny patients' unreasonable demands for fear of "weakening their patient's confidence in them and because of the fear of litigation if something goes wrong."[44]

In an area such as genetics, with the concomitant complex social, medical and ethical issues, these concerns seem particularly relevant. In relation to genetic testing, Dykes (1996) has noted:

> As more disease susceptibility genes, and their causative mutations, are identified, the pressure from some concerned individuals to have such tests performed will increase. Advertisements offering genetic testing directly to private individuals have already appeared in the national press in the U.K. and U.S.A. (p. 693).

In Britain, the new Advisory Committee on Genetic Testing (ACGT) (1997) has recommended that companies wishing to sell genetic tests through the mail be restricted to carrier testing. However, this permissive approach to carrier testing has been criticized (see King, 1997, p. 7, who calls for further formal regulation).

Even if direct-to-consumer advertising is not accepted in Canada, it must be recognized that Canadians are exposed to American campaigns via our close ties to the American media. In addition, biotech and pharmaceutical companies have devised creative ways of conveying the same information without the use of overt mass media advertising schemes (De Lisser, 1995; Spaulding & Joseph, 1992; Spurgeon, 1993). And, of course, biotech companies also advertise directly to physicians. Milunsky (1993) observes that "despite the obvious complexities and required interpretations of cytogenetic, molecular genetic and pregnancy screening tests, mega-commercial laboratories have sought to deliver such services directly to physicians' offices" (p. 627); see also Milunsky (1992). Office presentations by pharmaceutical detailers, the furnishing of drug samples and office supplies, biotech sponsored CME (continuing medical education) courses (Natowicz & Ard, forthcoming) and corporate sponsorship of post marketing clinical trials all seem to influence physician behaviour (Lexchin, 1993, p. 1405; Woollard, 1991, p. 932). Given that at least some studies have shown that "physicians tend to prescribe drugs according to advertised guidelines rather than published scientific literature," the impact of *direct-to-physician* advertising should not be underestimated (Rosner, 1992, p. 266).

As noted earlier, physicians will obviously play a critical role in the delivery of genetic services to Canadian consumers. They will have the dual role of acting as gatekeepers to genetic resources and educating their patients about the benefits, limits, risks, and ramifications associated with genetic testing. However, many studies seem to demonstrate the low level of understanding that many professionals have concerning human genetics. Andrews (1997, p. 40) summarizes a variety of studies on this issue. In one study, 35% of the physicians surveyed "incorrectly believed one needed have an affected relative in order to have a child with cystic fibrosis; 43% believed that cystic fibrosis affected children of one sex." See also Stephenson (1995, p. 1161) and Milunsky (1992, p. 644), who argue that most physicians who request genetic services straight from commercial laboratories will not be able to interpret results. Research by Wertz (1996) reveals a consistently low understanding among primary care physicians concerning basic genetics.

This indicates that direct-to-physician advertising may hold many of the same concerns as direct-to-patient advertising (Boer, 1993; Chren & Landefeld, 1994). In addition, if physicians have such a poor level of understanding, how will they adequately inform patients of their options vis-à-vis genetic testing? Clearly, further education of Canadian physicians is required – particularly general practitioners as they will be the first point of contact and will receive a large portion of the marketing material on genetic services. Because genetics is a field which is moving forward at an incredibly fast pace, this education process will not be easy.

Skewing the Direction of University Based Research

> The fact that the universities must function within the marketplace culture should not trick us into thinking that the generation of profit is the only relevant cultural value to steer our enterprise. . . . What industry is out to do is to take control of our futures. Accordingly, our prime political objective must be to get industry off the campuses and out of our laboratories.[45]

One of the most common criticisms of the commercialization process in the realm of human genetics is that it will erode the quality of university based research. Although this paper is focused on the more direct potential impacts of commercialization for Canadian consumers, the concerns associated with the growing link between academia and industry are worth noting – both because of the intensity of the debate and because the issues are ultimately relevant to the consumers of genetic services.

Although university researchers have always been under a variety of less than ideal pressures which have affected the direction and "purity" of research (e.g., pressures to produce militarily relevant innovations), industry influence has probably never been stronger – particularly in the biotech arena. Alliances between universities and industry are becoming increasingly common (Murray, 1996, p. 28; Natowicz & Ard, forthcoming). Schultz (1996) found that from the decade of 1960–69 to the decade of 1980–89 the number of university-industry research centres went from 61 to 284. At the University of British Columbia, between 1985 and 1995 "the university-industry related funds have gone up 10-fold" (Miller, 1995, p. 91S). It is also reported that the number of technologies licensed increased from 28 to 200 in the same period. Similarly, see Blumenthal et al. (1996, p. 369), where it is reported that 90% of the U.S. life sciences companies surveyed had a "relationship with academia." Later, the authors report: "We found that life science firms were significantly more likely to support academic research in 1994 than in 1984" (p. 371).

In fact, obtaining industry funding is generally encouraged by both academic institutions and federal funding agencies such as the Natural Sciences and Engineering Research Council and the Medical Research Council (see, e.g., Edwards, 1993, p. 18). These arrangements hold numerous advantages for both parties. They inject essential capital into expensive genomic research, allow the private sector to benefit from the university research infrastructure (Haber, 1996, p. 442), and increase research capacity generally. In fact, such arrangements are often perceived as essential; see, for example, Patterson and Emanuel (1996) who argue:

> In a society in which private industry is the only mechanism by which we can bring new knowledge to bear on practical application, relationships between industries and academics and health care institutions are essential. But they need to be conducted with clear rules, vigilance, and good judgement (p. 317).

See also the recommendations regarding "Genetic Science and Industry" issued by the Science and Technology Committee of the British House of Commons (1995).

However, these arrangements also create concerns for both industry (e.g. lack of control of research, fear of early dissemination, and loss of intellectual property rights) and the academic community (Greely, 1995; Korenman, 1993; Medical Research Council, 1996; Rothman, 1991; Tri-Council Working Group, 1996).

Some of the more often stated concerns and conflicts in the University research setting are as follows:

- Restriction of the academic community's ability to collaborate with other researchers;[46]
- Loss of academic freedom and ability to control the direction of research;
- Loss of objectivity;[47]
- Distortion of the priorities of the university toward research with commercial potential;[48]
- Exploitation of students;[49]
- Pressure (both internal and external) on researchers to spend less time on non-commercial activities such as teaching;[50] and, when clinical trials are involved,
- Abuse of the research/physician-patient relationship.

Although such concerns may not have obvious implications for the Canadian consumer of biotechnology, if commercialization adversely affects the quality of the long term benefits (which Canadians hope to reap from scientific inquiry), then it is the Canadian citizen who ultimately loses.

However, there may also be more immediate ramifications. Scientific inquiry which is removed from profit motivated commercial initiatives can act as an important scientific "watch dog." Can biotechnology deliver the results promised by industry? Are there associated dangers of which consumers should be aware? These questions are answered with the most persuasive force by researchers who do not have a financial stake in the expansion of the industry which produces the product. As argued by science historian Charles Weiner a decade ago: "[T]he dual roles played by many leading biologists have begun to impair the credibility of scientists when they provide advice on matters of public concern relating to their research" (Weiner, 1988, pp. 32–33). So, in the end, the commercialization of university-based research may rob Canadian consumers of a valued and unique quality control mechanism – the independent academic researcher.

Other Concerns Relevant to the Canadian Consumer

Liability concerns. Will commercial pressures create artificially high expectations among patients, thus inviting increased litigation (e.g., failure to provide, inform about, or interpret a particular genetic test)?[51] Will this medical legal environment encourage an over-utilization of genetic services?[52] The high profile which wrongful birth and wrongful life law suits receive will undoubtedly foster liability concerns among those health care providers who are involved in the provision of genetic services (see, for example, Capen, 1995; Southam Newspaper, 1996). The possibility certainly exists that these fears will lead to the inappropriate use of diagnostic tests (Malinowski & O'Rourke, 1996, p. 239). In addition, the complexity of the information which needs to be conveyed in the consent process may make this area of clinical practice particularly susceptible to informed consent litigation. However, as with all informed consent litigation, a patient/plaintiff may have difficulty in proving that the physician's non-disclosure "caused" (in the legal sense) the injury being litigated. For example, would a child not have been born with a disability "but for" the failure of the physician to inform the parents of a particular genetic test? Would the parent(s) have acted on the information? As we have seen from the recent Supreme Court of Canada case, *Arndt v. Smith*,[53] it may be difficult for plaintiffs to satisfy these causation issues. Nevertheless, as noted by Andrews (1997), "people may undertake genetic testing without sufficient consideration of its potential ramifications" (p. 39). Physicians should therefore carefully adhere to emerging disclosure and counselling protocols.

Cost to the health care system. Will the introduction of these services become a burden on an already strained health care system? Some have argued that the emerging genetic technology will have just such an effect; see, e.g., Vaillancourt-Rosenau (1994) and Suhr et al. (1995). The latter authors state:

> [T]his burgeoning technology is not inexpensive. With the prevailing recognition that health care resources are finite and/or diminishing, there are major implications for the entire health care community (including patients, health care professionals, and hospital and government administrators, as well as the general public) (p. 765).

Interestingly, issues of health care economics also seem to be recognized by industry (Social and Ethical Conerns Committee, 1996). A study of biotech companies done by KPMG shows "More than half [of those interviewed] understood that health economics/cost effectiveness will play an increasing role in a product's commercial success" (Nadeau, 1996, p. 41).

Such economic concerns are particularly relevant to Canada where provincial health care plans will be called upon to pay for many of the genetic services (Rowland, 1993, p. 188). In fact, Wertz's survey of Canadian genetic professionals found (1995, p. 115) that, in general, if a professional thought a genetic service should be available, they also felt it should be paid for by the public health care system. Given the influence that health care professionals continue to have over what services are deemed "medically necessary" (Caulfield, 1996), and therefore covered by the medicare system, such statistics highlight the likelihood that clinical genetic services will first be implemented through the public system.

In addition, we must recognize that providing genetics in an appropriate manner (e.g., comprehensive counselling services) will be time consuming and expensive. As summarized by Suhr et al. (1995, p. 766): "studies [have] concluded that clinical genetic services [are] time-consuming, labour-intensive, and not financially self-supporting." This study also notes that problems may arise if we expect Canadian family physicians to simply assume a primary role in the allocation of genetic services. An "average" family physician visit has been estimated at seven minutes. An average "direct genetic service" has been observed to be 47 minutes. As a result, this research group concludes that: "the issue of adequate [family physician] remuneration for cognitive activities, as well as the average physician's knowledge of molecular genetics, must be addressed." As a result, provincial health care schemes must consider not only the initial fee for the genetic test but also total aggregate costs.[54]

It should be noted that the economic impact of the commercialization of human genetic research may be less obvious than a potential over-utilization of genetic tests. Because much of the genetic financial investment is in the realm of pharmaceuticals, biotech corporations may feel compelled to push their genetically derived drugs in order to recoup early research costs.[55] Given that the pharmaceutical sector is the fastest growing component of the health care sector, this concern seems particularly relevant.

SOME PRELIMINARY RECOMMENDATIONS

> Public policy responds to perception and appearance as much as it does to reality, and the values and norms that shape expectations and influence perceptions are central to our understanding of conflict in the policy arena . . .[56]

As noted earlier in this paper, many of the concerns which are heard in relation to the commercialization process remain highly speculative (e.g., the eugenic effect of commercial pressure). Others are not (e.g., the impact on the sharing of scientific information). Nevertheless, given what is at stake, and given the speed and momentum of industry involvement, a consideration of possible responses is, at least, warranted.

Further Research

First and foremost, as pointed out by, e.g., Blumenthal (1992, p. 3349), more research into the actual benefits and possible negative ramifications of commercialization is required. For a process that is part of the fibre of our culture, there is surprisingly little information available. Policy should be built on a sound understanding of what is actually occurring (and what reasonably might occur) and not on unjustified fears nor overly optimistic visions of the future.

Obviously, even if relevant data can be collected, the policy-maker's job will still not be a clearly defined or easy one. What one person considers a vice another may consider a virtue. A difficult balancing of social needs and values will be required. Many of the prevailing values are strongly held – such as a belief in the free market or a fear of the commodification of life – and compromises will undoubtedly be required.

Examples of general questions which may warrant further research include:

- How to improve the regulatory environment (including a re-examination of the patent regime)?
- Is a new legal mechanism necessary for protection against "genetic discrimination"?
- What is the best way to effectively educate both professionals and the general public?

- How can we ensure meaningful, ongoing, public consultation?
- Who should be offered a given genetic service (e.g., what population would benefit from its availability)?
- What is the impact of marketing strategies on practice patterns, standard of care, consumer perceptions, etc.?
- How should we integrate this emerging technology into our health care system to ensure equitable access to genetic services?

A Consideration of Comprehensive Legislation or Regulation

As illustrated by the criticism the federal government received over its proposed *Reproductive and Genetic Technology Act*, Bill C-47, it is difficult to respond in a simple and comprehensive manner to the dilemmas which surround these emerging technologies. The state only has so many ready-made "regulatory arrows" in its quiver and the use of the most all encompassing – that is, criminal legislation – seems to have far too many negative implications to be sufficient (Caulfield, Hirtle, & LeBris, 1997). However, the federal government may wish to consider developing a more flexible mechanism to address a number of the issues raised above. For instance, this could be a component of the regulatory scheme suggested in the material which accompanied Bill C-47 (this would be the "third phase" of the federal government's proposed new reproductive and genetic technologies policy, Minister of Health, 1996). In particular, an overview body could be established to monitor the introduction of genetic technologies in Canada. Such an entity could, *inter alia*, review whether there is sufficient evidence (e.g., consider outcomes research, ethical and legal findings, etc.) to justify making the test commercially available; ensure that genetic tests are only provided as part of a comprehensive genetic program which includes appropriate counselling services and follow-up; and help to disseminate the findings of ongoing research initiatives.[57]

Education of the Public and Health Care Professionals

There is clearly a need for the increased education of clinicians in the area of human genetics (Riccardi, 1997). This education process should not only include continued updates on the scientific facts surrounding genetic testing, but also information on the legal, social and ethical implications. Effective education of both the public and the general practitioners may be one of the most meaningful regulatory mechanisms available. An educated physician population can more effectively monitor the claims of biotech companies and inform their patients of the true benefits and risks. There is no reason that the federal and provincial governments cannot become involved in this process.[58]

Monitoring of Marketing Strategies

A comprehensive policy for the monitoring of biotech marketing strategies could be considered. As noted by one commentator: "The pharmaceutical industry has too much at stake to monitor its own promotional activities in a reliable manner. Government is in a better position to represent the consumers' interests, to allocate the necessary resources to monitor evolving marketing strategies and to crack down in a meaningful way on offenders" (Shapiro, 1997, p. 361).[59] Given the expanding nature of the "information highway," this may be a difficult task (Pines, 1997):

> [The] expanding communications technologies in medical care pose new challenges to the regulatory process. . . . How the government responds in this period of new challenges will determine promotional policies for medical products for years to come (p. 65).

Other Regulatory Mechanisms

Other regulatory mechanisms of which the government should be aware, and which are relevant to the protection of consumer interests, include:

- University conflict-of-interest guidelines;
- The informed consent process;
- Private law (e.g., negligence and fiduciary law);
- Public law (e.g., The Canadian Charter of Rights and Freedoms);
- Professional practice guidelines and accreditation mechanisms (both for the health care providers and laboratories).

CONCLUSION

> **Some people think that there is a deep opposition between "government intervention" and "free markets." But the opposition is too simple. No one is really opposed to "government intervention." Markets depend for their existence on law, which is necessary to establish property rights and to set out the rules governing contracts and torts. Short of anarchy, a system of legal entitlements, set by the state, is inevitable. If we are asking whether the state is an important part of the solution to current social problems, there can be only one answer: It had better be.[60]**

This paper began by noting that any comprehensive policy governing the commercialization of genetic research will necessarily involve a careful balancing of many factors and social forces. However, action is clearly needed to ensure the healthy and appropriate integration of genetic technology into our health care system. In fact, the creation of an environment where there is an ongoing monitoring of medical, scientific, legal, ethical, and social issues may, in fact, help to foster a strong biotech sector by building consumer confidence.

The task will clearly be challenging. Can we structure the research and commercialization environment in a manner that will ensure both the future of basic research and the effective and efficient participation of industry? Can we maintain the free flow of information among scientists while still reaping the benefits of a competitive biotech market? And perhaps most importantly, can we regulate the potentially adverse impact of market forces while still allowing the rapid dissemination of genetic innovations? It is hoped that this overview document will assist the government in its efforts to address these difficult questions.

I close with one last question. Can this expensive research be supported by the private sector without an artificially created, inappropriately large, consumer market? In other words, would a market determined independent of commercial pressure be sufficient to sustain future genetic research? From the perspective of commercialization policy this may, in the end, be the ultimate dilemma.

NOTES

[1] J. Tam, Director of Scientific and Professional Affairs for the Canadian Drug Manufacturers Association, in Allentuch (1996).
[2] The Canadian initiative, a federal project entitled the Canadian Genome Analysis and Technology Program, first received funding in June of 1992 (Worton, 1993, p. 1).
[3] Much has been written recently on this topic. For example, see Ducor (1977); Heagerty (1997); Kurland (1997); Malinowski and O'Rourke (1996); Natowicz and Ard (forthcoming). Svatos (1996).
[4] Kevin Kinsella, CEO, Sequana Therapeutics, quoted in Cohen (1997, p. 775).
[5] See, for example, C. Anderson (1993); Culotta (1993); Dodet (1993); Goudey and Nath (1997, p. ix.). See also Science and Technology Committee (1995, para. 149), where it is noted that in Britain the number of biotech companies grew from 386 in 1994 to 485 in 1995.
[6] See, for example, Damsell (1996); Day (1995); Lehrman (1997). See also the editorial, Capitalizing on the genome (1996), where the impressive performance of genomic companies is noted. For instance, since their initial offering, the value of Genome Therapeutics stocks has increased 381%, Human Genome Sciences has increased 208%, and Incyte Pharmaceuticals has increased 227%.
[7] See Cohen (1997, p. 772), where it is estimated the pharmaceutical investment in genomics has topped $1 billion. See also Erramouspe (1996), where it is noted that "SmithKline Beecham committed $125 million last year [1995] for the rights to genes sequenced by Human Genome Sciences. In March 1994, Hoffman-La Roche agreed to invest more than $70 million in Millennium Pharmaceuticals for the right to exploit genes relating to obesity and adult-onset diabetes" (p. 995).
[8] See Natowicz and Ard (forthcoming), where it was found, for example, that of the U.S. companies surveyed 85% of the "consultants for these companies have dual affiliations with academia."
[9] Thomas, Davies, Birtwistle, Crowther, and Burke (1996, p. 387). The vast majority of these patents are privately held (76%). Seventeen percent of the patents are held by public entities and 7% are held by individuals. The authors also note that a significant proportion of the patents are held by companies in the U.S. and Japan.
[10] See, for example, Prouty (1996). According to Haber (1996, p. 442), "In 1994, U.S. pharmaceutical companies funded 79 [academic-industry] alliances for a total expenditure of $993 million. Swiss and Japanese pharmaceutical companies funded 10 alliances at $357 million and 19 alliances at $319 million, respectively. See also Murray (1996), for a discussion of the programs necessary to "help accelerate technology commercialization" in Canadian universities.
[11] The following data are a summary of information from Executive summary (1996, pp. 4–6); Goudey and Nath (1997); Heller (1996, pp. 9–11).
[12] The exact revenue generated by the Canadian biotech industry varies between reports. For example, the Goudey and Nath report (1997, p. iv) puts the value at $1.141 billion, an increase of 45% since 1994. Much of the difference may be attributed to how "biotechnology" is defined. Suffice it to say, there has been healthy growth in this sector over the past few years.
[13] Companies such as Human Genome Sciences and Incyte Pharmaceuticals make millions of dollars selling little pieces of genetic information (expressed sequence tags or ESTs) to pharmaceutical companies.
[14] It should be noted, however, that recent research indicates that interest in genetic testing seems to be high. See Julian-Rynier et al. (1997), where it was found that 87.7%

of those surveyed said they would ask for the breast cancer gene test if it became available. Whether such statistics will actually translate into sales is another question.

[15] A. Anderson (1996, p. 4).

[16] See, for example, Prouty (1996, p. 956), where it is noted that Quebec offers corporations "a 40% refundable tax credit on all eligible R&D expenditures when incurred on the company's behalf by a recognized research centre." See also the Science and Technology Committee (1995) report in Britain, and Alberta Science and Research Authority (1996), which focus on how to strengthen the respective regions' biotech sector.

[17] Bill C-47, *An Act respecting human reproduction technologies and commercial transactions relating to human reproduction*, 2nd Session, 35th Parliament, 45 Elizabeth II, 1996. This Bill never made it through the enactment process before the federal election was called in the Spring of 1997.

[18] Bill C-47, see Note 17.

[19] Gitlin (1995, p. 53).

[20] See, for example, *Ciararliello v. Schacter*, [1993] 2 S.C.R. 119 at 135: "The concept of individual autonomy is fundamental to the common law and is the basis for the requirement that disclosure be made to the patients." See generally, *Fleming v. Reid* (1991), 4 O.R. (3d) 74 (C.A.); *Re Baby R.* (1988), 53 D.L.R. (4th) 69 (B.C.S.C.), *rev'g* (1987), 9 R.F.L. (2d) 415 (Prov. Ct.); *Walker v. Region 2 Hospital Corp.* (1994), 116 D.L.R. (4th) 477 (N.B.C.A.); *Malette v. Shulman* (1990), 67 D.L.R. (4th) 321 (Ont. C.A.). See also, Royal Commission on New Reproductive Technologies (1993, p. 93).

[21] *B.(R) v. Children's Aid Society of Metropolitan Toronto* [1995] 1 S.C.R. 315 at 317.

[22] Leslie Alexandre, vice-president of Oncor Med. Inc, quoted in Papp (1996).

[23] See also Frankel (1995), who notes: "[G]enetic research and technology represent powerful forces for change. The law, meanwhile, is a mechanism, for maintaining stability. When the two forces interact, there can be considerable tension. . . .".

[24] For example, one issue which I do not cover is the need to ensure equitable access to genetic services; see World Medical Association, *Declaration on the Human Genome Project* (44th World Medical Assembly, Marbella, Spain, 1992), cited in *International Digest of Health Legislation*, 1993, p. 151. Likewise, I do not discuss whether and which genetic services should be considered "medically necessary" and thus covered by provincial health care schemes (see generally, Caulfield, 1996).

[25] Rosenberg (1996, p. 393).

[26] As the reader is undoubtedly aware, there has been a large degree of controversy around the very propriety of patenting of human genetic material. This paper is not the forum to revisit this ongoing debate, however, there has been a great deal written on this subject. See, for example, Caulfield, Cherniawsky, and Nelson (1996); Churchill (1994); Danish Council of Ethics (1994); Ducor (1997); McGoodwin (1996); Straus (1996).

[27] A number of commentators have suggested that a portion of the concerns associated with the patent regime could be mitigated through relatively minor adjustments. For example, Ducor (1997) advocates an expanded "experimental use" exemption. In addition, some commentators have argued that the test for utility, a test which states that an invention must have industrial application, should be applied with particular strictness in the context of human genetic sequences. By doing so, only true "genetic inventions" would receive patent protection. This would avoid the "hoarding" of genetic data and may lesson the need for secrecy in association with a broader range of genetic research. For a discussion of this issue, see Caskey, Eisenberg,

and Straus (1995); Caskey and Williamson (1996); Kiley (1992). The Science and Technology Committee (1995, para. 204) concludes: "Only a combination of a gene and a known utility which is novel and not obvious should be patentable in the context of that utility; and a combination of the same gene and a further novel utility should also be patentable."

28 Commenting of the U.S. situation, Stix (1996) notes: "[T]he rush to test outside the research environment has started. . . . Two companies – Myriad Genetics and OncorMed – have developed more comprehensive tests that look for a broader range of mutations in both the known breast cancer genes, *BRCA1 and BRCA2*. These tests are expected to come into routine clinical use in a few years" (p. 107).

29 As argued by Ed Rygiel, president of MDS Health Venture Capital Corp., quoted in Lanthier (1996b, p. 62): "[The biotech] industry is facing enormous challenges. Cutting the time it takes to get to market will become increasingly important . . .".

30 However, see Lerman et al. (1997), who found that while both an education and counselling approach increased patients knowledge regarding the limits of BRCA1 testing, "neither approach produced changes in intentions to have BRCA1 testing" (p. 148 [abstract]).

31 See Roy, Williams, and Dickens (1994, p. 187): "There is no need for a state-inspired and state-organized and, by implication, coercive eugenics programme, if voluntary parental uptake and utilization of prenatal diagnosis, with selective abortion of fetuses found to be defective, will, for all practical purposes, achieve the same result."

32 In Ledley (1994, p. 158).

33 S.A. 1928, c. 37, as amended.

34 See, for example, Keller (1992, p. 299): "There is no question that eugenics has become a vastly more realizable prospect than it was in the earlier part of the century, and it must be granted that, in many ways, the very notion remains as disturbing as it was in 1945."

35 Renaud, et al. (1993, p. 298, Table 4.10), where it is noted that 51% of physicians surveyed thought that "PND makes disorders out of conditions hitherto considered normal" and 49% agreed that "PND increases intolerance toward anomalies."

36 I have discussed this issue elsewhere; see Caulfield and Robertson (1996).

37 See also McGee (1997, pp. 59–62) for an interesting critique of the notion of genetic determinism.

38 Shapiro (1997, p. 359). The author goes on to note that "the $12.3 billion annual budget for drug advertising in the United States exceeds that for all undergraduate and postgraduate medical education and comes close to the entire budget of the National Institutes of Health."

39 While there are many examples of this phenomenon, see the following Websites: Genzyme Corporation, www.genzyme.com/ir/genz/press/970424.html; and Mitokor, www.mitokor.com. For another example see an October 15, 1996, press release by the Montreal based Algene Biotechologies Corporation announcing that they were close to finding another Alzheimer predisposition gene. The press release goes on to note, "Dr. Gauvreau is confident that these results should allow Algene to attract partners at favourable terms in the near future." Two other scientific press releases by Algene conclude thus: "The Class B Subordinate Shares of Algene are listed on the Montreal Exchange under the symbol AGN.B." It is interesting to note that the location of an actual gene has yet to be reported in a peer reviewed publication.

40 See, for example, Walholz (1997).

41 The potential aggressiveness of firms should not be underestimated. For example see Tanner (1997), where it is reported that a syringe company, Becton Dickinson,

"asks customers to pass along the names of friends who also purchase syringes" for the purpose of formulating mailing lists.

[42] See Websites mentioned in Note 39.

[43] See Allentuch (1996) and Wysong (1996), where a PMAC position paper is quoted: "By reducing the restrictions on advertising which currently exist, more and better information can be made available to patients, resulting in a more fully informed consumer able to exercise health care choices effectively" (p. 32).

[44] These studies are discussed in more detail elsewhere, Caulfield (1995, p. 224).

[45] Kurland (1997, p. 762).

[46] See, for example Blumenthal et al. (1996, p. 372); Wadman (1996).

[47] Maatz (1993, p. 147).

[48] See, for example, Natowicz and Ard (forthcoming), and Blumenthal (1992): "Among the respondents to the Harvard project's faculty survey, 30% of biotechnology faculty with industrial support (compared with 7% without it) said that their choice of research topics had been influenced by the likelihood that the results would have commercial application" (p. 3346).

[49] Lewontin (1992) notes that molecular biologists involved with industry may face serious conflicts of interest. "In some cases graduate students working under entrepreneurial professors are restricted in their scientific interchanges, in case they might give away potential trade secrets" (p. 74).

[50] Maddox (1994, p. 555).

[51] Milunsky (1993, p. 627). For Canadian case law see, *Freeman v. Sutter*, [1996] 4 W.W.R. 748 (Man. C.A.); *Arndt v. Smith* (1997) S.C.C; and *Cherry v. Borsman* (1992), 94 D.L.R. (4th) 487 (B.C.C.A.). See also Andrews (1992) and Burk (1994, p. 806).

[52] Renaud et al. (1992, p. 289), report that 62% of Alberta physicians surveyed thought that lawsuits would lead to more prenatal diagnosis than is medically required.

[53] See Note 51.

[54] Of course, the use of genetic services may also be justified as a cost saving mechanism – a concept that brings its own concerns; see Clarke et al. (1990, p. 1145); Science and Technology Committee (1995, para. 139).

[55] Vaillancourt-Rosenau (1994, p. 555) notes that 10–20% of all drugs are "genetically designed." See also Malinowski (1996, p. 343) where it is noted that in addition to the over 40 genetic drugs in commerce, "approximately 120 genetics-based drugs are in Phase III advanced clinical trials, 494 other such products are in human clinical trials, approximately 2000 proposed biologically derived drugs are in developmental stages, and genetics-related investigative drug applications (INDs) now represent forty-five percent of all INDs."

[56] Frankel (1996, p. 1298).

[57] Of course, any such entity would need to be integrated with existing regulatory mechanisms. See, for example, *Food and Drugs Act* R.S.C. 1985, c. F-27; and the *Medical Devices Regulations*, C.R.C, c. 871. *The Food and Drugs Act* could be amended to more directly address the issues associated with DNA-derived drugs and genetic tests. It has been suggested, however, that because the Act is focused on clinical safety and efficacy, it is not an appropriate forum for the type of ethical and social considerations genetic technologies demand (telephone interviews with Dr. D. Freeland of the Health Protection Branch, Health Canada, November 4, 1996 and February 5, 1997). See also Health Canada (1993). In addition, the Act does not catch genetic testing done as part of an existing laboratory's services – likely to be a large portion of all initial genetic testing. This regulatory hole was recently recognized by a U.S. Task Force which was struck to examine issues of genetic testing (NIH-

DOE Working Group on Ethical, Legal and Social Implications of Human Genome Research, *Proposed Recommendations of the Task Force on Genetic Testing* found at http://www2.med.jhu.edu/tfgtelsi/fedregister/index-RECOMMEN.html.).

58 See, for example, WHO (1996), Recommendation 2 (noting the need to develop "appropriate health education material for the public"); Recommendation 5 ("Genetic screening should not be imposed on an uninformed public, and means should therefore be found to improve the community's understanding of human heredity"); and Recommendation 18 ("Information and educational materials about the health implications of the Human Genome Project should be made widely available in forms that can be freely used and copied)."

59 The author argues that fines are an insufficient deterrent to inappropriate advertising, as the pharmaceutical industry will simply view them as the cost of doing business.

60 Sunstein (1997, p. 108).

REFERENCES

Advisory Committee on Genetic Testing (ACGT) (1997). New guidelines for postal genetic testing. *GenEthics News*, *15*, 3.

Alberta Science and Research Authority (1996). *The commercialization of biotechnology in Alberta*. Edmonton: ASRA.

Allentuch, A. (1996). Drug ads: A prescription for trouble. *Globe and Mail*, 29 October, p. C7.

American College of Medical Genetics/American Society of Human Genetics Working Group on ApoE and Alzheimer Disease (1995). Statement on use of apoliprotein E testing for Alzheimer disease. *Journal of the American Medical Association*, *274*, 1627–1629.

Anderson, A. (1996). Commercialization of genetic screening. The genetics revolution: Myth or reality? http://www.uclan.ac.uk/ facs/ethics/euroiss5.htm

Anderson, C. (1993). Genome project goes commercial. *Science*, *259*, 300–302.

Andrews, L. B. (1992). Torts and the double helix: Malpractice liability for failure to warn of genetic risks. *Houston Law Review*, *29*, 149–184.

Andrews, L. B. (1997). Compromised consent: Deficiencies in the consent process for genetic testing. *Journal of the American Medical Women's Association*, *52*, 39–42.

Andrykowski, M., Munn, R., & Studts, S. (1996). Interest in learning of personal genetic risk for cancer: A general population survey. *Preventive Medicine*, *25*, 527–536.

Angus, D. E., Aver, L., Clautier, J., & Albert, T. (1995). *Sustainable health care for Canada*. Ottawa: University of Ottawa.

Armstrong, P., & Armstrong, H. (1996). *Wasting away*. Toronto: Oxford University Press.

Banks, D. (1996). The economic attributes of medical care: Implications for rationing choices in the United States and United Kingdom. *Cambridge Quarterly of Healthcare Ethics*, *4*, 546–558.

Bekker, H., Moddell M., Denniss, G., Silver, A., Mathew, C., Bobrow, M., & Marteau, T. (1993). Uptake of cystic fibrosis testing in primary care: Supply push or demand pull?. *British Medical Journal*, *306*, 1584–1586.

Blumenthal, D. (1992). Academic-industry relationships in the life sciences. *Journal of the American Medical Association*, *268*, 3344–3349.

Blumenthal, D. (1996). Ethics issues in academic-industry relationships in the life sciences: The continuing debate. *Academic Medicine*, *71*, 1291–1296.

Blumenthal, D., Causino, N., Campbell, E. G., & Louis, K. S. (1996). Relationships between academic institutions and industry in the life sciences – An industry survey. *New England Journal of Medicine*, *334*, 368–373.

Blumenthal, D., Campbell, E. G., Anderson, M. S., Causino, N., & Louis, K. S. (1997). Withholding research results in academic life science: Evidence from a national survey of faculty. *Journal of the American Medical Association*, *277*, 1224–1228.

Boer, A. (1993). Pharmaceutical advertising in medical journals. *Journal of the American Medical Association*, *269*, 147.

Burk, D. L. (1994). Biotechnology and tort liability: A strategic industry at risk. *University of Pittsburgh Law Review*, *55*, 791–864.

Capen, K. (1995). New prenatal screening procedures raise spectre of more "wrongful-birth" claims. *Canadian Medical Association Journal*, *152*, 734–737.

Capitalizing on the genome (1996). Editorial. *Nature Genetics*, *13*, 1–5.

Caskey, T., Eisenberg, R., & Straus, J. (1995). HUGO statement on patenting of DNA sequences. *Genome Digest* (April), 6–10.

Caskey, T., & Williamson, A. (1996). Merck, SmithKline and patents. *Nature*, *381*, 360.

Caulfield, T. (1995). The allocation of genetic services: Economics, expectations, ethics, and the law. *Health Law Journal*, *3*, 213–234.

Caulfield, T. (1996). Wishful thinking: Defining "medically necessary" in Canada. *Health Law Journal*, *4*, 63–85.

Caulfield, T., Cherniawsky, K., & Nelson, E. (1996). Patent law and human DNA: Current practice. In: B. Knoppers, T. Caulfield, & D. Kinsella (Eds.), *Legal rights and human genetic material*, pp. 117–148. Toronto: Emond Montgomery.

Caulfield, T., Hirtle, M., & LeBris, S. (1997). NRTs: Is criminalization the solution for Canada?. *Health Law in Canada*, *18*, 3–14.

Caulfield, T., & Robertson, G. (1996). Eugenic policies in Alberta: From the systematic to the systemic? *Alberta Law Review*, *35*, 59–79.

Chaloner-Larsson, G., Haynes, F., & Merritt, C. (1993). The role of the biotechnology industry in the development of clinical diagnostic materials for prenatal diagnosis. In: *Royal Commission on New Reproductive Technology, Research Volume 5*, pp. 235–289. Ottawa: Canada Communications Group.

Chren, M., & Landefeld, C. S. (1994). Physicians' behavior and their interaction with drug companies. *Journal of the American Medical Association*, *271*, 684–689.

Churchill, J. (1994). Patenting humanity: The development of property in the human body and the subsequent evolution of patentability of living things. *Intellectual Property Journal*, *8*, 249–284.

Clarke, A. (1990). Genetics, ethics, and audit. *Lancet*, *336*, 120.

Cohen, J. (1997). The genomics gamble. *Science*, *275*, 767–772.

Cole, D. E., Gallinger, S., McCready, D. R., Rosen, B., Engel, J., & Malkin, D. (1996). Genetic counselling and testing for susceptibility to breast, ovarian, and colon cancer: Where are we today? *Canadian Medical Association Journal*, *154*, 149–155.

Committee of the Health Council of the Netherlands (1994). *Proper use of human tissue*. The Hague: Health Council of the Netherlands.

Couch, F. J., DeShano, M. L., Blackwood, M. A., Calzone, K., Stopfer, J., Campeau, L., Ganguly, A., Rebbeck, T., & Weber, B. L. (1997). BRCA1 mutations in women attending clinics that evaluate the risk of breast cancer. *New England Journal of Medicine*, *336*, 1409–1415.

Council of Europe (1994). *Recommendation 1240 on the Protection and Patentability of Material of Human Origin*. Text adopted by the Assembly of the Council of Europe on 14 April.

Crossley, M. (1996). Choice, conscience, and context. *Hastings Law Journal, 47*, 1223–1239.

Culotta, E. (1993). New startups move in as gene therapy goes commercial. *Science, 260*, 914–915.

Cuttler, L., Silvers, J. B., Singh, J., Marrero, U., Finkelstein, B., Tannin, G., & Neuhauser, D. (1996). Short stature and growth hormone therapy: A national study of physician recommendation patterns. *Journal of the American Medical Association, 276*, 531–537.

Damsell, K. (1996). Young biotechs mingle in UBC's research halls. *Financial Post*, 16 August, p. 24.

Daniels, N. (1997). The Human Genome Project and the distribution of scarce medical resources. In: T. Murray, M. Rothstein, & R. Murray (Eds.), *The Human Genome Project and the future of health care*, pp. 173–195. Bloomington, IN: Indiana University Press.

Danish Council of Ethics (1994). *Patenting human genes: A report*. Copenhagen: The Danish Council of Ethics.

Day, K. (1995). The rights to a medical dream: Local firm to commercialize gene therapy. *Washington Post*, 22 March, p. A1.

De Lisser, E. (1995). Medicine's shortsighted approach to profit. *Globe and Mail*, March 3, p. A11.

Dodet, B. (1993). Commercial prospects for gene therapy – A company survey. *Biotech, 11*, 182.

Donaldson, C., & Gerard, K. (1993). *Economics of health care*. London: Macmillan.

Ducor, P. (1997). Are patents and research compatible? *Nature, 387*, 13–14.

Dykes, C. W. (1996). Genes, disease, and medicine. *British Journal of Clinical Pharmacology, 42*, 683–695.

Edwards, L. (1993). Discovery, community, and profit: An overview of the science and technology system. In: *Royal Commission on New Reproductive Technologies, Research Volume 5*, pp. 1–70. Ottawa: Canada Communications Group.

Emanuel, E., & Dubler, N. N. (1995). Preserving the physician-patient relationship in the era of managed care. *Journal of the American Medical Association, 273*, 323–329.

Emson, H. E. (1995). Rights, duties, and limits of autonomy. *Cambridge Quarterly of Healthcare, 4*, 7–11.

Erramouspe, M. (1996). Staking patent claims on the human blueprint: Rewards and rent-dissipating races. *University of California at Los Angeles Law Review, 43*, 961–968.

Executive summary (1996). In: *Canadian Biotechnology 96*, pp. 4–6. Georgetown, Ontario: Contact International.

Fletcher, J., & Wertz, D. (1994). *Patients' ethical views on autonomy, new reproductive technology and prenatal diagnosis: A pilot study*. Poster presented at the 44th Annual Meeting of the American Society of Human Genetics, Montreal, October 21.

Frankel, M. S. (1995). Man's freedom and the human genome: Genetic knowledge and the legal system. *The Human Genome Project: Legal Issues, 1*, 255–259.

Frankel, M. S. (1996). Perception, reality, and the political context of conflict of interest in university-industry relationships. *Academic Medicine, 71*, 1297–1304.

Galas, D., & Eisenberg, R. (1996). Presentations at First International Conference:

DNA sampling: Human genetic research: Ethical, legal and policy aspects, Montreal, September 8.

Gene tests "need research protocols" (1996). *Nature*, *384*, 202.

Gitlin, T. (1995). *The twilight of common dreams*. New York: Henry Holt and Co.

Goudey J., & Nath, D. (1997) *Canadian Biotechnology 97: Coming of Age*. Canada: Ernst and Young.

Greely, H. (1995). Conflicts in the biotechnology industry. *Journal of Law, Medicine and Ethics*, *23*, 354–359.

Haber, E. (1996). Industry and the university. *Nature Biotechnology*, *14*, 441–442.

Heagerty, A. (1997). Industry-sponsored research. *Lancet*, *349*, 588.

Health Canada (1993). *A strategic direction for change: A review of the regulations under the Foods and Drugs Act*. Volume 1. Ottawa: Supply and Services Canada.

Healy, B. (1992). On gene patenting. *New England Journal of Medicine*, *327*, 664–668.

Healy, B. (1997). BRCA genes – Bookmaking, fortunetelling, and medical care (1997). *New England Journal of Medicine*, *336*, 1448–1449.

Heller, J. (1996). Highlights from the recent background economic study of the Canadian biotechnology industry. In: *Canadian Biotechnology 96*, pp. 9–12. Georgetown, Ontario: Contact International.

Ingersoll, B. (1997). FDA to clear the way for blitz of TC drug ads. *Wall Street Journal*, August 8, p. B1.

Julian-Reynier, C., Eisinger, R., Vennin, P., Chabal, F., Aurran, Y., Nogu'es, C., Bignon, Y. J., Macheland-Roumagnac, M., Maugard-Louboutin, C., Serin, D., Blanc, B., Orsoni, B., & Sobol, H. (1997). Attitudes toward cancer predictive testing and transmission of information to the family. *Journal of Medical Genetics*, *33*, 731–736.

Keller, E. F. (1992). Nature, nurture, and the Human Genome Project. In: D. Kevles & L. Hood (Eds.), *The code of codes*, pp. 281–289. Cambridge, MA: Harvard University Press.

Kessler, D. A., Rose, J. L., Temple, R. J., Schapiro, R., & Griffin, J. P. (1994). Therapeutic-class wars – Drug promotion in a competitive marketplace. *New England Journal of Medicine*, *331*, 1350–1353.

Kiley, T. (1992). Patents on random complementary DNA fragments? *Science*, *257*, 915.

King, D. (1997). Genetic testing by post: How should it be controlled? *GenEthics News*, *15*, 6–7.

Kitcher, P. (1996). *The lives to come: The genetic revolution and human possibilities*. Toronto: Simon and Schuster.

Knoppers, B. M., & Robertson G. (1995). Genetic testing and health care: Points to consider. *Health Law Journal*, *3*, 323–328.

Kolata, G. (1995). Tests to assess risks for cancer raising questions: Wide marketing debated. *New York Times*, 27 March, p. 1.

Korenman, S. (1993). Conflicts of interest and commercialization of research. *Academic Medicine*, *68*, S18–S22.

Krainer, M., Silva-Arrieta, S., FitzGerald, M.G., Shimada, A., Ishoka, C., Kanamaru, R., MacDonald, D. J., Unsal, H., Finkelstein, D. M., Bowcock, A., Isselbacher, P. J., & Haber, D. A. (1997). Differential contributions of BRCA1 and BRCA2 to early onset breast cancer. *New England Journal of Medicine*, *336*, 1416–1421.

Kurland, C. (1997). Beating scientists into plowshares. *Science*, *276*, 761–762.

Langley, G. R., MacLellan, A. M., Sutherland, H. J., & Till, J. E. (1992). Effect of nonmedical factors on family physicians' decisions about referral for consultation. *Canadian Medical Association Journal*, *147*, 659–666.

Lanthier, J. (1996a). Is the biotech sector a victim of its good fortune? *Financial Post*, 6 July, p. 6.

Lanthier, J. (1996b). Visionary without clout. *Financial Post*, September 17, p. 62.

Ledley, F. (1994). Distinguishing genetics and eugenics on the basis of fairness. *Journal of Medical Ethics*, *20*, 157–164.

Lehrman, S. (1997). Investment analysts paint rosy prospect for biotech. *Nature*, *385*, 189.

Leopold, M. (1993). The commercialization of biotechnology. *Annals of New York Academy of Sciences*, *700*, 214–231.

Lerman, C., Biesecker, B., Benkendorf, J., Kerner, J., Gomez-Caminero, A., Hughes, C., & Reed, M. (1997). Controlled trial of pretest education approaches to enhance informed decision making for BRCA1 gene testing. Abstract. *Journal of the National Cancer Institute*, *89*, 148–157.

Lewontin, R. C. (1992). *Biology as ideology: The doctrine of DNA*. New York: Harpers Perennial.

Lexchin, J. (1993). Interaction between physicians and the pharmaceutical industry: What does the literature say? *Canadian Medical Association Journal*, *149*, 1401–1407.

Loader, S., Caldwell, P., Kozyra, A., Levenkron, J. C., Boehm, C. D., Kazazian, H. H., Rowley, P. T. (1996). Cystic fibrosis carrier population screening in the primary care setting. *American Journal of Human Genetics*, *59*, 234–247.

Lowry, F. (1996). Larger private-sector role in health care needed now, think tank warns. *Canadian Medical Association Journal*, *154*, 549–551.

Maatz, C. (1993). University physician-researcher conflicts of interest: The inadequacy of current controls and proposed reforms. *High Technology Law Journal*, *7*, 137–188.

Maddox, J. (1994). Academics in business mix poorly. *Nature*, *371*, 555.

Malinowski, M. (1996). Capitation, advances in medical technology, and the advent of a new era in medical ethics. *American Journal of Law and Medicine*, *22*, 331–360.

Malinowski, M., & O'Rourke, M. (1996). A false start? The impact of federal policy on the genotechnology industry. *Yale Journal of Regulation*, *13*, 163–254.

Marshall, E. (1997). Gene tests get tested. *Science*, *275*, 782.

McGee, G. (1997). *The perfect baby: A pragmatic approach to genetics*. New York: Rowman and Littlefield.

McGoodwin, W. (1996). Council for Responsible Genetics. *International Legal Practitioner* (June), 57–67.

McIlroy, A. (1996). Ottawa set to regulate fertility industry. *Globe and Mail* (6 August), A3.

Medical Research Council (1996). *Draft document: Guidelines for the commercialization of medical research*. Ottawa: Medical Research Council of Canada.

Miller, R. C, Jr. (1995). University-industry collaboration. *American Journal of Medicine*, *99*, 90S–92S.

Milunsky, A. (1992). Threatened survival of academic-based genetic laboratories. *American Journal of Human Genetics*, *50*, 643–645.

Milunsky, A. (1993). Commercialization of clinical genetic laboratory services: In whose best interest? *Obstretics and Gynecology*, *81*, 627–629.

Minister of Health (1996). *New reproductive and genetechnologies: Setting boundaries, enhancing health*. Ottawa: Supply and Services Canada.

Moran, N. (1996). U.K. spurns proposed genetics regulatory commission. *Nature Medicine*, *2*, 139.

Morton, O. (1995). Biotechnology and genetics: Breaking nature's limits. *The Economist* (25 February), 59–74.

Motulsky, A. (1994). Predictive genetic diagnosis. *American Journal of Human Genetics*, *55*, 603–605.

Moysa, M. (1996). Angel of mercy or mercenary? *Edmonton Journal* (14 July), E1.

Murray, J. (1996). Technology transfer from a Canadian perspective. In: *Canadian Biotechnology 96*, pp. 28–30. Georgetown, Ontario: Contact International.

Nadeau, R. (1996). The biotechnology industry: Seeking a blueprint for growth. In: *Canadian Biotechnology 96*, pp. 39–42. Georgetown, Ontario: Contact International.

National Institute on Aging/Alzheimer' Association Working Group, Consensus Statement (1996). Apolipoprotein E genotyping in Alzheimer's disease. *The Lancet*, *347*, 1091–1095.

Natowicz, M., & Ard, C. (forthcoming). The commercialization of clinical genetics: An analysis of interrelations between academic centers and for-profit clinical genetics diagnostic companies. *Journal of Genetic Counselling*.

Nuffield Council on Bioethics (1995). *Human tissue: Ethical and legal issues*. London: Nuffield Council on Bioethics.

Papp, L. (1996). Health Canada to investigate U.S. retailing of genetic testing. *Toronto Star* (2 March), L7.

Patterson, W., & Emanuel, E. (1996). Physician-drug company conflict of interest. *Journal of Clinical Oncology*, *14*, 316–320.

Percy, S. (1997). Research conflict: Doctor may be in legal battle after reporting negative findings of drug company study. *Medical Post* (January 21), 1.

Physicians say direct-to-consumer advertising affects patient behavior (1993). *American Journal of Hospital Pharmacy*, *50*, 1329.

Pines, W. (1997). New challenges for medical product promotion and its regulation. *Food and Drug Law Journal*, *52*, 61–65.

Post, S., Whitehouse, P. J., Binstock, R. H., Bird, T. D., Eckert, S. K., Farrer, L. A., Fleck, L. M., Gaines, A. D., Juengst, E. T., Karlinsky, H., Miles, S., Murray, T. H., Quaid, K. A., Relkin, N. R., Roses, A. D., St. George-Hyslop, P. H., Sachs, G. A., Steinbock, B., Truschke, E. F., & Zinn, A. B. (1997). The clinical introduction of genetic testing for Alzheimer disease. *Journal of the American Medical Association*, *277*, 832–836.

Postman, N. (1993). *Technopoly: The surrender of culture to technology*. New York: First Vintage Books.

Prouty, A. F. (1996). Quebec biotechnology: The quiet revolution. *Nature Biotechnology*, *14*, 955–956.

Relkin, N., Kwon, Y., Tsai, J., & Gandy, S. (1996). The National Institute on Aging/Alzheimer's Association recommendations on the application of apolipoprotein E genotyping to Alzheimer disease. *Annals New York Academy of Sciences*, *802*, 149–176.

Renaud, M., Bouchard, L., Bisson, J., Labadie, J. F., Dallaire, L., & Kischchuk, N., (1993). Canadian physicians and prenatal diagnosis: Prudence and ambivalence. In: *Royal Commission on New Reproductive Technologies, Research Volume 13*, pp. 235–507. Ottawa: Canadian Communications Group.

Reproductive technology bill could make couples criminals. *Ottawa Citizen* (15 June), A2.

Riccardi, V. (1997). Educating clinicians about genetics. In: T. Murray, M. Rothstein, & R. Murray (Eds.), *The Human Genome Project and the future of health care*, pp. 21–38. Bloomington, IN: Indiana University Press.

Rosenberg, S. (1996). Secrecy in medical research. *New England Journal of Medicine, 334*, 392–394.
Rosner, F. (1992). Ethical relationships between drug companies and the medical profession. *Chest, 102*, 266–269.
Rothman, K. J. (1991). The ethics of research sponsorship. *Journal of Clinical Epidemology, 44*, 25S–28S.
Rowland, J. (1993). Commercial involvement in new reproductive technologies: An overview. In: *Royal Commission on New Reproductive Technologies, Research Volume 5*, pp. 141–233. Ottawa: Canada Communications Group.
Roy, D., Williams, J., & Dickens, B. (1994). *Bioethics in Canada*. Scarborough: Prentice Hall.
Royal Commission on New Reproductive Technologies (1993). *Proceed with care*. Ottawa: Canada Communications Group.
Salloum, S., & Franssen, E. (1993). Laboratory investigation in general practice. *Canadian Family Physician, 39*, 1055–1061.
Schneiderman, L., & Jecker, N. (1995). *Wrong medicine*. Baltimore: The Johns Hopkins University Press.
Schrag, D., Kuntz, K. M., Garber, J. E., & Weeks, J. C. (1997). Decision analysis – Effects of prophylactic mastectomy and oophorectomy on life expectancy among women with BRCA1 and BRCA2 gene mutations. *New England Journal of Medicine, 336*, 1465–1471.
Schultz, J. (1996). Interactions between universities and industry. In: F. Rudolph & L. McIntire (Eds.), *Biotechnology*, pp. 131–146. Washington, D.C.: Joseph Henry Press.
Science and Technology Committee (1995). *Third report: Human genetics: The science and its consequences*. London: House of Commons. Session 1994–95, July.
Selick, K. (1996). Restricting reproductive freedom is the real indignity. *Canadian Lawyer* (October), 46.
Shapiro, M. (1997). Regulating pharmaceutical advertising: What will work? *Canadian Medical Association Journal, 156*, 359–361.
Skolnick, M. (1996). *The role of the biotechnology industry*. Paper presented at the III World Congress of Bioethics, San Fransisco, November 23.
Social and Ethical Concerns Committee of the New England Regional Genetics Group (1996). *Cost-benefit/Cost-effectiveness analysis in genetics: Reading material*. Cambridge MA: New England Regional Genetics.
Southam Newspaper (1996). MDs must pay in "wrongful birth" suit. *Edmonton Journal* (12 June), A4.
Spaulding, C., & Joseph, J. (1992). Is Rx couponing a viable direct-marketing technique? *American Pharmacy, NS32*, 67–71.
Spurgeon, D. (1993). Mail-order sales of drugs without prescriptions under investigation by U.S., Canadian agencies. *Canadian Medical Association Journal, 149*, 1700–1701.
Stephenson, J. (1995). Questions on genetic testing services. *Journal of the American Medical Association, 274*, 1661–1662.
Stix, G. (1996). Is genetic testing premature? *Scientific American* (September), 107.
Straus, J. (1996). Intellectual property issues in genetic research. *Genome Digest* (July), 1–3.
Suhr, L. C., Wright, P. G., Cappelli, M., Kasaboski, A, Hasting, V. A., & Hunter, A. G. (1995). Delivery of molecular genetic services within a health system: Time analysis of the clinical workload. *American Journal of Human Genetics, 56*, 760–768.

Sunstein, C. (1997). *Free markets and social justice*. New York: Oxford University Press.
Svatos, M. (1996). Biotechnology and the utilitarian argument for patents. *Social Philosophy and Policy, 13*, 113–144.
Tanner, A. (1997). Diabetic upset firm was given his private medical information. *Edmonton Journal* (5 February), B9.
Testart, J. (1995). The new eugenics and medicalized reproduction. *Cambridge Quarterly of Healthcare Ethics, 4*, 304–312.
Thomas, S. M., Davies, A. R., Butwistle, S. M., Crowther, S. M., & Burke, J. F. (1996). Ownership of the human genome. *Nature, 380*, 387–388.
Tri-Council Working Group (1996). *Draft document: Code of Conduct for Research Involving Humans*. Ottawa: Medical Research Council of Canada.
Vaillancourt-Rosenau, P. (1994). Reflection on the cost consequences of the new genetic technology for health policy. *International Journal of Technology Assessment in Healthcare, 10*, 546–561.
Veatch, R. (1991). Allocating health resources ethically: New roles for administrators and clinicians. *Frontiers of Health Service Management, 1*, 3–29.
Wadman, M. (1996). Commercialization interests delay publication. *Nature, 374*, 574.
Walholz, M. (1997). Team closes in on identifying Alzheimer gene. *Wall Street Journal* (2 May), B1.
Weiner, C. (1988). Genetic engineering: When and where do we draw the line. *Marine Biology Laboratory (MLB) Science, 3*, 30–34.
Wertz, D. (1995). Professional perspectives: A survey of Canadian providers. *Health Law Journal, 3*, 59–130.
Wertz, D. (1996). *Testing children: An international survey*. Paper presented at the British Medical Association and Genetic Group Conference on Genetic Testing of Children, London, June 5.
WHO (1996). *Control of heriditary diseases: Report of a WHO Scientific Group*. Geneva: World Health Organization.
Whyte, J. (1993). Direct consumer advertising of prescription drugs. *Journal of the American Medical Association, 269*, 146, 150.
Williams, J., & Beresford, E. (1991). Physicians, ethics, and the allocation of health care resources. *Annals Royal College of Physicians and Surgeons of Canada, 24*, 305–309.
Woollard, R. F. (1991). Snake oil and Caeser salad: The ethics of physician and pharmaceutical relationships. *Canadian Medical Association Journal, 145*, 931–933.
Working Group of the Stanford Program in Genomics, Ethics, and Society (1996). *Draft report*. Stanford, CA: Stanford University Center for Biomedical Ethics.
Worton, R. (1993). CGAT – The first year. *Genexpress, 1*, 1–3.
Wysong, P. (1996). Direct to consumer drug ads spark debate. *Medical Post* (6 August), 32.

THE AUTHOR

Timothy Caulfield is Assistant Professor and Research Director, Health Law Institute, Faculty of Law, University of Alberta, 4th Floor, Law Centre, University of Alberta, Edmonton, Alberta, T6G 2H5. E-mail: tcaulfield@law.ualberta.ca. He is also a Fellow with Stanford University's Program in Genomics, Ethics and Society.

The author would like to thank Colin Feasby, Jill Frizzley, Lori Goldberg, Nina Hawkins, Marie Hirtle, Bartha Maria Knoppers, Erin Nelson, Dorothy Wertz, my PGES colleagues, and the entire MELSI committee for their assistance. Portions of this paper are modified from a paper prepared for the Canadian Genome Analysis and Technology Program (CGAT) in early 1997. Finally, the author would like to thank the Alberta Heritage Foundation for Medical Research for their continued support.

Zhiqi Chen and Alison McDermott

International Comparisons of Biotechnology Policies

ABSTRACT. Biotechnology policies are in many respects similar across the countries studied and are of three types: industrial policies, regulatory frameworks, and protection of intellectual property. Government-funded research, seed money, and tax credits are used within all of these countries, with the majority of support concentrated in R&D and capital investment, reflecting the knowledge and capital intensity of the industry. Tariffs, quotas, and other non-tariff barriers to trade are not widely used.

While the European Union enacted *process-driven* legislation to regulate genetically modified organisms, biotechnology-derived products in the U.S. and Japan are regulated under existing *product-driven* legislation. Regulatory stringency in each country appears to reflect public attitudes towards biotechnology. Labelling of genetically modified products and patentability of transgenic animals remain controversial issues that divide the U.S. and the EU.

As a country with a small domestic market for biotechnology, Canada has an interest in reducing trade obstacles caused by differential national policies.

Biotechnology poses new challenges to national government policies. On the one hand, regulations have to be imposed to ensure the health and safety of humans, animals, and the environment. On the other hand, policies have to be formulated to assist or stimulate research and commercialization in this promising new area. National governments in the developed countries have been struggling to strike a balance between these two objectives ever since the U.S. National Institute of Health issued the world's first guidelines on rDNA research in 1976.

In this paper, we study and compare biotechnology policies in the U.S. and Japan as well as the European Union and three of its member countries (France, Germany, and the United Kingdom). We divide biotechnology policies into three categories: industrial policies, regulatory frameworks, and protection of intellectual property. In the following three sections, we review policies in these three areas. Thereafter, we discuss the international policy initiatives that Canada should be cognizant of.

Journal of Consumer Policy **21**: 527–550, 1998.
Bartha M. Knoppers & Alan D. Mathios (eds), Biotechnology and the Consumer.

INDUSTRIAL POLICIES

Biotechnology is recognized in many countries as a key sector worthy of government support. Countries differ, however, in their approaches to the promotion of biotechnology. Japan is an example of a country which consciously targets biotechnology, with comprehensive government policies that specifically promote the growth of biotechnology industry. The United States, in contrast, has no industrial policy pertaining to biotechnology *per se*, but rather a series of policies formulated by various agencies to encourage growth, innovation, and capital formation in all high-technology industries. The approaches taken by countries in the European Union lie somewhere in between those of Japan and the U.S. In the European Union, programs aimed at promoting biotechnology also exist, but they are not as prominent as in Japan.

Despite their differences in approach, countries use similar policy instruments in their support of biotechnology. Because the biotechnology industry is both knowledge intensive and capital intensive, its development depends on two key factors: research and development (R&D) and capital investment. These are precisely the two areas in which most of government support is concentrated.

Government Assistance in Research and Development

Government funded research is the most common and most prominent form of government support for biotechnology. In both Europe and Japan, governments have launched a large number of biotechnology research projects since the 1980s. For example, in 1981, Japan's Ministry of International Trade and Industry launched the Next Generation basic technology research projects, in which biotechnology was allocated 26 billion yen over ten years. The actual amount was smaller owing to the rising government budget deficit in the mid-eighties (Brock, 1989, p. 101). Another high profile project in Japan is the Protein Engineering Research Institute (PERI) project which from 1986 to 1996 received a funding of 17.1 billion yen, 70% of which is provided by the Japanese Government (Nature, 1992). In Europe, Germany created the world's first national biotechnology program, the German Society for Chemical Equipment, Chemical Technology and Biotechnology, in 1974. Since 1982, the European Commission launched a number of high profile research initiatives, including the Biomolecular Engineering Programme (BEP) which dispensed ECU15 million from 1982 to 1986; the four-year Biotechnology Action Programme (BAP) launched in 1986 with an annual budget of nearly ECU14 million; and Biotechnology Research and Innovation for Development and Growth in Europe (BRIDGE), launched in 1989 (Sorj, Cantley, & Simpson, 1989, p. 17). In the United States, although there is no program specifically targeted at biotechnology, the U.S. government is the single most important source of research funding. The Office of Technology Assessment (OTA) estimates that the U.S. Federal Government funds more than half of total biotechnology-related research (OTA, 1991, p. 163). In 1990, the National Institute of Health alone provided an estimated $2.9 billion for biotechnological research (OTA, 1991, p. 249).

Government assistance in R&D may also take other forms. For example, through specific tax privileges, conditional loans, and direct grants, the Japanese government sponsored research associations or research consortia which coordinated projects carried out by researchers at different companies. In fact, the Next Generation projects were conducted through research associations (Brock, 1989, p. 101; Scheidegger, 1988, p. 9). In the United States, Congress and the executive branch provided stimulus to biotechnology R&D by allowing universities and businesses to retain title to patents arising out of federally funded research (OTA, 1991, p. 209). Similar measures are under consideration in France (BWI, 1996a).

Government Assistance in Capital Investment

Government assistance in investment can take the direct form of grants to firms, such as seed money for start-up companies, and government guaranteed loans. For example, as part of its Crusade for Biotechnology initiative, the U.K. government made up to 5 million pounds available over three years starting 1996 to set up "business incubators" for biotechnology companies. Grants covering up to 50% of costs are awarded to cover facilities, business coaching, and laboratory units (BWI, 1996a). In France, the Caisse des Depots et Consignations, a finance organization controlled by the French government, created a venture capital fund worth US$ 79 million as seed capital for start-ups (BWI, 1996a). In Germany, both federal and some state governments (e.g., Bavaria) set up risk financing funds that can be invested in biotechnology companies (Althaus, 1996). In

Japan, small firms can have 80% of their private loans guaranteed by the Venture Enterprise Centre, a part of the Japan Development Bank (Scheidegger, 1988, p. 10).

Government assistance in investment can also take the indirect forms of R&D investment tax credits and special depreciation allowances designed to stimulate activities in the biotechnology industry. All countries in this study allow more generous depreciation rates for R&D capital than for conventional investment. In addition, France, Japan, the U.K., and the U.S. all have R&D tax credits (OECD, 1996, p. 114).

Two Observations

1. Trade restrictions such as tariffs, quotas, and other non-tariff barriers do not appear to be used as tools to promote domestic biotechnology industries. This is true even of Japan, whose post-war success in promoting its targeted industries is often attributed to the use of trade restricting policies which would effectively reserve the domestic market for Japanese firms. This is not to say that trade frictions over biotechnology products have not existed. In fact, the high degree of strategic intervention to which trade in agricultural products has been subject for years, appears to be making its way into the ag-biotech sphere. The recent dispute between the U.S. and the EU over the exports of Ciba-Geigy's genetically modified Bt-corn to the European market, is one such example. In this case, the EU's reluctance to allow the imports of the Bt-corn does not appear to be motivated by a desire to promote its own biotechnology firms, and may be an entirely genuine response to public concerns over health and the environment. At the same time, the potential influence of some strategic considerations, such as the vocal opposition of groups in non-biotechnology sectors who felt threatened by the product (e.g., the Corn Producers Organization in France) would be difficult to deny.

2. Efforts to promote biotechnology are not always successful. In fact, a study done by the Office of Technology Assessment found that targeting of biotechnology for special support was one of the least significant factors affecting competitiveness in biotechnology. "Government targeting efforts everywhere, including Japan, seem to have had marginal impact, at best" (OTA, 1991, p. 151). Indeed, despite earlier predictions that the targeting of biotechnology by the Japanese government would again successfully transform Japan into

a world leader in biotechnology, Japan's biotechnology industry today is still far behind the United States. Imports still provide 80% of Japan's demand for biotechnology (Economist, 1995). Japan, which successfully used industrial policies to shape its major industries (such as steel and consumer electronics) into world leaders, has not been able to duplicate the success in the biotechnology industry. The existence of a complex myriad of factors which might influence the development of a such an industry (if biotechnology can in fact be characterised as a distinct industry), with each present in varying degrees in the different countries, makes the separate impact of any promotional efforts on the part of governments very hard to sort out.

REGULATORY FRAMEWORKS

Countries do differ significantly with regard to their regulatory frameworks. Such differences exist both in terms of stringency and in terms of general approach.

The United States

Oversight of biotechnology in the United States began in June 1976 when the National Institute of Health (NIH) issued comprehensive guidelines designed to permit recombinant DNA research to proceed while maintaining safety for workers and the environment. A key feature of the NIH guidelines was the linking of the degree of containment with the amount of hypothetical risk. Over time as the risks associated with many organisms and experiments were better understood, the Guidelines were revised and relaxed. As a result, the current NIH Guidelines are much less restrictive than the original 1976 version. The NIH Guidelines had a significant influence on global biotechnology regulation because they were widely copied by many countries.

The authority to regulate biotechnology-based products is divided among several federal agencies, including the Food and Drug Administration (FDA), the U.S. Department of Agriculture (USDA), and the Environmental Protection Agency (EPA). To better coordinate the regulatory activities of these agencies, in 1985 the President's Office of Science and Technology Policy (OSTP) established a

Biotechnology Science Coordination Committee (BSCC). In 1990 the BSCC was replaced by Biotechnology Research Subcommittee. In June 1986 the OSTP published the document "Coordinated Framework for Regulation of Biotechnology," which established that biotechnology-derived products would be regulated under the existing legislation, and which outlined the jurisdictions of various agencies. In 1991, the President's Council on Competitiveness recommended four principles of regulatory review, which later were endorsed by then President George Bush. These principles confirmed the long-standing federal policy that regulatory oversight should focus on the characteristics and risks of the biotechnology product, not the process by which it is created. They also proclaimed, for the first time, the principle that performance-based standards are generally preferred over design standards.

The three agencies with principal regulatory authority are the FDA, the USDA, and the EPA. Among the three agencies, the FDA has had the greatest influence on commercial biotechnology since many products sold have been health care products. It is also believed to be the agency that has the clearest and most consistent policy regarding the regulation of biotechnology (Coombs & Campbell, 1991, p. 220).

One interesting aspect of the U.S. regulation is that in addition to federal government, local governments play an important role in regulating the environment and health aspects of biotechnology. In the United States, health and environmental regulations can be enacted by different levels of government. Local governments seem to be free to enact their own laws to address environmental questions posed by biotechnology-derived products (Coombs & Campbell, 1991, p. 210). In fact, several states have enacted legislation regulating field tests of genetically modified organisms or the use of certain products developed using biotechnology (OTA, 1991, p. 188). For example, Hawaii, Illinois, and Wisconsin require notification before the release of genetically engineered organisms into the environment, while Minnesota and North Carolina have more formal permit systems for field tests.

The OECD

Before we discuss the regulatory frameworks in Japan and Europe, we must mention an important report by the Organization for Economic Cooperation and Development (OECD). In 1983, OECD member countries set up a committee of experts to examine safety issues associated with the use of genetically modified organisms (GMOs) in large-scale industrial applications and agricultural and environmental applications. Recommendations on contained use were issued in 1986. The report's conceptual framework is similar to the NIH Guidelines. It describes containment requirements based on the level of estimated risk. The OECD report also recommends a case-by-case review of environmental and agricultural applications of biotechnology. A stepwise progression of experiments was recommended so that experiences could be gained and safety evaluated. The OECD recommendations formed the basis of biotechnology regulations in many countries, including Japan and the European Union.

Japan

The interest of Japanese government in biotechnology dates back to 1971, when the Science and Technology Council identified the life sciences as an area worthy of special government and private assistance (Brock, 1989, p. 62; Saxonhouse, 1986, p. 98). Despite its early attention to biotechnology, Japan had a late start in adopting rDNA regulatory policies. The first version of Japan's guidelines for rDNA experiments, which was almost a direct copy of the 1976 version of the NIH Guidelines (Brock, 1989, p. 85), was published by the Ministry of Education, Science, and Culture (MESC) on March 30, 1979, just one day before the American NIH guidelines underwent their first revision. In August 1979, Japan's Council for Science and Technology published "Guidelines for Recombinant DNA Experiments," amending the MESC guidelines to include research conducted at institutions that are unaffiliated with the MESC.

In June 1986, the Ministry of International Trade and Industry (MITI) published "Guidelines for Industrial Application of Recombinant DNA Technology," which was generally consistent with the OECD recommendations. In December of the same year, the Ministry of Health and Welfare issued process guidelines for pharmaceuticals manufactured using Recombinant DNA Technology. In 1989, the Environment Agency published its views on the release of GMOs, and the Ministry of Agriculture, Forestry and Fishery brought into effect the first guidelines for the use of recombinant DNA in industries under its jurisdiction.

Like the United States, regulation of biotechnology in Japan is based on existing legislation governing drugs, worker health and safety, agriculture, and environmental protection (OTA, 1991). In contrast to the United States, however, in Japan there is no agency mandated to coordinate the different ministries involved. It has been suggested that guidelines issued by some ministries were motivated more by concerns about protecting their jurisdictions than by necessity (Brock, 1989, p. 88). While bureaucratic turf wars are not unique to Japan, the absence of a coordinating body in Japan has led to overlapping initiatives launched by competing agencies. For example, after the MITI established the Bioindustries Development Centre (BIDEC) in 1983, the Ministry of Health and Welfare introduced a similar organization in 1986 (Brock, 1989, p. 98).

The European Union

In addition to ensuring human, animal, and environmental health and safety, the regulatory process in the European Union has the added task of developing an internal biotechnology market within the Union. The two most important directives that deal with biotechnology regulation in the EU are Directive 90/219 and Directive 90/220. Other directives that are related to biotechnology include Directives 90/679/EEC and 93/88/EEC which set minimum standards of worker health and safety protection against exposure to biological agents. Both directives are based on the 1986 OECD recommendations. Directive 90/219 covers the contained uses of genetically-modified microorganisms for commercial and research purposes. Directive 90/220 regulates the deliberate release into the environment of genetically modified microorganisms. Initially intended to address environmental safety, Directive 90/220 had until recently been interpreted broadly to cover the approval of food products. This has changed, however, as the Novel Food Regulations (Regulation No. 258/97), which harmonize the approval and labelling of imports of genetically modified food, came into force in all EU member countries in May 1997 (EC food regulations published, 1977; Johnson, 1997). Directive 90/220 itself was expected to be modified in 1997 (BWI, 1996c).

While member countries are required to bring their national laws in harmony with these EC directives, national authorities have considerable discretion in regulating biotechnology. Depending on how national laws are written and implemented, the actual stringency of regulations varies from country to country. Regulatory guidance and harmonization between regions have not been fully achieved, as the directives continue to be interpreted differently by different EU states (BWI, 1996b, p. 2). For example, under 90/220, several steps are required to obtain authorization for field trials of genetically modified organisms. The ease with which these steps can be accomplished varies greatly between countries, with France, Belgium, and the UK being relatively open, and Germany traditionally more hesitant.

The approach to regulation in the EU is significantly different from that of the U.S. First, instead of relying on existing legislation, the EU has enacted new legislation that specifically regulates genetically modified organisms. Second, the EU regulation is explicitly based on the method by which the organism has been produced (process-driven oversight), rather than on the intended use of the product (product-driven oversight). This is an important distinction, which is often cited as a reason for the strength of the American industry.

Member Countries of the EU

France. Regulatory procedures in France can be characterized as simple and voluntary. They are based on existing laws, with a limited number of guidelines and regulations. There are no specific regulations and guidelines in the field of deliberate releases. The regulatory approach has been to employ industry-sponsored voluntary standards to as great a degree as possible in supplementing existing legislation. Notification of rDNA experiments, for example, is voluntary. In recent years, however, there are signs that France is tightening its regulation of biotechnology. The most recent example is the ban on the cultivation of the Bt-corn (Hodgson, 1997).

Germany. Responding to public pressure, Germany enacted new legislation imposing tight restrictions in 1990. The comprehensive Genetic Technology Law covers recombinant micro-organisms, viruses, cells, plants, and animals, in addition to plasmid vectors. The law outlines specific requirements for both research and commercial production. It specifies conditions for building and operating production facilities, releasing GMOs into the environment, transporting GMOs, and marketing of products containing GMOs.

Two features of the German law are worth noting. First, despite its stringency, the regulation is risk-based. The law divides work with rDNA into four safety levels, with activities considered more risky subject to more stringent requirements. For example, commercial manipulation of organisms in class one and academic research in all four classes need only be notified to the authorities – no permission is needed – but commercial work in classes two to four requires permission (MacKenzie, 1990b).

Second, in contrast to the situation in most countries where regulation of biotechnology is administered mainly by the national or federal government, German regulations are enforced by the State governments. The Federal Ministry of Health plays only an advisory role, with one exception. The Ministry retains control over the approval of licensing of any release of GMOs and the marketing of products containing them (MacKenzie, 1990a).

Since 1993, however, the climate for biotechnology in Germany has changed substantially, to become more favourable to the promotion of the industry. Germany has streamlined its laws regulating genetic engineering helping expedite product approvals (Althaus, 1996).

The United Kingdom. In the United Kingdom, most biotechnology-derived products are regulated under existing legislation, with some special procedures and data requirements for regulating products of genetic engineering. For instance, the Health and Safety (Genetic Manipulation) Regulations of 1978 were issued under the general authority of the Health and Safety at Work Act. It requires that the Health and Safety Executive be notified in advance of any activities involving genetic manipulation (Yuan, 1987, p. 146). The deliberate release and marketing of GMOs are controlled by Part VI of the Environmental Protection Act 1990, and by the 1992 and 1995 Deliberate Release Regulations made under the Act, requiring environmental risk assessment, notification, and consent from the Secretary of State prior to the release or marketing of GMOs. An advisory committee of scientific and medical experts advises the Secretary of State on applications for consent.

Comparative Analysis of Regulatory Policies

Risk-based oversights are common to all countries in this study. The approach of linking the degree of containment with the estimated risks was first proposed in the NIH Guidelines. The same approach was taken by the 1986 OECD recommendations, which formed the basis of regulatory oversight in Japan and the EU. Even the stringent 1990 German law adopts a risk-based approach.

The principles behind regulatory oversight are different for different countries. The United States, for example, has a clearly stated principle that biotechnology oversight should be product-driven. In practice, however, the temptation to move away from product-driven oversight is sometimes hard to resist. Even the FDA, the agency credited with having the most consistent biotechnology policy, contemplated a registration requirement targeted specifically at foods made with rDNA techniques in 1993. The proposal was withdrawn in 1995 after objections from scientific and professional groups (Miller, 1997). Germany, on the other hand, has a process-driven regulatory regime that requires all rDNA-derived products to meet certain regulatory hurdles.

Some countries drafted new legislation specifically for biotechnology, while others base their regulatory oversight on existing or amended legislation. The first group includes the EU and Germany, while the second group includes the U.S., France, and Japan.

The stringency of regulation appears to be closely related to public attitudes towards biotechnology. For example, until the early 90s, Germany and France were at the two extremes of regulatory stringency. At one extreme, Germany, where the Green Party's anti-biotechnology stance was particularly influential, had comprehensive tight restrictions on biotechnology. At the other extreme, France, where little public concern existed about the use of biotechnology (OTA, 1991, p. 193), had simple regulatory procedures. Similarly, the ease of regulation in Germany since 1993 has been accompanied by waning public opposition to biotechnology (Althaus, 1996).

On the controversial issue of product labelling, countries remain divided. On the one side is the European Union whose Novel Food Regulations require that genetically modified foods be labelled if the modifications render a novel food or food ingredient no longer equivalent (EC food regulations published, 1997). On the other side is the United States which is reluctant to endorse special labels out of concern because labelling could be used as a trade barrier (Fox, 1997).

INTELLECTUAL PROPERTY RIGHTS

Biotechnology-derived products are knowledge-intensive. Protection of intellectual property rights is, therefore, of crucial importance to the growth of the biotechnology industry. Among various categories of intellectual property protection, patents and plant breeders' rights are particularly relevant to the biotechnology industry. In theory biotechnology inventions can also be protected through trade secrets. However, a number of drawbacks associated with trade secrets make it an unattractive option for biotechnology inventors (see Olson, 1986, for details).

All countries in this study have existing intellectual property legislation that applies, to at least some extent, to biotechnology-derived products. They are also signatories to all major international intellectual property agreements, which include the Paris Union Convention, the Patent Cooperation Treaty, the International Union for the Protection of New Varieties of Plants (the UPOV Convention), and the Budapest Treaty. These treaties provide a certain measure of protection for biotechnological innovations; however, biotechnology also poses new challenges to intellectual property rights. Countries differ in their ways of meeting these new challenges.

The United States

The most prominent feature of the U.S. patent law in relation to biotechnology is that there is no statutory exclusion for specific subject matters. In the U.S., a patent may be obtained for "any new and useful process, machine, manufacture, or composition of matter, or any new and useful improvement thereof" (35 U.S.C. 101). In particular, living organisms can be patented. In the case *Diamond v. Chakrabarty*, the Supreme Court held that a live, human-made micro-organism is patentable subject matter under section 101 as a "manufacture" or "a composition of matter." After *Chakrabarty*, the patenting of micro-organisms became common place in the United States. In 1987, the Patent and Trademark Office (PTO) announced that it would consider non-naturally occurring, non-human, multi-cellular organisms (including animals) to be patentable subject matter. In April 1988, the first patent on a non-human animal was issued to Harvard University for transgenic non-human mammals genetically engineered to contain a cancer-causing gene (the "Harvard Mouse"

Patent). An important new law for biotechnology, the Biotechnology Process Patent Act, was enacted in 1995. It overturned the so-called *Durden Doctrine* and makes it easier to obtain biotechnology process patent protection. In a 1985 decision known as *In re Durden*, the U.S. Court of Appeals for the Federal Circuit ruled that merely using a new and patentable starting material in, or producing a new and patentable product using, an old process did not necessarily make the process itself new and patentable. Since then the PTO repeatedly used the *Durden* decision to reject biotechnology process inventions (Sorell & Seide, 1996).

Plant breeders' rights are protected under the Plant Patent Act (PPA), which extends patent protection to most new and distinct asexually propagated varieties, and the Plant Variety Protection Act (PPVA), which provides patent-like protection for certain types of new, sexually reproduced plant species (OTA, 1991, p. 204). Furthermore, since the *Hibberd* ruling in 1985, ordinary patents (utility patents) have been granted on plants, even though some protection was already available under the PPA or PVPA. In *Ex parte Hibberd*, the PTO Board of Appeals and Interferences ruled that a corn plant containing an increased level of tryptophan, an amino acid, was patentable subject matter under 35 U.S.C. 101 (OTA, 1989, p. 11).

Japan

Japan's patent system is similar to the U.S. law regarding biotechnological inventions (OTA, 1989, p. 163). In Japan, a patentable invention must utilize a law of nature in the highly advanced creation of technical ideas. In 1975, the patent law was revised to recognize patents on living organisms. A patent will be approved if a microorganism itself can be considered as an "accumulation of technological know-how," and depositing such a microorganism involves disclosing certain trade secrets (Coombs & Campbell, 1991, pp. 98–99). Furthermore, transgenic animals are patentable in Japan (Industry Canada, 1996, p. 75).

Plant breeders' rights are protected through the "Seed and Seedling Law" enacted in 1978, which protects intellectual property of new plant varieties, specifically seed and seedlings for agricultural use. By joining the UPOV in 1982, Japan established a system for the protection of new plant varieties that conforms to international standards. Furthermore, since 1975, patents can be obtained for new plant variety inventions that possess reproducibility (Coombs & Campbell, 1991, p. 99).

The European Union

Most EU states have joined the European Patent Convention (EPC), which is a system of law, common to all of the member countries, established for the granting of the so-called European patents. Primarily, the Convention establishes a single supranational European Patent Office (EPO) with a uniform procedural system for the centralized filing, searching, examination, and opposition with respect to a single European patent application. If granted, a patent matures into a bundle of individual European patents, one for each of the countries designated by the applicant. This European granting system and the resulting European patents exist in parallel with the conventional national granting procedures and the resulting national patents.

Article 52(1) of the EPC defines patentable subject matter as inventions which are susceptible to industrial application, which are new, and which involve an inventive step. While this definition is very broad, the EPC explicitly specifies a number of exclusions. One such exclusion that is important to biotechnology is Article 53(b), which stipulates that European Patents will not be issued for plant or animal varieties and essentially biological processes for the production of plants and animals (with the exception of microbiological processes or the products thereof). The European Patent Office's view on patenting living materials is based strictly on the provisions of the European Patent Convention, which permit patenting of certain life forms if they are novel, inventive, and industrially applicable, if the invention is not contrary to public order and morality, and does not cover plant or animal varieties *per se*. It should be noted that although plant varieties are excluded, no general exclusion for plants is contained in the EPC. While animal patents have been allowed, the patentability of animals is still a hotly debated subject in Europe.

The EPC reduces the costs of obtaining patent protection because it eliminates duplications in filing, searching, and examination and it minimizes the number of translations that must be made (OTA, 1989, p. 157). In theory, patents issued through the European Patent Office provide one-step patent protection throughout the EU. Unfortunately, the lack of coordination between countries, such as the absence of a common definition of patentable material and of unified enforcement mechanisms, has compromised the level of effective protection in Europe.

Because of these weaknesses in the EPC, the European Union has been attempting to set up a harmonized patent system for biotechnology since 1988. But in March 1995, after a sudden reversal by members of the socialist group, the European Parliament rejected a draft directive on the Legal Protection of Biotechnological Inventions. A new version of the proposed directive was launched in January 1996 and is being discussed by the legal affairs committee of the European Parliament. Among the controversial items is the potential to patent human genes, when they are isolated from the human body.

The EU has had greater success, however, in setting up a system of plant breeders' rights. In April 1994, the Council Regulation 2100/94, which implements the full provisions of the revised 1991 UPOV Convention, came into effect. Relative to the existing national systems, the EU system strengthens breeders' rights considerably by enabling them to control almost any use of their variety (Ardley & Hoptroff, 1996). However, the above-mentioned proposed directive on biotechnological inventions would extend "farmer's privilege" to animals, which would permit livestock farmers to reproduce biological inventions in order to replenish animal numbers.

The national patent laws of EU member states generally complement the EPC provisions. Microorganisms are patentable in all three European countries in this study. Both France and Germany permit patenting of plant varieties that are not the subject matter of the specific plant variety law, but Britain does not (OTA, 1989, p. 161).

Comparative Analysis of Intellectual Property Protection

Intellectual property laws in the U.S., Europe, and Japan differ in many aspects. In this analysis we list a few that are most relevant to biotechnology.

The patentability of life forms. This is probably the most controversial patent issue in biotechnology. While all countries in this study have accepted the patentability of micro-organisms, not all of them are prepared to grant patents on transgenic animals. As discussed earlier, animals are patentable in the U.S. and Japan but not in the EU.

The relevance of social considerations to patent decisions. The European Patent Convention, as well as the corresponding national patent laws, stipulate that a patent shall not be granted if the exploitation of the patent would be "contrary to public order or morality" (Crespi, 1997). Similarly, Japan's patent law has a statutory exclusion of inventions detrimental to public order, morality, or health (OTA, 1989, p. 163). In the United States, however, these social factors are not considered under the patent law, and instead, fall under the purview of Federal, State, and local laws that regulate the development and use of commercial products.

Protection of plant inventions. The systems of plant variety rights are similar across countries in this study. Plant varieties are protected under plant breeders' rights with some exclusions such as the farmer's exemption. Furthermore, ordinary patents (utility patents) may be obtained for plant inventions. The U.S. system is somewhat unique in having a special kind of patent (called a plant patent) for asexually reproduced varieties. Although they are termed patents, the protection afforded by plant patents is less broad than that offered by ordinary patents (see OTA, 1991, p. 12, for a comparison between utility patents and plant patents).

INTERNATIONAL POLICY INITIATIVES

Before we discuss the international policy initiatives that Canada should be cognizant of, it is useful to review a number of facts about the biotechnology industry and their policy implications.

Analysis of Biotechnology Industry

1. Biotechnology is a knowledge intensive industry. Strong learning effects exist for both firms and government regulatory agencies. In the biotechnology industry, a firm's success depends critically on its research and development efforts. Other things being equal, the more experience a firm has in research and development, the more competitive a firm is. In addition, regulatory agencies are able to improve efficiencies as they accumulate more experience in handling biotechnology-derived products. A country's competitiveness in the

biotechnology industry thus improves as its firms and regulatory agencies accumulate experience.

Economic theory (for example, Chen, forthcoming) suggests that in an industry characterized by such learning effects, a national government can help its domestic firms gain a competitive advantage by closing its domestic market temporarily to imports (through tariffs, quotas, or other non-tariff barriers). Such a move will allow domestic firms to move down their learning curves more quickly because they have the whole domestic market to themselves (as opposed to sharing it with foreign competitors). As an industry matures, and enough firms have moved sufficiently down the learning curve to have gained a competitive advantage over foreign rivals, the country can open its market again. It has been argued that it was this kind of strategy that helped Japan gain competitive advantage in its targeted industries after World War II (e.g., Borrus, 1983).

In the case of biotechnology, a government interested in pursuing this kind of strategy has at its disposal a convenient tool for restricting trade: regulatory policies. A government can use instruments such as product standards and testing procedures to erect trade barriers for the purpose of helping domestic firms acquire a competitive advantage. So far, however, there are no indications of any country having pursued such a strategy explicitly. After Japan's Ministry of International Trade and Industry announced that it would target biotechnology industry as one of the priority industries in 1981, there were concerns in the West that Japan would use unfair trade practices to help its biotechnology firms (Saxonhouse, 1986, p. 97). This, it turned out, did not happen. As can be seen from earlier discussion, Japan's strategy for promoting biotechnology is just like other major industrialized countries, through instruments such as research funding and R&D tax credits.

Trade frictions in biotechnology, such as the dispute over the export of Ciba-Geigy's Bt-corn to the EU, did occur, but they were the result of public concerns over health and environment and political pressure from non-biotechnology sectors (such as farmers using traditional technology), and were not motivated by the desire to protect and promote domestic biotechnology industries.

This does not mean that in the future no country will attempt to acquire competitive advantage through trade-restricting policies. At this stage, biotechnology is still a very young industry. Biotechnology firms, domestic or foreign, face a common set of obstacles to their growth: public resistance to biotechnology-derived products, burdensome government regulations, and shortage of capital. They have a great deal of common interest to defend. However, as biotechnology industry matures and public concerns and regulatory burden ease, biotechnology firms in some countries may turn their attention to foreign competitors and there may be political pressure to erect trade barriers in the name of promoting an industry of strategic importance.

2. To commercialize a biotechnology product, a firm typically has to incur large research and investment start-up costs and to undergo costly regulatory processes.

This creates incentives for firms to seek regulatory approval and to commercialize their products first in a large market, because commercialization in a large market will generate much needed cash to finance further product development and commercialization in other smaller markets. If a firm is to commercialize its product first in a large market, it is also likely to conduct research and product testing and to locate production facilities in this market as well.

This puts Canada's biotechnology industry at a potential disadvantage because, relative to the U.S., Japan, and the European Union, the biotechnology market in Canada is small. Both of the features of the biotechnology industry as described above (knowledge intensity, and large research/start-up costs) present special challenges to Canada. Harmonization presents itself as a potential solution.

International Policy Initiatives

Harmonization will address both issues discussed above. First, harmonization will limit a foreign government's ability to use regulatory policies as trade barriers, should it attempt to do so. Second, harmonization will offset, to at least some extent, the limitations of Canada's small domestic market. It would place Canada on an even footing in the battle to attract firms to the country, and would assure access to larger foreign markets, which will create more incentives for them to do research, product testing and commercialization in Canada.

While we should pursue harmonization with all of our trading partners through international institutions such as the World Trade Organization, we would argue that Canada has a special interest in negotiating bilateral harmonization agreements with the United States.

There are at least two rationales for Canada-U.S. harmonization agreements:

1. The U.S. is the dominant leader in biotechnology and it has one of the world's largest markets for biotechnology. The easier access to the U.S. market provided by harmonization agreements will both stimulate investment within Canada and attract foreign investment from outside North America in our biotechnology industry.

2. The regulatory frameworks in the U.S. and Canada are already similar in many respects. Comparable legislation, similar agencies with similar jurisdiction, and basic similarities in philosophy and principles (KPMG, 1995, pp. 45–46) would make harmonization with the U.S. much easier than with other jurisdictions such as the European Union.

Canada-U.S. harmonization agreements could cover a full range of issues from the contained use to the deliberate release of GMOs. The most important issues are those related to the commercialization of biotechnology products, such as product approvals and protection of intellectual property. On these issues, harmonization could be achieved at three possible levels. While here the focus of this discussion is on product approvals, harmonization of intellectual property rights can also be discussed in terms of the same three levels:

1. At a minimum, Canada and the U.S. should reduce or eliminate the procedural differences in regulatory approvals of biotechnology products. They would include acceptance of each other's test data and harmonization of test protocols and product standards. While a firm would still have to go through two separate regulatory processes in the two countries, standard date packages and submissions would eliminate the need to prepare two sets of different application materials.

2. A more ambitious goal would be to set up a central filing system for product approvals. Under this system, an application for marketing a product would be sent to a common agency established by both countries to make an initial assessment. The assessment would then be sent to national agencies for approval in each country. Under such a system, the decisions to approve a product would be made independently in the two countries, but a firm needs to file only one set of applications, rather than two separate sets. The two countries might even agree upon the methods used in making the assessment, and on the format of the assessment reports. This would constitute mutual acceptance of the scientific part of assessment reports.

3. The highest level of harmonization would be a form of "free trade" in biotechnology, where the approval for marketing a biotechnology product would be granted for the whole North American market either through mutual acceptance of product approval or through the creation of a bi-national agency that grants product approval under a common regulatory framework. Despite certain similarities of philosophy and procedure between the two countries, mutual acceptance of product standards would require a degree of coordination and mutual confidence that would not be arrived at without considerable effort and negotiation. Since biotechnology is an enabling technology that can be used in many industries, harmonization with the U.S. would undoubtedly carry impacts that extend beyond biotechnology itself. Given the principle of product-driven oversight adopted in both Canada and the U.S., these harmonization agreements will have to cover products derived using advanced biotechnology as well as products produced conventionally. Efforts to negotiate such agreements would be likely to trigger public debates in Canada, and to raise a host of issues that extend beyond the traditional purview of regulatory policy. The eventual level of harmonization will to a great extent depend on the outcome of such debates.

For many, harmonization carries the implication of a lowering of standards, or an abandonment of Canadian regulatory processes and approaches in favour of the American ones. A more healthy way to look at harmonization is as a way to achieve efficiencies through greater understanding, and confidence in each other's processes. It goes without saying that Canada will want to adopt only those approaches which serve its interest in the long-term.

Global Institutional Framework for Biotechnology

Harmonization of biotechnology policies can also be carried out within the framework of international organizations. Involved in various facets of setting up an international framework for biotechnology are institutions such as the Convention for Biological Diversity (CBD), the Food and Agriculture Organization (FAO), the United Nations Environment Program (UNEP), the United Nations Industrial Development Organization (UNIDO), the World Health Organization (WHO), and the World Trade Organization (WTO). Among them, the CBD and the WTO will probably have the most important impact on international harmonization of regulation and international trade in biotechnology.

The CBD. The Convention on Biological Diversity was signed in 1992 by 160 countries prior to the UN Earth Summit in Rio de Janeiro. Article 19(3) of CBD specifies that the Parties shall consider the need for and modalities of a protocol setting out appropriate procedures, including, in particular, advance informed agreement, in the field of safe transfer, handling, and use of any GMOs that may have an adverse effect on the conservation and sustainable use of biological diversity. Since then the CBD has sponsored consultations to determine the need for an international biosafety protocol. In November 1995, the second conference of the parties to the CBD agreed to formulate an international protocol regulating the transfer of GMOs between states (Nature, 1995, p. 326). An ad hoc working group of experts is currently working on a biosafety protocol and is expected to submit its proposed text for the protocol in October 1997.

Setting up an international biosafety protocol has its obvious benefits, but also creates potential concerns among some parties that the protocol may lead to a rigid and unscientific framework that will impede development and international trade in biotechnology. Indeed, earlier works by the CBD have been criticized by some for endorsing a nonrisk-based, process-driven framework (Miller, 1996). More recently, 15 African countries led by Ethiopia tabled a proposal calling for tough international regulations governing the handling, transfer, use, and release of all "living modified organisms" (Nature, 1997, p. 326). As a country with a small market and a product-driven regulatory system, Canada may stand to lose from a rigid and unscientific international framework for biotechnology. In its efforts with the CBD to develop a biosafety protocol, Canada should recognize the legitimate concerns of other countries, and at the same time insist on an international framework that is conducive to the development and trade in biotechnology.

The WTO. The World Trade Organization was established in 1995 as the successor to the General Agreement on Tariffs and Trade (GATT). Two WTO agreements are particularly relevant to biotechnology. First, the Agreement on Technical Barriers to Trade seeks to ensure that technical regulations and standards, as well as testing and certification procedures, do not create unnecessary barriers to trade. Currently, negotiations are under way for the international harmonization of test protocols and acceptance of test data (Industry Canada, 1996, p. 11). As discussed earlier, countries with small domestic markets such as Canada will benefit from such a harmonization agreement. Second, the WTO Agreement on Trade-Related Aspects of Intellectual Property Rights (TRIPs) requires that patent protection be available for all inventions, whether of products or processes, in almost all fields of technology. It also specifies a number of permissible exclusions from patentability, including plants, animals (other than microorganisms), and inventions whose commercial exploitation is prohibited for reasons of public order or morality. This agreement so far avoids the controversial patent issues in biotechnology, such as patentability of life forms.

CONCLUSIONS

There are a large number of similarities in the biotechnology policies of the countries in this study. All countries actively promote their biotechnology industry through instruments such as government funded research, seed money for start-up companies, and generous tax credits and depreciation allowances. Regulatory oversights in all these countries are more or less risk-based, linking the degree of containment with the estimated risk. Most biotechnology inventions are protected by patents and/or plant breeders' rights.

The differences in biotechnology policies, however, are also striking. Some countries (e.g., the U.S.) adopt a product-driven approach and base their regulatory oversight on existing legislations, while others (e.g., Germany) opt for a process-driven approach and pass new laws specifically for biotechnology. Patentability of transgenic animals remains a contentious issue that divides the U.S. and the EU.

As a country with a small domestic market for biotechnology, Canada has a strong interest in, and will likely continue its efforts to achieve, regulatory efficiencies through cooperation with other countries. Canada would appear to have a particular interest in pursuing harmonization agreements with the United States. Canada's efforts should also be extended to the international community: to the development of a scientific, product-driven biosafety protocol, and to the reduction of trade obstacles caused by differential national biotechnology policies.

REFERENCES

Althaus, S. (1996). Survey – Biotechnology: Public opposition now on the wane. *Financial Times*, November 26, p. 2.

Ardley, J., & Hoptroff, C. G. M. (1996). Protecting plant "invention": The role of plant-variety rights and patents. *Trends in Biotechnology, 14*, 67–69.

Borrus, M. (1983). The politics of competitive erosion in the U.S. steel industry. In: J. Zysman & L. Tyson (Eds.), *American industry in international competition*, pp. 60–105. Ithaca, NY: Cornell University Press.

Brock, M.V. (1989). *Biotechnology in Japan*. London: Routledge.

BWI (1996a). *Bio-World International*, August 14, p. 4.

BWI (1996b). *Bio-World International*, November 20, p. 2.

BWI (1996c). *Bio-World International*, December 18, p. 1.

Chen, Z. (forthcoming). Adoption of new technology by a lagging country: Leapfrogging or no leapfrogging? *Pacific Economic Review*.

Coombs, J., & P. N. Campbell, P. N. (1991). *Biotechnology worldwide*. Newbury: CPL Press.

Crespi, R. S. (1997). Biotechnology patents and morality. *Trends in Biotechnology, 15*, 123–129.

EC food regulations published (1997). *Nature Biotechnology, 15*, 300.

The Economist (1995). November 18, p. 91.

Fox, J. (1997). Biotech food labelling: Polite hedging or loggerheads? *Nature Biotechnology, 15*, 308.

Hodgson, J. (1997). UK's Ciba maize decision political? *Nature Biotechnology, 15*, 308–310.

Industry Canada (1996). *Sector competitiveness frameworks – Bio-industries*. Ottawa: Industry Canada.

Johnson, E. (1997). European politics pass food regulations lacking meat. *Nature Biotechnology, 15*, 210.

KPMG (1995). *Improving Canadian biotechnology regulation – A study of the U.S. experience*. Toronto: KMPG.

MacKenzie, D. (1990a). Germany bows to pressure over gene research. *New Scientist, 125* (March 10), p. 28.

MacKenzie, D. (1990b). West Germany's gene law weakens role of public opinion. *New Scientist, 126* (April 14), p. 17.

Miller, H. I. (1996). Biotechnology and the UN: New challenges, new failures. *Nature Biotechnology, 14*, 831–834.

Miller, H. I. (1997). The end of big government? *Nature Biotechnology, 15*, 111.

Nature (1992). *359* (October 15), p. 577.

Nature (1995). *378* (November 23), p. 326.

Nature (1997). *387* (May 22), p. 326.

OECD (1996). *Science, technology and industry outlook*. Paris: Organisation for Economic Co-Operation and Development.

Olson, S. (1986) *Biotechnology: An industry comes of age*. Washington, DC: National Academy Press.

OTA (1989). *New developments in biotechnology: Patenting life*. Washington, DC: Office of Technology Assessment, Congress of the United States.

OTA (1991). *Biotechnology in a global economy*. Washington, DC: Office of Technology Assessment, Congress of the United States.

Saxonhouse, G.R. (1986). Industrial policy and factor markets: Biotechnology in Japan and the United States. In: H. Patrick (Ed.), *Japan's high technology industries: Lessons and limitations of industrial policy*, pp. 97–135. Seattle, WA: University of Washington Press.

Scheidegger, A. (1988). Biotechnology in Japan: A lesson in logistics? *Trends in Biotechnology, 6*, 7–15, 47–53.

Sorell, L.S., & Seide, R. K. (1996). Patenting biotechnology process inventions. *Biotechnology, 14*, 158–159.

Sorj, B., Cantley, M., & Simpson, K. (Eds.) (1989). *Biotechnology in Europe and Latin America*. Dordrecht: Kluwer Academic Publishers.

Yuan, R.T. (1987). *Biotechnology in Western Europe*. Washington, D.C.: U.S. International Trade Administration.

THE AUTHORS

Zhiqi Chen is an associate professor at Carleton University. His address is: Department of Economics, Carleton University, 1125 Colonel By Drive, Ottawa, Ontario K1S 5B6. E-mail: zchen@ccs.carleton.ca. Alison McDermott is an economist at Industry Canada. Her address is: 1801 Hollis St, P.O. Box 940, Station "M", 4th Floor, Halifax, Nova Scotia B3J 2V9. E-mail: mcdermott.alison@ic.gc.ca.

Gillian K. Hadfield and David Thomson
An Information-Based Approach to Labeling Biotechnology Consumer Products

ABSTRACT. Using an economic framework this paper explores the need for labeling of biotechnological consumer products. In particular, we assess the impact of labeling on information problems faced by consumers and regulators. Using information analysis, we propose an approach to labeling biotechnology products that attempts to respect both the real nature of consumer information-processing capacities and approaches and the environment of uncertainty in which any regulatory policy for biotechnology will operate. We conclude that the fact of uncertainty on the scientific front and the nature of consumer concerns in this area gives rise to a need for some type of labeling. Using labels to convey substantive information, however, is likely to be of limited value to consumers. A comprehensive approach to information policy for consumers in this field should aim instead to use labeling requirements to harness the incentives of producers and other private entities to effectively convey to consumers what they want and need to know. We therefore recommend that governments require a simple alert label on biotechnology consumer goods that will prompt consumers to assess their information needs and producers or others to supply those needs. Government regulation in this scheme would consist of basic health and safety regulation and direct or indirect monitoring and regulation of the content of the information ultimately conveyed to consumers by producers and others.

From an information economist's perspective, the central problem of consumer protection is an information problem: Are consumers getting what they intended and expected to get from a transaction in the marketplace? This is evident in consumer protection policy's core concerns: fraud and misrepresentation. In a complex and rapidly changing world, however, the relationship between what consumers expect and what they get in a transaction is a more difficult problem than the core attention to fraud and misrepresentation suggests. For one thing, consumer expectations are the product of complex and subtle psychological coping mechanisms, the consumer's way of dealing with extraordinary amounts of information. For another, the process of collecting, absorbing, interpreting, and applying information so as to bring expectations into line with realizations is a costly one. What this means is that whereas the problems of fraud and misrepresentation can be addressed in a direct fashion – thou shalt not deceive or misrepresent – the problem of consumer information

Journal of Consumer Policy **21**: 551–578, 1998.
Bartha M. Knoppers & Alan D. Mathios (eds), Biotechnology and the Consumer.

is a subtle and difficult one. It is no answer to the problem of consumer information to simply shower consumers with information: Consumers are underinformed or misinformed to begin with precisely because of the high cost of becoming informed relative to the perceived value of information. Unless an information policy starts with this basic premise it runs the risk of being at best useless, at worst counter-productive.

Taking information analysis as the starting point for a study of labeling for biotechnology products is particularly apt because of the extraordinary information challenges presented by this rapidly developing industry. A biotechnology product is not just another complex consumer good about which consumers want and need substantial information in order to determine their purchasing behavior. Rather, biotechnology, at the forefront of scientific development, presents the scientists who work in the field and the regulators who must determine appropriate policy in the field with extraordinary information needs of their own. Those who produce and those who must regulate the science involved are themselves at places uncertain about the match between what they expect or intend and what they will realize. Addressing the problem of consumer protection in the context of biotechnology requires a hefty dose of hubris and a recognition that, just as consumers must figure out how to both take account of and act in spite of uncertainty, so must regulators.

In this paper we propose an approach to labeling biotechnology products that attempts to respect both the real nature of consumer information-processing capacities and approaches and the environment of uncertainty in which any regulatory policy for biotechnology will operate. We conclude that the fact of uncertainty on the scientific front and the nature of consumer concerns in this area gives rise to a need for some type of labeling of products. Using labels themselves to actually convey substantive information, however, is likely to be of limited value to consumers. What a comprehensive approach to information policy for consumers in this field should aim to do instead is to use labeling requirements as a way of harnessing the incentives of producers and other private entities to effectively convey to consumers what they want and need to know. Government regulation in this scheme would consist of basic health and safety regulation, simple labeling requirements, and direct or indirect monitoring and regulation of the content of the information ultimately conveyed to consumers by producers and others.

PREMISES: BIOTECHNOLOGY AND CONSUMER BEHAVIOR

In this section we set out the scientific and consumer behavioural premises that underlie the analysis of labeling that follows.

Modern biotechnology offers the prospect of dramatically changing the characteristics of the plants and animals on which human beings depend for food and other needs. In this paper, "modern biotechnology" is used interchangeably with "genetic engineering" and should be distinguished from "traditional" biotechnology, the techniques for selectively breeding plants and animals that have been used for centuries. With a precision unknown to traditional methods, modern biotechnology permits identified genes to be isolated, duplicated, and inserted into a different organism. In this manner, desired traits can be transferred between organisms without any barriers other than those imposed by the genetic material itself. For instance, a mouse gene encoding metallothionein (a protein that binds heavy metals and prevents them from having a toxic effect) has been transferred to plants, enabling the transgenic crop to grow in soils with high concentrations of contaminants (Pan et al., 1994). Clearly, an exchange of genetic material between mice and plants is unlikely to occur naturally; nor could traditional plant breeding programs match the speed and specificity of genetic engineering in developing a crop plant with the desired trait.

The power of modern biotechnology methods lies in the fact that the transfer of just a single gene can have profound effects on the characteristics of a host organism. For instance, the introduction to tomato plants of a gene that produces the coat protein of the tobacco mosaic virus has allowed these transgenic plants to resist infections by the virus (Abel et al., 1986). A bacterial gene that produces an insecticidal protein has been transferred to tomato, potato, and cotton plants, rendering them more resistant to caterpillars (Fischhoff et al., 1987). As well, transgenic plants have been given a gene that produces an enzyme that depletes glyphosate, a common herbicide (Shah et al., 1986). Such single-gene improvements are expected to increase yields and decrease reliance on agrochemicals, which should produce a saving in costs that will be passed along to consumers. Consumers will also benefit from the enhanced quality characteristics of transgenic crop plants. A familiar example is the tomato that contains a gene blocking the synthesis of polygalacturonase, an enzyme that degrades pectin and causes softening of the fruit (Schuch, 1994).

This tomato need not be picked while green but can ripen on the vine and still withstand shipping and handling. Other transgenic plants have enhanced nutritional characteristics, such as transformed rapeseed that produces a canola oil low in saturated fats (Hildebrand, 1992) and modified, high-starch potatoes that contain less water and that absorb less fat during frying (Stark, Timmerman, Barry, Preiss, & Kishore, 1992).

The traits of animals can also be affected by manipulations of single genes. Transgenic salmon have been developed with greater resistance to frigid water and enhanced growth rates (Du et al., 1992). Work is currently underway to produce transgenic animals that express the traits of disease resistance or altered nutritional characteristics, such as leaner meat (Pursel et al., 1989). Other efforts are aimed at the development of transgenic lactating animals with altered milk components. In one such project, the aim is to develop a milk that will contain human milk proteins such as lactoferrin, which could be consumed as a human milk substitute or infant formula (Yom & Bremel, 1993). As techniques continue to develop, it will become possible to transfer traits controlled by multiple genes, extending both the sophistication and the complexity of the recombinant organisms.

The impact of biotechnology is not restricted to the whole food level: A substantial impact of biotechnology will be at the level of food additives and food processing. Perhaps the best example of this application of biotechnology is the engineering of the gene for calf renin (or chymosin, a milk-curdling enzyme) into fungi (Cullen et al., 1987). The use of microbial chymosin in the production of cheese was recently approved by Health Canada in 1995. Other examples include cyclodextrins (flavour and odour enhancers), recently engineered into potatoes (Oakes, Shewmaker, & Stalker, 1991), and recombinant monellin, a protein 100,000 times sweeter than sugar (Kim et al., 1989). However, with the exception of meat tenderizers, the direct use of enzymes by consumers is rare. Thus, for the purposes of this paper, we will focus only on the concerns of consumers regarding genetically modified whole food, particularly plants.

As with any new plant variety developed by a breeding program, transgenic plants must be tested to evaluate (a) their agronomic characteristics, such as growth performance, stress and pest resistance, seed production, phenotypic stability, and environmental effects, and (b) quality characteristics, wholesomeness, nutritional content, texture, flavour, appearance, and processing suitability (Barefoot, Beachy, & Lilburn, 1994). This evaluation process eliminates many plants with undesired traits. As well, the safety of a new plant variety as a food source must be considered.

Allergicity

For 2–4% of children and up to 2% of adults, ingestion of certain foods or food additives elicits an allergic response (Sampson & Metcalfe, 1992; Young, Stoneham, Petruckevitch, Barton, & Rona, 1994). All known food allergens are proteins; genes encode proteins. Therefore, there is a risk that introducing new genetic material to a host organism will cause the host to acquire new allergenicity. Genetic engineering, although more precise than traditional breeding techniques in terms of the quantity of genetic material that is inserted, is nonetheless vulnerable to this criticism because it has the potential to introduce entirely novel genes (encoding proteins for which there is no previous experience with allergenicity) into common food sources.

The ability of a food allergen to be transferred from one plant to another by genetic engineering was recently confirmed. The gene encoding the Brazil nut 2S albumin storage protein was engineered into soybean to increase the methionine content. The 2S albumin protein is a rich source of methionine and its introduction to soybeans would have reduced the need for soy-based animal feeds to be supplemented with sulphur-containing amino acids. However, it was subsequently discovered that 2S albumin is a major Brazil nut allergen and that the transgenic soybeans could elicit immune responses in hypersensitive subjects (Nordlee, Taylor, Townsend, Thomas, & Bush, 1996). Plans to market the transgenic soybeans were discontinued, avoiding a situation in which these soybeans might have inadvertently been consumed by humans as infant formula, meat extenders, baked goods, or dairy substitutes (Nestle, 1996).

The high-methionine soybean may be considered an easy case. The donor organism was one in which there was considerable experience both with its use as food and with its allergenicity. But other transgenic plants will be more difficult to assess because very few food allergens have actually been identified and characterized. When the allergenicity of the donor organism is undetermined, the safety assessment of the novel food can only rely on a comparison of the expressed protein to known allergenic and nonallergenic proteins.

One basis for comparison is the degree of homology between the amino acid sequence of the expressed protein and that of known allergens (Astwood & Fuchs, 1996). This comparison has the potential to reveal that the introduced gene encodes a known allergen. The value of such information is qualified by the fact that substantial sequence similarities can exist between allergenic and nonallergenic molecules. For instance, a major shrimp allergen, tropomyosin (Leung et al., 1994), shows at least a 60% homology in amino acid sequence to nonallergenic tropomyosin in beef, pork and chicken (Lehrer, Horner, & Reese, 1996). A more important use of the amino acid sequence of an expressed protein may be to reveal the presence of allergenic epitopes – small (12–18 amino acid) sequences in the protein that interact with antibodies or with cells in the immune system. A match of just 6–8 contiguous amino acids can indicate that the introduced protein shares an epitope with a known allergen (Astwood & Fuchs, 1996). However, epitopes are not always linear sequences; discontinuous epitopes consist of two separate amino acid sequences that are only brought into alignment by post-translational folding of the protein (Lehrer et al., 1996). Because discontinuous epitopes arise from the tertiary structure of a protein, it would be difficult to detect them in the amino acid sequence of an expressed protein.

The physiochemical properties of the expressed protein provide a second basis of comparison to known allergens. Although exceptions exist, most allergens are glycosylated, stable to proteolysis and acid conditions, and resist processing and cooking (Astwood & Fuchs, 1996; Lehrer et al., 1996). Proteins that exhibit these characteristics can be viewed as putative allergens. The potential for allergenicity can also be explored in gastric and intestinal models, which reveal to some degree the digestibility of proteins engineered into plants. Common food allergens appear to be stable within simulated gastric fluid whereas common nonallergenic plant proteins are not (Astwood & Fuchs, 1996). Another approach is to measure the ability of the engineered protein to stimulate an antibody response in animal models or in an in vitro assay of human antibodies (Lehrer et al., 1996).

The use of these various strategies for assessing and predicting the allergenicity of novel proteins will help to detect potentially serious allergic reactions. In addition, it should not be overlooked that modern biotechnology has the potential to eliminate known allergens from edible material or to create "edible vaccines" that deliberately exploit an immunogenic response to engineered epitopes (Haq, Mason, Clements, & Arntzen, 1995). However, the possibility remains that gene transfer could inadvertently insert a new and undesired allergen into a food plant. At present, "there is no definitive way to conclusively establish whether [novel] proteins are allergenic" (Lehrer et al., 1996; see also Kessler, Taylor, Maryanski, Flamm, & Kahl, 1992). This gap in scientific knowledge has important ramifications for consumer protection policy, which will be explored in detail in a later section. For now, it is sufficient to state that a policy that requires allergenicity to be assessed only on the basis of the histories of host and donor organisms (which is the approach recommended by Health Canada, 1994, p. 19), and that does not explicitly require any further assessment of allergenicity when the donor organism is not known to be allergenic, can be criticized as being less than thorough (e.g., Bereano & Wilker, 1992).

Toxicity

Allergenicity is not the only form of adverse reaction to food. Indeed, much of the world's biomass is directly toxic to human beings. Natural toxicants appear to be a greater food hazard than environmental contaminants (Ames, Profet, & Gold, 1990), food additives, and pesticide residues (Hall, 1992). For this reason, we consume only a small percentage of the total species of plants and animals (Stewart, 1992), selecting as food those organisms that contain toxicants and antinutritional factors at levels posing no risk to human health. Furthermore, there are some foods in which proper preparation such as soaking or cooking is required in order to render them safe to eat. An example is cassava root (or tapioca), an important source of carbohydrate for 300 million people in the tropics. This crop must be processed by any of a number of methods (boiling, roasting, frying, sun-drying, soaking, or fermenting) to reduce the content of hydrogen cyanide. Other cyanogenic plants include bitter almonds and certain varieties of lima beans (Conn, 1981). Because genetically engineered organisms may contain gene products and secondary metabolites that have not previously existed in material considered to be edible, a comprehensive safety evaluation of novel foods is required to assess their potential toxicity.

In some cases, the potential risks will be relatively easy to identify and evaluate. Some genetic modifications are intended to strengthen an organism's defences – in effect, making it more toxic to natural enemies. The presence of such novel gene products raises immediate questions about the suitability of the organism as human food. For instance, insecticidal proteins (such as enzyme inhibitors and lectins – sugar-binding proteins) expressed in transgenic plants could conceivably have toxic effects on human beings (Boulter, 1993).

This possibility can be assessed by studying the effectiveness of the enzyme inhibitors on human enzymes or by determining the stability of the novel proteins under simulated gut conditions. The latter approach was taken to evaluate the safety of the neomycin phosphotransferase II protein (NPTII), the product of a marker gene that is often used to identify plant cells that have incorporated new genetic material into their genome. Specifically, the transformed cells are the ones that grow in the presence of antibiotic. Because this protein confers resistance to kanamycin and neomycin, however, survival of the protein inside the digestive tract might compromise the effectiveness of these antibiotics as therapeutic agents, even if the protein itself was not toxic (Fuchs et al., 1993). The results showed that the protein was rapidly degraded under simulated gut conditions. The toxicity of NPTII was also evaluated conventionally by feeding large quantities of the compound to mice, and it was found that a millionfold greater dose than the average daily amount caused no adverse effects (Fuchs et al., 1993).

In certain cases, the concern will focus more on the metabolic by-products of a novel gene product than on the protein itself. For example, an enzyme that breaks down herbicide or a protein that binds heavy metals could produce high levels of toxic substances in the consumable portion of the plant. When the presence of such compounds is suspected, given the nature of the genetic modification, a conventional toxicity analysis can be performed to evaluate the safety of the food.

In other cases, however, the potential risks will be more difficult to detect because the putative toxicant(s) will be unknown or unexpected. Current techniques do not permit precise control over the integration of new genetic material into the host genome. New DNA is inserted more or less randomly into the chromosome, often in multiple copies, and sometimes at different sites on the chromosome. This pattern of insertion can affect the expression of the novel gene. For instance, multiple copies of a gene encoding a particular trait can sometimes produce suppression of that trait, instead of expression (Courtney-Gutterson et al., 1994; Flipse, Straatman-Engelen, Kuipers, Jacobsen, & Visser, 1996). The expression of other genes on the chromosome can be affected as well: Normally active metabolic pathways may be suppressed by the integration of novel DNA into a region of the chromosome that codes for an essential protein. Because much of the genome is non-coding, this particular risk may actually be quite low. However, non-coding regions of DNA may contain normally silent metabolic pathways that could be activated by the insertion of the novel DNA (which usually contains regulatory sequences along with the gene that confers the desired trait). The consequences of such a perturbation to normal cell function could be the production of natural toxicants or antinutritional factors that were not previously associated with that food. Presently, our knowledge of the control mechanisms for naturally occurring toxic materials in plants is "very limited" (Stewart, 1992), suggesting that we cannot confidently predict the effect that gene transfer will have on the rest of the genome.

Uncertainty in Predicting the Effect of Gene Introduction

There is also a degree of uncertainty in predicting the effect of introducing a gene product into a new environment. One possibility is that the protein will have unanticipated effects on the organism. For example, experiments with transgenic pigs that expressed high levels of bovine growth hormone exhibited the desired traits of significant weight gain and reduction in subcutaneous fat; however, the animals also suffered a high incidence of gastric ulcers, arthritis, and other complications (Pursel et al., 1989). This example is perhaps unfair. Domesticated animals are complex organisms and chronically elevated hormone levels would be expected to have pathological long-term effects. After all, any substance is toxic at high enough dosages. But the example does serve to illustrate the point that homeostasis can be altered dramatically by a single gene product. There is no reason to believe that the effects produced by a protein must be dramatic in order to be deleterious. For example, altered levels of enzyme production could modify the nutritional content of foods. An indigestible compound that normally occurs at low levels may be produced in high concentrations, or a normally digestible compound may be converted to an indigestible form (Kessler et al., 1992).

Another possibility with which genetic engineers must contend is that a protein will be expressed in an altered form in the new environment of the host organism. Post-translational modification appears to depend on host-specific factors (Roitsch & Lehle, 1989), which means that a modification such as glycosylation may not occur identically in both host and donor organisms. Such a difference could affect the digestibility or allergenicity of the expressed protein. Alternatively, if the expressed protein is an enzyme, host-specific modification could affect its biological activity, which again raises nutritional questions.

A further uncertainty is that the techniques themselves may not be genetically neutral. In experiments with yeast cells, methods involving physical or chemical promotion of DNA uptake had unpredicted genetic consequences for the host cell (a slow growth phenotype) whereas cytoduction – transfer of the same plasmid DNA from one nucleus to another in a mated pair of yeast cells – did not produce this phenotype (Danhash, Gardner, & Oliver, 1991). The significance of this observation is not yet clear, but the implication is that the choice of technique for transferring genetic material can affect the traits expressed by the recombinant organism.

The Need to Develop New Methods of Assessment

Thus, modern biotechnology has not (yet) achieved complete control over the expression of novel and endogenous traits in recombinant organisms. But this admission is by no means equivalent to a statement that genetically engineered foods must a priori be considered toxic. Rather, the recent advances in biotechnology have revealed the inherent shortcomings of the science of food toxicology, deficits with which human cultures have traditionally coped through trial and error. It is becoming clear that toxicological assessment, based on a pharmacology paradigm (i.e., testing specific ingredients or food additives for levels of safe exposure), is inappropriate for the evaluation of whole foods; a univariate model that examines the effect of one compound on homeostasis fails to represent adequately the multivariate nature of living systems (Miller, 1992). Moreover, the standard methodology of feeding laboratory animals large quantities of test substances to determine safety factors simply cannot be adapted to whole foods because nutritional and metabolic disturbances interfere with the interpretation of the results (Miller, 1992).

An alternative is to assess novel foods using analytical biochemistry and analytical chemistry methods instead of traditional toxicological methods (Stewart, 1992). These approaches yield information on the composition of the genetically modified food, which can then be compared to the same food in its traditional form. A change in the amounts of major constituents or in levels of inherent toxicants can be detected, and more detailed toxicity analyses performed if necessary. This approach goes a long way toward ensuring that genetically modified food is not "less safe" than the traditional variety. The concept of substantial equivalence to analogous conventional foods was proposed by the OECD (1993) as the most practical way to assess the safety of new foods. Similarly, Health Canada (1994, p. 16) recommends, inter alia, a comparison of the composition of the novel food to that of the unmodified host, which may encompass an examination of the principal chemical characteristics, significant nutrient constituents, and non-nutrients including natural toxicants typically associated with the food. But because it is not possible to analyze every single component in a given food, the possibility remains that this method could overlook toxicants that were not previously recognized as a component of the food (Stewart, 1992).

Consumer Assessment of Risks

To some extent, the risks associated with genetically engineered foods are merely extensions of the same risks that consumers face with respect to conventional foods. The nutritional qualities of conventional foods are uncertain; studies are reported on a continual basis that specific foods are now considered carcinogenic, are apparently responsible for excessive weight gain, or are the source of certain antinutritional factors. As for allergenicity, individuals develop in early childhood a knowledge of their own particular sensitivities to different conventional foods. It is simply a feature of the world in which consumers live that there is some doubt about the long-term or hidden safety of foods. Consumers cope with this uncertainty by assuming that government regulation and producer self-interest (mediated through reputation and legal liability) combine to ensure that food products are basically safe – or, at least, that immediate risks are absent. In a recent survey, 86% of respondents were either concerned (74%) or unsure (12%) about the safety of using supplemental recombinant bovine growth hormone (rbST) in the production of milk. However, when asked how they would feel if they learned that rbST was approved by Health Canada, 46% of these respondents said they would be a little less concerned and 17% said they would be a lot less concerned (Optima Consultants, 1994, p. 14). See also International Centre for Agricultural Science and Technology (1995), which found that consumers generally take food safety for granted and assume that a product is not on the shelves unless it is safe.

This assumption, along with other simple rules and heuristics, is the sort of device that consumers routinely use to solve information problems (Slovic, 1987). In fact, a scientific assessment of risk may be beyond the capability of most consumers because the information required is complex, is costly to obtain and to process, and may not be within their grasp. As a result, consumer assessments of risk rarely coincide with scientific assessments of risk.

The introduction of genetically engineered foods only compounds these latent uncertainties that consumers have about food products. The food market will change dramatically as new plant and animal varieties are created and rapidly brought to the commercialization stage. Consumers' beliefs about the nutritional quality, toxicity, or allergenicity of certain foods will be challenged, and it is unlikely that they will be able to respond adequately using the simple devices employed in the past. Moreover, there is incomplete scientific knowledge in several key areas, including the genetics and biochemistry of natural toxicants (Stewart, 1992) and gut immunology (Castro & Arntzen, 1993). Consumers' concerns also extend to the environmental and ecological risks of genetically engineered organisms (Environment Canada, 1996) and to the morality of manipulating the genetic makeup of plants and animals. Consumer preferences like these already operate in the market; as evidenced by the avoidance by some consumers of products, like veal, believed to be produced under unethical conditions. In addition, one recent survey found that 32% of respondents often purchased organic foods and another 30% occasionally made such purchases (Optima Consultants, 1994). Any solution to the consumer information problems posed by biotechnology should build on what is known about the source of consumer information problems (not an absence of information per se but rather a limited and costly capacity to absorb and apply information) and about the ways in which consumers can exploit less costly and more manageable ways of obtaining and applying information.

A recent development in the scientific assessment of food safety is the recognition by some experts that human studies will sometimes be required to evaluate fully the safety of new products (Curtis & Cichoracki, 1994; WHO, 1991). Clinical trials, performed under carefully controlled conditions, have the advantage of allowing foods to be tested in their intended forms, and can expose many of the hidden risks that face the consumer population. Other authorities have stopped short of recommending clinical trials, but urge nonetheless that extended observation and experience prior to market introduction must be an important part of the development cycle (Lindemann, 1990). Whether human subjects are studied using full double-blind trials or using less formal protocols, it is becoming clear that some degree of monitoring of genetically engineered foods is required in order to obtain a comprehensive understanding of the risks they pose. Scientists must be able to match the experiences of human subjects to the food products they consume, if gaps in the knowledge of food safety are to be narrowed. In the next section, we argue that an effective consumer protection policy must have the same goal for consumers: helping them to align as much as possible their knowledge and expectations for a particular type of food to their actual experiences with novel foods.

ANALYTICAL FRAMEWORK: MATCHING CONSUMER EXPECTATIONS AND REALIZATIONS

With the above premises in mind, we turn to an analytical framework that can take these premises and from them work out a set of insights for how regulation of biotechnology might best be structured.

The analytical framework we rely on is introduced in Hadfield, Howse, and Trebilcock (1998). The central concept is that of the "bad deal," namely the consumer transaction in which what the consumer gets out of the transaction deviates from what the consumer intended and expected to get. In the ordinary consumer context, the notion of a "bad deal" is often trivial from a public perspective: the inexpensive gizmo that breaks after three uses or fails to live up to its infomercial promises. In many settings however, and importantly in the case of food products, the "bad deal" is far from trivial: food that is expected to be nutritious and safe but is in fact toxic or unhealthy is a matter of public concern. We use the concept of a "bad deal" to capture the essential way in which this analysis of consumer protection is grounded in the consumer's expectations and desires, rather than public conceptions of acceptable risks.

Thus this analysis begins with an idea of deference to consumer assessments in the market rather than paternalism. This is of particular significance in the biotechnology setting because of the multiple discrepancies between scientific and consumer assessments of the risks associated with biotechnology. What we are terming a "paternalistic" approach to regulation would begin with the concept that the goal of regulation is to induce consumer behavior that accords with what consumers would choose were their beliefs in line with scientific risk assessments. This is the approach taken by Viscusi (1994):

> The task for hazard warnings is to create an awareness such that consumers will make the decisions that they would have made if they were fully informed. For both the purchase decision and the product use decision, the critical issue is whether expected consumer welfare has been enhanced, based on the true risks associated with the product. These probabilities will not necessarily be those perceived by the consumer. Moreover, it may be the case that the consumer's expected welfare has been raised by a product even though the product lowers the consumer's welfare based on the true probabilities. What matters from the policy standpoint is the consumer's expected welfare using the actual risks associated with the product. The fact that consumers might gladly purchase a product while ignorant of the associated risks does not make a purchase worthwhile (p. 326).

There is a complex set of ideas in this approach. First, there will be little dispute in the area of consumer hazards that the fact that consumers might act against their own interest by "gladly" purchasing a product they mistakenly believe to be safe argues for a paternalistic form of regulation. Clearly, this gap between perceptions and scientific assessments should be closed in favor of the scientific assessment. The more difficult issue arises with respect to those cases in which consumers' risk assessments exceed those of the scientific community. Here, identifying a consumer protection problem solely by reference to scientific assessments of risk ignores both the reality of how consumers manage uncertainty and the reality that scientific assessments are themselves subject to substantial uncertainty and are reached only through an exercise of judgment. In addition, in the context of biotechnology, it misses the point that consumers' concerns are not coincident with scientific concerns; (some) consumers care not only about nutritional and health risks, but also about environmental risks and ethical issues.

By focusing on the divergence between what consumers actually expect and what they get from a transaction we can, we believe, both capture the impulse to protect consumers against unsuspected risks and at the same time defer to the divergent ways in which consumers on the one hand and scientists and regulators on the other assess risk. The way this works is as follows. We assume that when consumers choose products, they do so on an expectation that the products do not pose undesired risks of injury or harm. Thus information that corrects consumers' mistakes about the safety of a product brings their expectations into closer alignment with what they in fact get. But so too does any information that allows consumers to bring their actions into line with their beliefs, even when those beliefs diverge from scientific beliefs. So, for example, if consumers perceive a risk that science does not, a non-paternalistic approach to consumer protection will nonetheless set as its goal conveying to consumers the information they need in order to act on their perception.

This consumer-centered approach to the problem of information in markets is grounded in the basic principle of a market economy, namely that consumers assess for themselves the costs and benefits of various transactions. The fact that consumer perceptions are not scientific is not a basis for substituting a principle that those transactions that science would make are those that should guide policy. All this is not to say, however, that policy has no role to play in addressing the deviations between subjective and scientific assessments of risk. First, policy can be directed at improving consumer assessments of risk; often this will be in fact what consumers desire. Second, since regulators must assess the benefits of intervention against the costs, scientific as opposed to consumer assessments of risk will play an important role in determining whether efforts to help consumers bring their actions into line with their subjective beliefs are warranted. Where intervention is relatively low-cost, it will not matter that consumers, from a scientific perspective, are overestimating risks. But where intervention is costly it may not be justifiable for government to expend resources protecting consumers from divergences between their subjective, but not scientific or public, beliefs and the facts.

It must be emphasized that this is an analytical approach and not, yet, a conclusion about appropriate regulatory policy. As we will see, what emerges from the consumer-centered framework set out above is policy that sees the problem of bringing consumer assessments of risk into line with scientific assessments as flowing out of a consumer demand for information. It is the divergence between what consumers expect, in fact, and what they get, in fact, that drives policy with respect to information. This is opposed to what we have called paternalistic policy which is governed by an ex ante determination of whether consumer perceptions about gaps between what they want and what they get are scientifically rational. To the extent that a consumer-centered approach to policy results in greater convergence between scientific and subjective risk assessments it is because such convergence in fact brings consumer expectations in line with their experiences in the market.

The Value and Cost of Information

Having set out the basic approach it is now important to be clearer about the framework in which consumer perceptions are addressed. Consumers' beliefs are a function of their existing state of knowledge and the information made available to them, mediated by the cost of acquiring and processing that information. In this framework, consumers are conceived of as purchasers of information. They assess the value of information and compare that value with the cost of acquiring it. What then determines the value of information?

The value of information is understood in information economics to refer to the benefits that can be derived by the consumer by making different choices in response to the information. The concept is an inherently tricky one. By hypothesis the consumer making an information purchase decision is uninformed about what it is he or she will learn. She must decide, therefore, based on what she expects the information to tell her. Consider for example the decision whether or not to research the repair record of an unfamiliar make of automobile. Information about repair records is potentially valuable if information could lead the consumer to alter her decision about whether to purchase the new make as opposed to a familiar make. The consumer's starting point is what he or she knows already about the frequency and cost of repairs of other makes, the variability among different makes with which she is familiar, and whatever she knows about the likelihood that the unfamiliar make is similar on this dimension to the familiar makes. If, for example, variability in her experience is low and she believes the new make is like the familiar makes, she will perceive the value of information to be low and so will be unwilling to expend significant resources obtaining information. She may in fact be quite wrong about the "true" value of information. The repair record of this new make may in fact be quite different from what she expects. But information has no value in a world in which there is no uncertainty and hence no potential for divergences between beliefs and facts, and so it misses the point and indeed is logically incoherent to begin by assuming the consumer knows the true value of information.

The second part of the analysis of the consumer's decision about whether to collect information is the cost of information. The cost of information relative to its perceived value will determine the consumer's decision. Hence even where information is perceived to be highly valuable, it may not be obtained if the cost is high. Conversely, even low value information will be obtained if it is low cost.

Once obtained, information economics supposes that consumers integrate the information in a systematic way, according to Bayes' theorem. Without going so far as to presume that consumers are mathematically adept in the application of this statistical theorem, the idea of Bayes' theorem is an important and intuitive one. It simply says that new information is evaluated relative to the prior information state. Learning does not happen on a clean slate; it happens relative to what is already known. Consumers who already expect products to be safe will not be much affected by learning more about safety. Consumers who have strong beliefs that products are dangerous will both be affected by contrary information about safety and slow to adjust their beliefs (Viscusi & Magat, 1987).

The intuition of Bayes' theorem is also inherent in the consumers' assessment of the value of information and the decision whether to pursue information. Where beliefs are strong, that is approaching certainty, consumers will require substantial contrary information in order to adjust their beliefs significantly so as to alter their behavior. Simply put, their starting point is the strong belief that they do not need further information. Hence the value of information will be perceived to be low. Correspondingly, the amount of information that would have to be acquired to potentially dispel the original belief will be perceived to be large and hence costly. Learning is then not likely to occur.

REGULATORY INSIGHTS: A ROLE FOR LABELING

With the premises about consumer behavior in relation to biotechnology and the above analytical framework in place, it is relatively straightforward to identify the implications for a regulatory approach to labeling for biotechnology consumer products.

The starting point for this regulatory analysis is the consumer's starting point: the purchase of, for example, a bioengineered tomato. In the absence of any labeling, the consumer will make this purchase decision in the same way that he or she makes the decision to purchase a conventional tomato, that is, against the backdrop of the general belief that foods offered for sale in Canadian markets are safe for consumption and the consumer's own experience with conventional tomatoes. The consumer will expect and intend that the purchase will result in the same outcome as a conventional purchase. The experience with the bioengineered tomato purchase, however, may in fact deviate from these expectations for several reasons: The tomato may contain a higher-than-conventional level of toxicants, may require different preparation for safe consumption, may contain a substance to which this consumer is allergic, may have a longer shelf life, may taste different, may contain an unknown carcinogen, may have lower or higher levels of vitamin C, may pose the consumer with an ethical issue (if the consumer has ethical views about biotechnology or the environment), and so on. Some of these deviations will be welcome, some not; either way, consumer choices will be mismatched to some degree with expectations and intentions.

We assume that the bounds on some of these deviations will be limited by standard health and safety regulation, which endeavors to ensure that the bioengineered tomato is not knowingly toxic, allergenic, carcinogenic, and so on. Such regulation in fact promotes the consumer belief in the general safety of foods, a belief that is both cost-reducing for consumers, and productive of the potential for error when the belief is overgeneralized. One of the premises we have emphasized above is the fact that health and safety regulation in this area cannot guarantee that a bioengineered food lacks unexpected attributes such as toxicity: The unpredictability of secondary effects with biotechnology and the limits of current toxicological approaches implies that even best practices health and safety regulation will result in some products getting to the market that possess unknown characteristics. This is most vivid with respect to long-term carcinogenic effects or idiosyncratic effects such as allergenicity. Until a wide variety of individuals have tried the new tomato over an extended period it cannot be known whether the altered genetic make-up of the tomato has produced unwelcome effects. It is these risks, those that escape best practices health and safety regulation, that labeling policy has to address.

One approach to the labeling issue at this point would be to adopt the position that if best-practices health and safety regulation clears the tomato, then there is no scientific basis for additional warning or labeling of the food. This approach, which we have earlier called paternalistic, misses, we believe, two points. First, it fails to address the idiosyncrasy of consumer demands. Allergic response is one form of idiosyncrasy. Another is individualized approaches to nutrition and diet, such as combining diets which recommend eating foods from one group – such as vegetables – only in combination with foods from specified other groups – such as grains. Yet another is represented by the array of ethical preferences with respect to food.

The second point missed by the paternalistic approach is that this approach treats a risk which is currently unidentified or unknown to producers and regulators as not a risk at all. But from the consumer's perspective, a risk exists when in fact there is the potential for any deviations between the experience with conventional tomatoes and that of the bioengineered tomato. In light of the scientific uncertainty, the fact that there are no known scientific risks is not a basis for concluding that there is no need to inform consumers about potential unknown risks.

Now consider the other extreme: labeling the tomato with the details of what is known and what is possible about the deviations between the bioengineered product and conventional tomatoes. This is clearly not a wise or indeed feasible approach to regulation. First, we know that consumers are not able to process complex scientific information about food. Second, the cost of doing so is very likely to exceed what consumers perceive to be the value of the information and so may be completely ignored, thus failing to address the information gap facing consumers. Third, an excess of poorly understood information cluttering food labels can have a negative impact on valuable information, such as nutritional data. Empirical research demonstrates that "clutter" on labels reduces the impact of all information conveyed (Viscusi, 1994).

Application of Our Framework to Information About Biotechnology Consumer Products

What then does follow from the analytical framework outlined earlier? The first point is that in the absence of any indication that a food has been bioengineered, there is no potential for consumers to connect their experiences with the food over time and hence no incentive or ability for consumers to resolve their information deficiencies over time. Suppose a particular consumer in fact has an allergic response to the bioengineered tomato but not conventional tomatoes. Unless the tomato has been labeled in such a way as to distinguish it from the conventional tomato, the consumer will be unable both to identify the source of the allergic response and, even if this is ultimately learned, unable to avoid the product at low cost in the future without avoiding tomatoes entirely.

There is a more subtle point to be made here. Consumers, in the analysis suggested earlier, decide what and how much to learn about a product based on their assessment of the value of information relative to its cost. If the bioengineered tomato is indistinguishable from the conventional tomato, consumers will perceive the value of information to be low; indeed, they might perceive it to be zero. Consumers will be prompted to invest in learning about the bioengineered tomato only if they first are alerted to the fact that there may be something to learn, that learning may have value.

The Role of An Alert Label

Against the backdrop of the general belief in the safety of foods on the market, and the accumulated repertoire of knowledge about conventional foods with which each of us enters the supermarket, the need to signal the value of information about bioengineered foods is a basis for at least minimal labeling alerting consumers to the fact that a product is not identical to the conventional product. Consumers will only search for information if they are aware that the search may be potentially valuable. Hence there would seem to be a role for simple alert labels that serve merely to distinguish biotechnology products from conventional products.

A simple alert label – a double helix on the can, for example – also serves the basic function of allowing consumers who wish to avoid biotechnology products a way of doing so. This directly serves the goal of consumer protection policy from a consumer-centered perspective as it allows consumers to match their intentions with their transactions. We emphasize again that there may be a variety of reasons as to why consumers have these preferences; they include environmental, ethical, and dietary approaches to life that it is not the place of consumer protection policy to judge in the absence of some indication that facilitation of these preferences is excessively costly or counterproductive. One of the virtues of the simple label is that it is relatively low cost on this score and, because it can be relatively uncluttered, would not impose a negative externality on other labeling information. There are some costs associated with even this simple label, however, and they include keeping differently-sourced inputs (such as grains) separate during production or labeling processed foods that contain a minute amount of a bioengineered foods. A labeling policy in practice would have to assess the magnitude of these costs and weigh them against consumer benefits.

The deeper virtue from the simple alert label, however, comes not from the information it conveys itself, but from the role it plays in structuring an information mechanism that is mediated through the market. Once consumers are alerted to the fact that information might be valuable in connection with the bioengineered tomato, the questions are raised, where will they get that information, what type of information will they get, what will be the cost of the information, and what, ultimately, will be the value of the information in terms of assisting them to make desired adjustments in their buying behaviour? The simple alert label presents itself as potentially the key to having these questions answered by the producers of biotechnology products and the market.

If consumers are alerted to the potential need for information, this creates a demand for information that the market will be prompted to satisfy. Particularly if consumers are inclined to overestimate the risks of biotechnology and/or display an aversion to these foods rooted in affect rather than information, as seems true at least at present states of knowledge (Zimmerman, Kendall, Stone, & Hoban, 1994), then producers will have a strong incentive to supply information to overcome these distortions. This information could be provided directly on labels, through point-of-sale signs or brochures, through advertising media, and so on. The key point is that the function of the regulatory label is not to provide information but rather to structure the market mechanism that will provide information.

Market Provision of Information

There is good reason to believe that a market mechanism for the provision of information in this area will dominate whatever could be achieved through conventional regulatory labels where the label itself is the source of information. Intuitively, if producers are faced with the prospect that consumers will, through lack of information, avoid biotechnology products, they will devote resources to providing information. More importantly, they will devote resources to understanding the complex nature of consumer beliefs, information-processing capacities, and so on, and devising effective methods for achieving the goal of conveying useful information that consumers can understand and apply.

The superior effectiveness of producer-originated information has been demonstrated by at least one empirical study comparing the effectiveness of producer advertising about the health benefits of eating fiber-rich foods with that of publicly-provided information. In this study it was discovered that producers were significantly more effective at conveying information to consumers about these benefits than government. Moreover, producers apparently tailored different forms of advertising to different segments of the market in order to respond to differences in information processing abilities, cost, and so on. This was demonstrated by a reduction in cross-sectional differences between consumers over time (Ippolito & Mathios, 1990). The results of this study deserve emphasis because consumers often report that government, health professionals, and independent organizations are more trusted sources of information than producers (Decima Research, 1996). The fiber study referred to above, however, makes the point that to the extent this is true (and its truth no doubt varies across different segments of the population), the market – that is, producers – will respond to this if it is in their interest to ensure that information is conveyed effectively to consumers. Statements by producers, if disbelieved, will be eschewed and endorsements from independent health organizations will be sought. Misleading statements will be contested by competing producers of conventional products. The point is that if there is an economic incentive to convey accurate and effective information, those incentives can be harnessed through a simple alert label that creates a market for information.

We expect, for example, that producers of the bioengineered tomato will waste no effort in communicating to consumers the positive attributes of their tomato compared to the conventional tomato: If it rots more slowly or has a higher Vitamin C content or is less likely to prompt an allergic response, consumers will be informed through advertising. We can expect also, however, that if there are negative attributes of the bioengineered tomato – lower Vitamin C content or increased risks of allergenicity or reduced nutritional value – they will not be advertised in the absence of any consumer demand for information. This then is the role of the alert label: to create a positive benefit for producers – protecting their markets from consumer misinformation or overestimation of risks.

The Role of Government

This returns us to a conventional role for consumer protection regulation – the monitoring of the truth of producer representations. The premise of this conventional regulation is that the market is not perfect in ensuring truth-in-advertising, but that a reasonably good approximation to reliable information in markets can be achieved through a combination of market incentives (reputation, competition) and oversight through regulation and legal liability. The simple labeling approach to information problems in biotechnology, then, draws on the conventional strength of consumer protection while harnessing the market in meeting consumers' information needs. The producers of the bioengineered tomato, facing an incentive to protect their market from misinformed consumer reactions, will produce information. Government oversight then operates to ensure that the information produced is in fact fair and accurate.

Thus the principal role for government policy that emerges from the information-based approach is a combination of a simple label to alert consumers to the fact of bioengineering in a product and conventional monitoring for truth-in-advertising. The goal of structuring the market for information about biotechnology, however, may give rise to a secondary role for policy. It is possible, perhaps probable, that consumer awareness of biotechnology is currently sufficiently primitive that consumers will not in fact be "alerted" to anything by the simplest of labels stating that the product contains bioengineered ingredients. Consumers may not recognize, for example, that bioengineering could mean that the product may elicit an allergic response that the conventional product does not or that the nutritional value of the bioengineered product could differ from that of the conventional product. There may then be a role either for government education directed towards conveying such generic ideas, or for a simple set of statements on food labels, such as "Check nutritional content" or "Introduce slowly." The point, it needs to be emphasized, is for government regulation to alert consumers to the value of information and so create a demand for information, not to satisfy that demand directly.

Possible Objections

There are two final points to explore, raised by objections that can be anticipated. The first is that labeling will stigmatize biotechnology foods by leading consumers to believe, irrationally, that the food is unsafe. Whereas this is an objection to a conventional approach to labeling, the very point of the market-based simple alert labeling approach suggested above is to create the possibility that consumers could form beliefs that are not warranted by the scientific evidence. This is precisely what will harness producer incentives to address the information gap, and to address it in a more cost-effective and tailored way than government can. There is every reason to believe that producer advertising, monitored for accuracy by regulation and legal liability, will dispel unwarranted – but only unwarranted – perceptions among consumers. The risk of uninformed stigma is precisely what does the work in this approach and the risk of stigmatization is a virtue not a vice in this setting. Indeed, this is why there is a role here for government, rather than reliance on voluntary labeling. Producers are not likely to label voluntarily because of a collective action problem: Unless *all* are labeling, no *one* producer will voluntarily stigmatize its products.

A second objection that can be anticipated is that a simple label will rapidly become ubiquitous and hence uninformative: If everything has an "alert" on it, no-one will pay any attention to it. This objection is well-grounded in the research on labels which indicates that if most products have "caution" warnings on them consumers disregard the warning (Viscusi, 1994). The objection in the biotechnology setting is, however, misplaced. If biotechnology products do in fact become so ubiquitous, then that will be testament to the experience of satisfaction for consumers that these products, in fact, are largely indistinguishable from conventional products in terms of meeting consumer expectations. The path that will have been traveled to the state in which the labels are ubiquitous will have been one of relatively informed decision-making if the label functions, as argued above, as an initial alert prompting consumer learning. The fact that the label is ubiquitous and therefore generally ignored will be a sign that this approach to consumer protection has been effective. Only those who have specific concerns – be they dietary, ethical, or environmental – will pay attention to the label.

CONCLUSION

The information-based approach to consumer protection policy with respect to biotechnology in food production begins with some basic premises about consumer behavior and the risks of biotechnology; and with an analytical framework sensitive to the nature of consumer information processing and demand. It then leads to the conclusion that the appropriate role for government in this area is to structure a mechanism whereby consumers reflect a demand for information which can then be met – in effective, low-cost, and targeted fashion – by the producers of biotechnology. The key to structuring this mechanism is a simple label that alerts consumers to the possibility that there is valuable information to be learned about these products. The remaining role for government is then the conventional one of ensuring truth-in-advertising through regulation and legal liability.

REFERENCES

Abel, P. P., Nelson, R. S., De, B., Hoffmann, N., Rogers, S. G., Fraley, R. T., & Beachy, R. N. (1986). Delay of disease development in transgenic plants that express the tobacco mosaic virus coat protein gene. *Science, 232*, 738–743.

Ames, B. N., Profet, M., & Wilker, N. L. (1990). Dietary pesticides (99.99% all natural). *Proceedings of the National Academy of Science, 87*, 7777–7781.

Astwood, J. D., & Fuchs, R. L. (1996). Allergenicity of foods derived from transgenic plants. *Monographs in Allergy, 32*, 105–120.

Barefoot, S. F., Beachy, R. N., & Lilburn, M. S. (1994). Labeling of food-plant biotechnology products. *Cereal Foods World, 39*, 760–765.

Bereano, P. L., & Wilker, N. L. (1992). Letter to the Editor: Regulations for genetically engineered foods. *Science, 258*, 1561–1562.

Boulter, D. (1993). Insect pest control by copying nature using genetically engineered crops. *Phytochemistry, 34*, 1453–1466.

Castro, G. A., & Arntzen, C. J. (1993). Immunophysiology of the gut: A research frontier for integrative studies of the common mucosolal immune system. *American Journal of Physiology, 265*, G599–G610.

Conn, E. E. (1981). Unwanted biological substances in foods: Cyanogenic glycosides. In: J. C. Ayres & J. C. Kirschman (Eds.), *Impact of toxicology on food processing*, pp. 105–121. Westport, CT: AVI Pub. Co.

Courtney-Gutterson, N., Napoli, C., Lemieux, C., Morgan, A., Firoozabady, E., & Robinson, K. E. P. (1994). Modification of flower color in florist's chrysanthemum: Production of a white-flowering variety through molecular genetics. *Bio/Technology, 12*, 268–271.

Cullen, D., Gray, G. L., Wilson, L. J., Hayenga, K. J., Lamsa, M. H., Rey, M. W., Norton, S., & Berka, R. M. (1987). Controlled expression and secretion of bovine chymosin in Apergillus nidulans. *Bio/Technology, 5*, 369–376.

Curtis, G. L., & Cichoracki, J. R. (1994). Food safety and health claims: The need for clinical research. *Food Technology, 48*(5), 92–96.

Danhash, N., Gardner, D. C. J., & Oliver, S. G. (1991). Heritable damage to yeast caused by transformation. *Bio/Technology, 9*, 179–182.

Decima Research (1996). *Report to the Canadian Institute of Biotechnology on attitudes towards biotechnology*. Unpublished manuscript. Ottawa: CIB.

Du, S. J., Gong, Z., Fletcher, G. L., Shears, M. A., King, M. J., Idler, D. R., & Hew, C. L. (1992). Growth enhancement in transgenic Atlantic salmon by the use of an "all fish" chimeric growth hormone gene construct. *Bio/Technology, 10*, 176–181.

Environment Canada (1996). *Environmental applications of biotechnology: Focus groups*. Ottawa: Environment Canada.

Fischhoff, D. A., Boudish, K. S., Perlak, F. J., Marrone, P. O., McCormick, S. M., Niedermeyer, J. O., Dean, D. A., Kusano-Kretzmer, K., Mayer, E. J., Rochester, D. E., Rogers, S. G., & Fraley, R. T. (1987). Insect tolerant transgenic tomato plants. *Bio/Technology, 5*, 807–813.

Flipse, E., Straatman-Engelen, I., Kuipers, A. G. J., Jacobsen, E., & Visser, R. G. F. (1996). GBSS T-DNA inserts giving partial complementation of the amylose-free potato mutant can also cause co-suppression of the endogenous GBSS gene in a wild-type background. *Plant Molecular Biology, 31*, 731–739.

Fuchs, R. L., Ream, J. E., Hammond, B. G., Naylor, M. W., Leimgruber, R. M., & Berberich, S. A. (1993). Safety assessment of the neomycin phosphotransferase II (NPTII) protein. *Bio/Technology, 11*, 1543–1547.

Hadfield, G. K., Howse, R., & Trebilcock, M. (1998). Information-based principles for rethinking consumer protection policy. *Journal of Consumer Policy, 21*, 131–169.

Hall, R. L. (1992). Toxicological burdens and the shifting burden of toxicology. *Food Technology, 46*(3), 109–112.

Haq, T. A., Mason, H. S., Clements, J. D., & Arntzen, C. J. (1995). Oral immunization with a recombinant bacterial antigen produced in transgenic plants. *Science, 268*, 714–716.

Health Canada (1994). *Guidelines for the safety assessment of novel foods*. Volume 2. Ottawa: Health Canada.

Hildebrand, D. F. (1992). Altering fatty acid metabolism in plants. *Food Technology, 46*, 71–74.

International Centre for Agricultural Science and Technology (1995). *Consumer acceptance of potatoes that have been genetically modified*. Saskatoon: International Centre for Agricultural Science and Technology.

Ippolito, P., & Mathios, A. D. (1990). Information, advertising and health choices: A study of the cereal market. *Rand Journal of Economics, 21*, 459–480.

Kessler, D. A., Taylor, M. R., Maryanski, J. H., Flamm, E. L., & Kahl, L. S. (1992). The safety of foods developed by biotechnology. *Science, 256*, 1747–1749, 1832.

Kim, S. H., Kang, C. H., Kim, R., Cho, J. M., Lee, Y. B., & Lee, T. K. (1989). Redesigning a sweet protein: Increased stability and renaturability. *Protein Engineering, 2*, 571–575.

Lehrer, S. B., Horner, W. E., & Reese, G. (1996). Why are some proteins allergenic? Implications for biotechnology. *Critical Reviews in Food Science and Nutrition, 36*, 553–564.

Leung, P. S. C., Chu, K. H., Chow, W. K., Asari, A., Bandea, C. I., Kwan, H. S., Nagy, S. M., & Gershwin, M. E. (1994). Cloning, expression, and primary structure of Metapenaeus ensis tropomyosin, the major heat-stable shrimp allergen. *Journal of Allergy and Clinical Immunology, 94*, 882–890.

Lindemann, J. (1990). Biotechnologies and food: A summary of major issues regarding safety assurance. *Regulatory Toxicology and Pharmacology, 12*, 96–104.

Miller, S. A. (1992). Novel foods: Safety and nutrition. *Food Technology, 46*, 114–117.

Nestle, M. (1996). Allergies to transgenic foods – Questions of policy. *New England Journal of Medicine, 334*, 726–728.

Nordlee, J. A., Taylor, S. L., Townsend, J. A., Thomas, L. A., & Bush, R. K. (1996). Identification of a Brazil-nut allergen in transgenic soybeans. *New England Journal of Medicine, 334*, 688–692.

Oakes, J. V., Shewmaker, C. K., & Stalker, D. M. (1991). Production of cyclodextrins, a novel carbohydrate, in the tubers of transgenic potato plants. *Bio/Technology, 9*, 982–986.

OECD, Group of National Experts on Safety in Biotechnology (1993). *Safety evaluation of foods derived by modern biotechnology: Concepts and principles*. Paris: Organisation for Economic Cooperation and Development.

Optima Consultants (1994). *Understanding the consumer interest in the new biotechnology industry: Study findings*. Ottawa: Office of Consumer Affairs.

Pan, A., Yang, M., Tie, F., Li, L., Chen, Z., & Ru, B. (1994). Expression of mouse metallothionein-I gene confers cadmium resistance in transgenic tobacco plants. *Plant Molecular Biology, 24*, 341–351.

Pursel, V. G., Pinkert, C. A., Miller, K. F., Bolt, D. J., Campbell, R. G., Palmiter, R. D., Brinster, R. L., & Hammer, R. E. (1989). Genetic engineering of livestock. *Science, 244*, 1281–1288.

Roitsch, T., & Lehle, L. (1989). Structural requirements for protein N-glycosylation: Influence of acceptor peptides on cotranslational glycosylation of yeast invertase and site-directed mutagenesis around a sequon sequence. *European Journal of Biochemistry, 181*, 525–529.

Sampson, H. A., & Metcalfe, D. D. (1992). Food allergies. *Journal of the American Medical Association, 268*, 2840–2844.

Schuch, W. (1994). Improving tomato quality through biotechnology. *Food Technology, 48*, 78–83.

Shah, D. M., Horsch, R. B., Klee, H. J., Kishmore, G. M., Winter, J. A., Turner, N. E., Hironaka, C. M., Sanders, P. R., Grassner, C. S., Aykent, S., Siegel, N. R., Rogers, S. G., & Fraley, R. T. (1986). Engineering herbicide tolerance in transgenic plants. *Science, 233*, 478–481.

Slovic, P. (1987). Perceptions of risk. *Science, 236*, 280–285.

Stark, D. M., Timmerman, K. P., Barry, G. F., Preiss, J., & Kishore, G. M. (1992). Regulation of the amount of starch in plant tissues by ADP glucose pyrophosphorylase. *Science, 258*, 287–292.

Stewart, K. K. (1992). Food composition and analysis in the assessment of the safety of food produced by biotechnology. *Food Technology, 46*, 103–107.

Viscusi, W. K. (1994). Efficacy of labeling of foods and pharmaceuticals. *Annual Review of Public Health, 15*, 325–343.

Viscusi, W. K., & Magat, W. (1987). *Learning about risk: Consumer and worker responses to hazard information*. Cambridge: Harvard University Press.

WHO (1991). *Strategies for assessing the safety of foods produced by biotechnology*. Geneva: World Health Organization.

Yom, H. C., & Bremel, R. D. (1993). Genetic engineering of milk composition: Modification of milk components in lactating transgenic animals. *American Journal of Clinical Nutrition, 58*(Suppl.), 299S–306S.

Young, E., Stoneham, M. D., Petruckevitch, A., Barton, J., & Rona, R. (1994). A population study of food intolerance. *Lancet, 343*, 1127–1130.

Zimmerman, L., Kendall, P., Stone, M., & Hoban, T. (1994). Consumer knowledge and concern about biotechnology and food safety. *Food Technology, 48*, 71–78.

THE AUTHORS

Gillian Hadfield is an Associate Professor, Faculty of Law, University of Toronto, 78 Queen's Park Cr., Toronto, Canada M5P 2C5. E-mail: gillian.hadfield@utoronto.ca. David Thomson is a third-year student at the same university. E-mail: david.thomson@utoronto.ca.

Alan D. Mathios

Economic Perspectives on the Dissemination of Science-Based Information to Consumers

ABSTRACT. The role of producers in disseminating information about biotechnology to consumers has been debated in policy circles. Prior to 1984 the U.S. prohibited firms from discussing any scientific relationships between diet and health in food advertising and labeling. The U.S. government changed its policy regarding health claims in advertising and labeling and after 1984 producer claims added to the flow of information linking diet and disease. This paper reviews the research Pauline Ippolito and the present author have published analysing the impact of this change in policy. Using a variety of U.S. data, the evidence shows that there were marked improvements in consumer knowledge of diet-disease relationships, healthier new food product introductions, and significant improvements in the composition of the typical U.S. diet during the 1985–1990 period. The benefit of producer diet-disease claim in changing knowledge and behavior provides a potentially important example as the Canadian government contemplates regulations governing the flow of biotechnology information to consumers. A review of the institutions that regulate advertising regulation is included as well as a discussion of their ability to deter false and deceptive advertising.

The gain to consumers from incorporating evolving scientific discoveries into basic decisions about products is likely to be large. Innovative new drugs and technologies present new opportunities for consumers to increase their welfare. The question of how best to get information about evolving technologies and scientific innovations to consumers is of central importance to policy makers, especially in the area of biotechnology. Accurate and reliable information about the benefits and risks of the use of biotechnology is of central concern to producers in the biotechnology industry and to consumers in the marketplace. Consumers express desires to be informed through a variety of mechanisms and are skeptical and confused about many of the claims that advocates and producers of biotechnology products advance. However, advertising and labeling are two of the major sources of information available to consumers and a potentially invaluable mechanism for turning relatively abstract scientific information into meaningful information that can be used by consumers to make improved choices.

The role of producers in disseminating scientific information to

Journal of Consumer Policy **21**, 1998.
Bartha M. Knoppers & Alan D. Mathios (eds), Biotechnology and the Consumer.

consumers has been widely debated in policy circles. Producers convey information to consumers via advertising and marketing efforts. There are widely varying views on whether this type of information provision leads to improved consumer choices and whether the dissemination of scientific information to consumers should be left to government and other less partial sources of information. This debate has surfaced over the last ten years as the U.S. has altered its regulations governing the use of science-based health claims for food products. For many years, the U.S. prohibited firms from discussing the relationship between nutrition and disease in advertising and labeling. This ban on producer health claims lasted for many years. In 1984, the U.S. government changed its policy regarding health claims in advertising and labeling and soon producer claims added to the flow of information linking diet and disease. This paper examines how these changes in regulations altered consumer knowledge of diet-disease links and changes in their consumption of fat, saturated fat, and cholesterol. Moreover, the paper focuses on how the empirical results in this paper can help guide developing policies regarding the regulations governing the flow of biotechnology information to consumers.

There are several *limitations* in this work that should be noted at the outset.

First, this study examines empirical evidence based *solely on U.S. consumers and solely in the area of nutrition messages*. The empirical work is not based on the Canadian marketplace and does not examine the effectiveness of information dissemination efforts in the area of biotechnology. Rather, the purpose of the empirical analysis is to examine, in very fine detail, the effectiveness of government/other sources of information and advertising/labeling in educating the public about certain types of scientific relationships. The advantages and disadvantages of each source of information are likely to carry over to a number of areas and is thus relevant to the regulation of information regarding biotechnology. Empirical evidence on the effectiveness of these different sources should be valuable to policy makers facing similar goals of educating the public about scientific findings and outcomes.

Second, this study is focused on the dissemination of information regarding the benefits of product use rather than the lack of danger of using a particular product. In the area of biotechnology it is important to recognize that dissemination of information regarding risk raises additional issues beyond those associated with the communication of information regarding benefits. This study does not address these additional issues.

Finally, in assessing the empirical evidence, it is important to recognize some caveats implicit in any study of this type. The type of evidence considered in this report cannot definitely prove or disprove that any two events (i.e., advertising of health benefits and changes in consumer behavior) are causally related. What we can do is to test hypotheses to determine which hypotheses are consistent with the available data, and thus, more likely to be true. Moreover, measurement of any consumer behavior as complex as food consumption is difficult, and every data set relevant to the issue has limitations. That is why three major data sets are considered in this paper. Finally, as in any study of the effects of a policy change, it is important to recognize other factors that could affect behavior. In particular in our case, public education efforts continued to evolve throughout the periods under study. Little quantitative information is available, however, to assess the effectiveness of these efforts and whether this effectiveness changes during the period under study. Consequently, the empirical evidence is best viewed as assessing the role of producer claims in an environment where the mass media and government continue to disseminate similar information.

SOURCES OF INFORMATION AND CONSUMER BEHAVIOR

Government and Other General Sources of Information

Government has some advantages as a source of information. As with all public goods, government is in a unique position to tax the population in order to fund the development and dissemination of information and thus avoid the complexities introduced by attempting to price information. Moreover, if the public interest theory of government is reasonably accurate in this arena (in which the government is assumed to maximize social welfare), government would be an unbiased and credible information source that would dispense information whenever the benefits of the information justified the costs of spreading it.

However, there are potential disadvantages to government provision of information, especially if private sources are constrained. For instance, if government is the sole or major source of such information, great power is concentrated in one body. This can be a significant problem if the process is susceptible to errors or if any of the other theories of government behavior apply. For instance, if the "capture" or "special interest" theories of government behavior are valid (Peltzman, 1976; Stigler, 1971), special interest groups might have undue influence on the types of information developed and disseminated, possibly blocking or slowing dissemination adverse to their interests. Similarly, if bureaucratic incentives influence government actions, these decisions may be excessively risk averse or otherwise unresponsive to changes in science and the marketplace (Niskanen, 1971).

Finally, the nature of government and the pressures to which it responds influence the way the information is likely to be dispensed. In the nutrition area, for instance, information is usually disseminated through the release of government studies or scientific panel recommendations. These releases are initially limited to one-time reports in the news media, though there is a second round dissemination through the popular press that reports nutrition information. A number of studies have found that the effects of information that is not repeated frequently can be short-lived (see Russo, Staelin, Nolan, Russell, & Metcalf, 1986, for instance, for such a finding on the effects of nutrition information in supermarkets). This information is highly concentrated in the news and print media, and therefore, likely to be absorbed disproportionately by those reached by these information channels and those most efficient at processing information. Feick, Herrman, and Warland (1986), for instance, find that more educated consumers are significantly more likely to acquire nutrition information from print media than their less educated counterparts. Moreover, the information is generally released in generic form and not in product-specific form. Generic information requires that consumers have other sources of information and a greater understanding of the issue to turn the information into behavior, again creating a potential bias towards those most efficient in processing information and those with better access to health information.

Producers as a Source of Information

Private sector firms are another potential source of information. Certain products have or can be formulated to have desirable characteristics that may not be well understood by potential consumers. If these potential consumers could be informed about these product features at a low enough cost, demand for the product would increase enough to create profit opportunities for producers. This mechanism creates an incentive for producers to attempt to provide the missing information to potential consumers. There are a host of issues related to producer provision of information that are beyond the scope of this paper but that are important for understanding these incentives and for designing policy in the area. For example, if the information is provided in generic form, other producers of similar products will simply "free-ride" on the information and reduce the benefits to the original producer. Thus, producers are unlikely to provide health information unless they can tie it directly to their particular product (Calfee & Pappalardo, 1989; Ippolito, 1986, 1988).

Producers have several advantages as providers of health information. First, they should be willing to devote substantial resources to information provision, if there are significant deficiencies in public knowledge and if there are products that can be sold profitably as a result of providing this new information. Thus, producers are capable of adding large amounts of some types of information to the market, when it is needed. Second, producers' incentives are to provide information in product-specific form. Thus, as compared with government information, producer-provided information is more directly tied to potential behavioral changes, making it easier to act upon. Finally, producers have strong incentives to find the best methods to communicate the information to those who do not have it and would use it if they had it. These considerations should improve consumer access to the information, especially for subgroups within the population that do not have the information, and should reduce the information processing requirements necessary to turn the information into behavior.

Producer-provided information also has potential disadvantages, however. One important issue is credibility. Since consumers cannot usually verify science based claims directly (especially for long term effects), there is the potential for deception. Unless the market or government has mechanisms to punish firms that lie, or consumers can verify the information in some way, consumers would be expected to be skeptical of producer-provided information, limiting food producers' incentives to make claims.

A second issue is the inherent bias of producer-provided information. Assuming they can be credible when they make claims, producers have strong incentives to provide information that is positive about their product, but they have no incentive to provide negative information. Despite this inherent bias at the individual firm level, economic theory suggests that in many cases competition among producers can eliminate this bias in the information provided by the market as a whole (Grossman, 1981). This "unfolding" theory suggests that despite firms' initial reluctance to highlight "bad" characteristics in their products, competition will often induce all but the worst firms to disclose the features of available products, if the market values the information. As long as consumers are skeptical of firms that do not disclose additional features, they would then be able to rank products on most key features.

Outline of Empirical Approach

Over the last decade Pauline Ippolito and I have focused our research interests on examining the issues of government/other sources of information and producer provided science information with respect to the dissemination of scientific information linking diet to health. This paper reviews this previously published research and then examines how the conclusions can provide insight into the regulation of advertising of biotechnology products. While the empirical research focuses on how different sources of information linking diet to health affect consumer food choices, the results are likely to have implications for the dissemination of biotechnology information.

The empirical work examines consumer choice within two distinct periods. First, consumer knowledge and behavior is examined during the years 1977–1985, when there was significant regulatory risk in advertising and a *de facto* ban in labeling of explicit health claims. Second, during the years 1985–1990, when truthful health claims were permitted and thus, potentially added to the continuing efforts of government and other general information sources to educate the public about diet-disease links. Content analyses establish that once science-based claims were permitted in advertising and labeling, firms were quick to seize on the opportunity to promote the health features of their products. Available evidence suggests that fiber-cancer claims began in 1985, whereas heart-health claims started slowly after 1985, but built to a more substantial level by 1987 for firms with products that could benefit from spreading the fat, heart-health information.

THE IMPACT OF SCIENCE-BASED HEALTH CLAIMS ON PRODUCER AND CONSUMER BEHAVIOR

Background on Regulations Governing the Dissemination of Diet-Disease Information

In any study designed to better understand how information spreads in markets, it is important first to examine when the information under study was available for dissemination. A review of the science linking diet to disease demonstrates that scientific research linking dietary consumption of fats, especially saturated fat and cholesterol to serum cholesterol levels, and, in turn, to heart disease was relatively well established by the 1950s and has gathered continued support since then. Numerous research studies can be found in the 1960s, 1970s, and 1980s documenting and refining scientific understanding of the links between dietary fats and cholesterol and serum cholesterol levels, and between serum cholesterol levels and heart disease. Reviews of the newer literature are available in U.S. Surgeon General (1988) and National Research Council (1989). Moreover, in the 1970s substantial epidemiological evidence developed supporting the relationship between fat intake and breast and colon cancer. Studies throughout the 1970s and 1980s have generally confirmed these results, though there is less support for the link between fat and breast cancer.

The dissemination of information by government and other sources. As scientific understanding of the role of fats and cholesterol increased, information began to flow to the public through several channels. Public health organizations, government, and the press were all significant sources of this information. By the early 1960s, many articles were appearing regularly in the popular press discussing the relationship between saturated and polyunsaturated fat and serum cholesterol, and their relationship to heart disease. By the early 1960s, the American Heart Association (AHA) assumed a leading role in advising consumers that they could reduce their risk of heart attack by limiting saturated fat and cholesterol consumption. Other groups soon followed.

The evidence also suggests that government played a role in disseminating this information, though considerably later than other sources. Prior to the 1970s, government dietary advice focused on the need to obtain sufficient energy and nutrients from foods to prevent nutritional deficiencies. By the 1970s, however, the focus of national nutrition policy began to shift to dietary components associated with chronic disease, though these moves often generated considerably controversy. The shift in emphasis can be seen in rules for voluntary and triggered nutrition labeling adopted in 1973, which allowed labeling of fat content by types, but the shift is most clearly reflected in a series of U.S. Senate hearings held between 1973 and 1977 on the role of diet in the development of chronic disease (U.S. Senate, 1974, 1977). Quantitative recommendations were issued for several aspects of food consumption, including recommendations to increase consumption of complex carbohydrates (55–60 percent of energy), and to decrease consumption of refined and processed sugar (less than 15 percent of energy), fat (less than 30 percent of energy), saturated fat (less than 10 percent of energy), cholesterol (300 mg/day), and salt (3 grams/day). In 1979, the Surgeon General also published dietary advice for the public, including recommendations that Americans consume fewer calories if overweight, less saturated fat and cholesterol, less salt, and relatively more complex carbohydrates such as whole grains, cereals, fruits and vegetables, etc. (U.S. Surgeon General, 1979). Similar public advice was given in 1990 and revised several times since in the *Dietary Guidelines for Americans* (U.S. Senate, 1981). In 1988 *The Surgeon General's Report on Nutrition and Health* provided another major review of the scientific literature on the relationship between diet and health and issued similar dietary guidelines. In 1989, the National Research Council also issued a major review with similar dietary recommendations in its report *Diet and Health: Implications for Reducing Chronic Disease Risk*. Thus, government and other general information sources appear to have made considerable efforts to communicate the growing scientific evidence linking diet to heart disease and cancer.

Regulatory constraints on producers. Food producers were also an information source throughout this period, though producers faced constraints in providing some types of information. Claims on food labels are primarily regulated by the Food and Drug Administration (FDA), and claims in advertising are primarily under the jurisdiction of the Federal Trade Commission (FTC). The types of claims allowed on labels or in advertising changed over time. For example, after producers reformulated margarine and cooking oil products to reduce saturated fat levels and began promoting their products' heart-related characteristics in the late 1950s, the FDA prohibited any label claims regarding cholesterol or fat content by type. During the early 1970s enforcement of the ban on labeling these characteristics appears to have ended, and by 1973 the labeling policy was explicitly changed to allow cholesterol and fat composition disclosures and simple nutrient claims about these characteristics on labels. For a discussion of the history of FDA regulation of cholesterol, see Calfee and Pappalardo (1989). The FTC allowed simple nutrient claims about fats and cholesterol in advertising throughout this period as long as the claim was not deceptive or misleading, though the agency did propose additional rules for nutrient content claims in its Food Rule, as described below.

Despite the policy change regarding fat and cholesterol labeling, manufacturers' claims linking these or any other dietary component to disease risks was explicitly prohibited on labels throughout the 1970s and into the 1980s. Thus, for instance, from 1973 to the mid 1980s a manufacturer could label the fiber, fat, saturated fat, and cholesterol content of a food product but could not on the label cite the health reasons why consumers should care about these characteristics, namely, the potential to reduce heart disease and cancer risks. From the early 1960s, this label prohibition appears to have been actively enforced. For example, Hutt (1986, pp. 32–34) reports that in 1964 the agency seized Nabisco Shredded Wheat, because the label contained information linking serum cholesterol to heart disease. Another example is in 1976, when the ITT Continental Baking Company was required to terminate a labeling campaign linking high fiber food to colon cancer, and in 1978 when Kellogg was required to stop a similar campaign. The labeling policy prohibiting health claims was restated by the agency at several occasions.

Rules for advertising never formally prohibited diet-disease claims or other more general health-related claims. Advertising cases are usually brought under the FTC's authority to pursue deceptive business practices. An assessment of what the agency considers deceptive must be determined from cases or other agency pronouncements during the period of interest. In the case of food claims, a variety of evidence suggests that diet-disease claims and other health related claims in advertising raised substantial legal risk at the FTC during the period from the mid 1970s to at least the beginning of 1983 for disease claims, and until 1980 for more general health-related claims. A primary indication of agency thinking at the time can be taken from the progress of the FTC's Food Rule, a broad rule-making undertaken in the mid 1970s to regulate food claims in advertising. In the original 1974 proposal for the Food Rule, the FTC staff recommended a ban on all diet-disease claims as inherently deceptive in advertising, as in the FDA labeling restrictions, and proposed to prohibit other general health-related claims, such as the term "health food."[1] The Commission itself did not propose to ban all health claims, instead reserving its options while soliciting comments on the issue in the rule's Federal Register Notice. By 1978 staff recommendations explicitly dropped the proposal to ban heart-health claims (and discussed criteria for such claims) and by 1982 the FTC began expressing reservations about many of the remaining features of the Food Rule, which ultimately led to a vote to terminate the rulemaking on December 17, 1982 and a formal public notice ending the rulemaking in May of 1983, in favor of case-by-case enforcement against deceptive food claims of all types.[2]

Thus, by May 1980 these FTC decisions in the Food Rule created considerably less legal risk for truthful claims in food advertising using general health-related claims, and by May 1983 less risk for claims that specifically referred to diet-disease risks, though no particular guidance was given on the form of such claims. A diet-disease claim in advertising continued to raise the risk of prosecution by the FDA, however. Under FDA law, a diet-disease claim in advertising allowed FDA to declare the product a "drug," and thus subject to drug law requirements.

The policy banning health claims on labels, with its implications for advertising, was effectively relaxed in 1985 following the introduction of Kellogg's highly publicized All-Bran advertising and labeling campaign, explicitly using the National Cancer Institute's statements on the potential relationship between fiber and cancer to promote its high fiber cereals. The FDA's decision not to challenge this prominent campaign, which was in direct violation of the prohibition of diet-disease claims on labels, presumably led firms to perceive a significantly reduced legal risk in using accurate and well founded health claims in advertising and labeling. FDA published a proposed rule to govern health claims on labels under a general deception standard in 1987,[3] but agency officials had publicly supported a change in policy earlier and had announced that well founded claims would not be prosecuted in the interim. FDA officials were often quoted in trade press articles during 1985 and 1986 as supporting valid health claims and promising an official change in policy to allow them (Hutt, 1986, pp. 19 and 49, citing speeches by FDA agency officials).

Thus regulatory events over this period suggest that possibly as early as 1983 in advertising and by sometime around 1985 in labeling producers faced considerably less regulatory risk in making truthful claims about well established diet-disease relationships in advertising and labeling. Initial claims focused on the relationship between fiber and cancer, but a number of food manufacturers soon began to promote the relationship between fats and cholesterol consumption and heart disease, as the policy debate continued on how best to regulate such claims. Thus, beginning in 1985, the regulatory environment changed to allow producers to become an additional source of consumer information linking diet to health. The regulatory uncertainties in the area seemed to induce firms to enter this field cautiously, but available evidence suggests that some claims were certainly made in 1986. For instance, Promise margarine, which was lower in saturated fat than margarines at the time, was introduced in 1986 with its "Heart smart" theme and focus on the role of saturated fat in coronary disease risks. The major TV campaign by Nabisco for Fleishmann's lower saturated fat margarine, featuring a 30-year-old man talking about his recent heart attack and discussing the role of diet in prevention, also ran in 1986. By 1987 a number of firms were making heart-health claims in major advertising campaigns.

Given these circumstances, the empirical work examines consumer choice within two distinct periods, the years 1977–1985, when there was significant regulatory risk in advertising and a *de facto* ban in labeling on producers' use of explicit health claims, but government and other general information sources provided diet-disease information, and the years 1985–1990, when truthful health claims were permitted and thus, potentially added to the continuing efforts of government and other general information sources. Again, while the empirical work described in the following sections is with respect to assimilation of diet-disease information, the evidence can be used to evaluate the likely effectiveness of policies governing the regulation of the dissemination of science-based information with respect to biotechnology, especially biotechnologies related to the health benefits of foods.

The Effect of Producer Health Claims in the Ready-to-Eat Cereal Market

Figure 1 provides the market-share-weighted fiber content of cereals for each year between 1978 and 1987. This figure is discussed in detail in Ippolito and Mathios (1989, 1991). The figure shows that despite growing evidence of the link between reduced cancer rates and high fiber diets during the years 1978–1984, a period before producer health claim advertising, there was no shift towards high fiber cereals. However, as soon as producer health claims in advertising and labeling began in late 1984, there was a significant increase in the market-share-weighted fiber content of cereals.

As demonstrated in Figure 1, during the health claims advertising and labeling years of 1985–1987, the market-share-weighted fiber content of cereals (grams of fiber/ounce) increased from 1.64 grams to 1.75 grams of fiber/ounce of cereal, an increase of approximately 7%. Under reasonable assumptions, it is estimated that health claims in advertising and labeling caused approximately 2 million more households to consumer high fiber cereals.

Cereal manufacturers, in response to the growing demand for high fiber cereals and knowing they could now advertise the health benefits

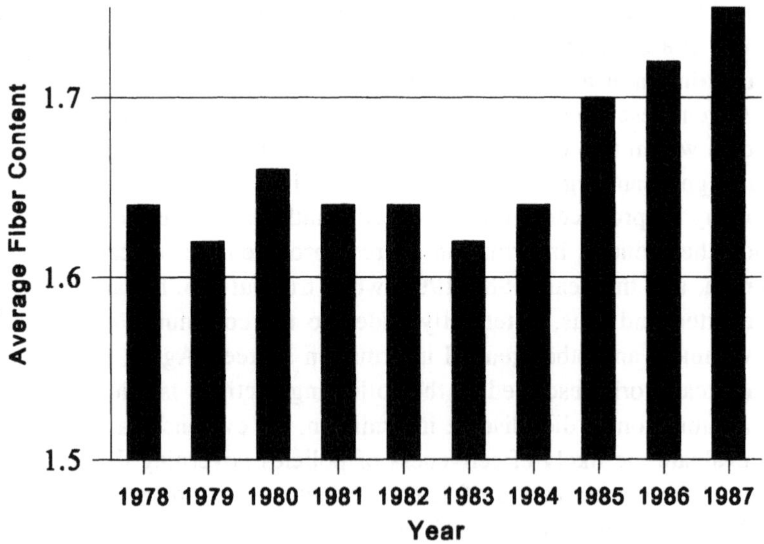

Figure 1. Average fiber from cereal 1978–1987.

of fiber, responded by developing new high fiber cereals. Excluding childrens cereals, cereals introduced between 1985 and 1987 averaged 3.59 grams of fiber per ounce compared to an average of 1.99 grams of fiber per ounce for cereals introduced between 1978–1984.

It is important to note that even prior to 1984 firms were permitted to disclose their fiber content on labels. Consequently, the dramatic effects on producer and consumer behavior are clearly linked to the use of the health claim rather than the ability to disclose fiber content. In other words, it is important to permit firms to explain the reasons why consumers should care about fiber.

The use of health claims in the ready-to-eat cereal market also had a profound effect on consumers' knowledge of the link between diet and disease. FDA survey data shows that consumer knowledge of the link between fiber and cancer was low and did not increase substantially during the six years prior to the introduction of health claims. (For more detail about these results see Ippolito & Mathios, 1991, and for a detailed description of the survey see Levy & Stephenson, 1990).

For the less than high school education group there was no gain in knowledge, and for high school graduates and those who attend some college there were some modest gains. The 1978 statistics are taken from Levy and Heimbach (1990), who combined the "Some College" and "College Graduate" categories. For example, in 1984, as in 1978, only 1 percent of those with less than a high school education knew the link between fiber and cancer. After the introduction of health claims, all education groups gained knowledge of the fiber-cancer link. For example, reported knowledge rose from 1.1 percent in 1984 to 18 percent in 1986 for those with less than a high school education. Knowledge levels increased dramatically for other education groups as well.

USDA consumption data allow for investigation of which types of consumers purchased high-fiber cereals immediately before and after the change in policy governing health claims. Table I provides the frequency distributions by a number of demographic characteristics.

The consumption data in Table I indicate that those groups that had relatively low levels of fiber cereal consumption in 1985 had the largest increases in high fiber cereal consumption by 1986. For example, the proportion of those who did not graduate from high school who ate fiber cereals increased from 1.9% to 4.8%. For college

TABLE I
Percent Eating Fiber Cereals,[a] Women 19–50 Years

	1985 (Before Health Claims) %	1986 (After Health Claims) %
Total	5.8	6.5
Education		
Not High School Graduate	1.9	4.8
High School Graduate	6.1	4.8
Some College	6.0	8.8
College Graduate	8.2**	8.6*
Income		
$0–$15,000	1.9	6.2
$15,000–$35,000	6.3	6.6
Over $35,000	8.5**	7.1
Race		
White	6.7	6.9
Non-White	0.7**	4.4
Smoking		
Non-smoker	7.2	7.1
Smoker	3.2**	5.4**
Head of Household Status		
Single	3.1	4.6
Not-single	6.6**	7.3*

Data: USDA Continuing Survey of Food Intakes by Individuals (CSFII), Women 19–50 Years, 1985 and 1986.
Notes:
[a] In this table, fiber cereals are defined as cereals containing at least two grams of fiber per ounce of cereal. The individual was coded as eating fiber cereal if he or she reported consumption of any quantity of a cereal of this type in the CSFII's in person, 24-hour recall of all food consumed the previous day.
** Indicates significance at the 95% level in a chi-square test of independence between the demographic characteristics and consumption of cereals with at least two grams of fiber per ounce.
* Indicates significance at the 90% level.
This table is reproduced from Ippolito & Mathios (1991).

graduates the increase was much smaller, from 8.2% to 8.6%. Similar results are found for other demographic characteristics as well. These results are dramatic enough so that some of the statistically significant differences in consumption that existed in 1985 no longer existed just one year later. These types of changes are found even when considered within a regression framework. Regression results indicate that many of the coefficients on the demographic characteristics are significant in explaining high fiber cereal consumption in 1985 and are insignificant in 1986.

In considering potential reasons why advertising had differential effects on various groups, several major differences between the information distribution methods used by government and private advertisers are worthy of mention. Government and general information is usually disseminated in generic form ("increased soluble fiber consumption may reduce risks of CHD") and this information is concentrated in news and print media reports about the latest scientific studies on diet and health. In contrast, most cereal advertising is distributed through television, with a smaller portion in print media. Moreover, health claim advertising and labeling is product specific so that advertising and labeling not only indicates the relationship between food characteristics and health, but also prominently features a product that contains these characteristics.

The Impact of Health Claims on Fat, Saturated Fat, and Cholesterol Consumption

In this section, the effects of information about the health effects of fat and cholesterol on consumption of these nutrients during the periods of the study are examined.

Changes in daily consumption of fats and cholesterol. Comprehensive data on food consumption can be found in a series of multistaged stratified probability samples representing the 48 contiguous states for the sex-age groups sampled, conducted by the U.S. Department of Agriculture (USDA). At ten year intervals, USDA conducts its Nationwide Food Consumption Surveys (NFCS), large scale, national probability surveys of food consumption by individuals. Since 1985, USDA has also conducted more frequent, smaller scale surveys of select populations, the Continuing Surveys of Food Intake by Individuals. In all of these surveys, detailed one-day recall data on all food consumed within the previous 24-hour period are collected and matched to nutrition data. For a more detailed description of the data see Ippolito and Mathios (1994).

Since one of the purposes of this section is to examine changes in consumption over time, it is important that the base of the analysis be consistent over time. For this reason, Ippolito and Mathios (1994,

1996) focused on data for sub-populations that were consistently sampled by the USDA in the years available and for which the basic variables of interest are available. In particular, these studies analysed those data from the USDA surveys that constitute 1-day recall data on food consumed in a 24-hour period in spring 1977, 1985, 1986, 1987/88, and 1989/90 for women, aged 19–50 years, who were heads of households and who consumed at least 300 calories on the interview day.[4] Since the focus of the analysis is on changes during the two periods of study, we analyze changes from 1977 to 1985, and from 1985 to 1990.

These data show that daily fat consumption for women fell during 1977–1985, and that the rate of decline accelerated during the health claims period from 1985 to 1989/90. Average fat consumption for women declined significantly by 3.7 grams in the eight years from spring 1977 to spring 1985 (from 73.3 to 69.6), and fell an additional 11.2 grams in the nearly 5 years between the 1985 and 1989/90 samples. The changes during the health claim period are statistically significant in simple means tests for both the 1985–1987/88 and 1985–1989/90 comparisons (with $t = -3.7$ and $t = -5.8$, respectively).

The changes in saturated fat consumption generally parallel changes in total fat consumption. As with total fat, the absolute reductions in saturated fat consumption were larger in the health claims period of 1985–1989/90 than in the pre-1985 period. Changes in average cholesterol consumption also show the same pattern of accelerated decline during the health claims period as found for fat and saturated fat. The magnitude of the acceleration is more pronounced, though a change in the cholesterol data used for eggs beginning in the 1987 data suggests the need for caution in interpreting this result. For example, for women, the daily consumption of cholesterol declined 40.4 milligrams (from 345.3 to 304.9) during the 8 years prior to 1985. From 1985-1989/90 average daily cholesterol consumption declined by 83.7 milligrams ($t = -7.4$).

Food sources of fat and cholesterol consumption. The data above show differential rates of decline in consumption of fat, saturated fat, and cholesterol during the two periods of study. One possible reason for the differential rates of decline has to do with the incentives the government faces in the dissemination of information, as opposed to private firms. This section establishes that these incentives are likely to lead to predictable differences in the types of improvements consumers are likely to make. The conclusion to this section will focus on the relevance of the empirical results to issues surrounding the dissemination of biotechnology information.

Government and general sources of information typically provide generic information about the effects of diet on disease risks and about the importance of eliminating or reducing foods that contain high levels of nutrients linked to the disease risk. For example, in providing information about saturated fat or cholesterol and heart disease, these sources typically stress the importance of reducing meat consumption and increasing foods that are very low in saturated fat and cholesterol, such as fruits and vegetables. For instance, the recommendations specifically mentioning foods in the *Dietary Goals for the United States* (U.S. Senate, 1977) were for consumers to reduce meats, increase poultry and fish, substitute nonfat milk for whole milk, increase consumption of fruits and vegetables, and reduce butterfat, eggs, sugar, and salt. This type of generic information provides a broad perspective on the major types of dietary changes that are desirable in reducing disease risks. These sources of information are unlikely to focus on particular brands of food, but rather on broad food categories where general statements can be made. Thus, food categories that have midrange levels of the nutrients at issue, or where simple generalizations are inappropriate, are unlikely to be the focus of these sources of information. Similarly, with respect to biotechnology the government is unlikely to focus on particular products or brands but instead on the generic information surrounding safety and risk features of biotechnology.

For these reasons, during the 1977–1985 period government and general sources of information would be expected to have their greatest effects in reducing fat, saturated fat, and cholesterol consumption in those food categories where this type of generic information about food categories can be given. Thus, during this period, reductions in these nutrients should arise primarily from reductions in the easily identified high fat or high cholesterol food categories, such as meat, dairy, and eggs. Similarly, the generic information should cause increases in the consumption of food categories that are quite "good" in these dimensions, specifically leading to increased consumption of simple grain products, fruits, and vegetables. We would expect reductions in the mid-range and higher fat food categories that have considerable variation within the category to be relatively limited, because these food categories are not likely to be the focus of information from these sources.

In contrast to government and general sources of information, producer-provided information is unlikely to contain generic information about food categories and is less likely to provide broad guidance on dietary choices. Instead, producer-provided information typically focuses on a particular brand of a particular food, identifying where the brand is superior to its competitors and why the consumer should consider this difference important enough to buy the product. Thus, producer-provided information is usually more specific than information from government and other general sources but more narrow in scope, though the competitive process itself fills in some of the information missing in individual claims. For instance, if a producer focuses on the low cholesterol content of its product and is gaining market share from the claims, a competing producer that has a low cholesterol and low fat product has the incentive to spread this information to regain the lost sales. This competitive process should provide a fuller picture of the range of products available in the market than is provided by any individual claim in isolation. Also, producer-provided information may be more prominent to many consumers because of the increased quantity of claims and the different mix of media used for making the claims.

If the added specificity and prominence of producer-provided information are important in helping consumers to make dietary improvements, and if context and broad dietary information are gathered from either the competitive process or from government and general sources, the removal of the ban on producer health claims would be expected to increase the rate of improvement across a broader range of product categories, compared to the period when such claims were prohibited. This would occur not only in food categories in which explicit health claims are made, but also in other food categories where the incentive to compete on nutrition dimensions is increased because of the greater prominence of the diet-health issue created by the added health claims information. In contrast, if producer-provided claims are deceptive or sufficiently incomplete to mislead consumers, the rate of dietary improvement would be expected to slow across food categories as consumers respond to the misleading information. We now turn to an assessment of the USDA consumption data to judge which of these hypotheses is most consistent with the available data.

Reductions in fat from various food sources. Table II gives the mean amount of fat consumed by women from a variety of product categories in the available years. We begin by focusing on the changes that occurred during the period of 1977–1985, when producers were not allowed to make health claims. The following section examines changes in the post-1985 period, when the policy was changed to allow producer health claims.

1. *Changes in dietary sources of fat between 1977 and 1985.* As shown in Table II, foods from the Meat/Mixtures category accounted for approximately 37 percent of the fat consumed per day by women. Clearly, consumption from the Meat category was the dominant source of fat in 1977. Foods from the Fruits/Vegetables category are the next largest source of fat for both sexes in 1977, contributing nearly

TABLE II
Average Fat Consumption Per Day by Food Category, Women 19–50 Years

Food Category (N)	1977 (1704)	1985 (1259)	1986 (1293)	1987/88 (889)	1989/90 (365)
Meat/Meat Mixtures	27.15	16.81†	15.56	15.18*	11.99*
Poultry/Poultry Mixtures	3.64	3.51	3.46	3.42	4.34*
Fish/Mixtures	1.83	2.27†	1.31*	1.51*	2.21
Grain/Mixtures/Soups	2.74	5.74†	6.28	4.90	7.50*
Egg/Mixtures	3.18	2.54†	2.67	2.33	1.81*
Breads	2.74	2.88	2.61	3.08	2.81
Sweet Bread	1.01	1.03	1.30	0.91	1.26
Fats and Oils	4.35	3.58†	3.77	4.23	3.49
Dressings/Sauces/Gravy	4.44	5.39†	4.73	4.24*	5.39
Milk	4.05	3.57†	3.73	3.67	2.83*
Cheese/Cream/Yogurt	3.88	4.48†	4.95	3.79*	4.30*
Desserts	4.70	7.23†	6.48	5.35*	3.87*
Snacks	2.15	4.18†	3.97	4.54	3.98*
Fruit/Vegetables	6.53	5.72†	5.18	6.27	5.87
Cereals	0.20	0.32†	0.45	0.41	0.39
Drinks/Other	0.11	0.32†	0.42	0.04*	0.08*
Totals	72.70	69.57†	66.80	63.86*	62.14*

Data: USDA National Food Consumption Surveys, Individual Intakes, 1-Day 1977,1987/88, and Continuing Survey of Food Intakes by Individuals, 1985,1986, and 1989/90. All means weighted.
Notes:
† Indicates differences from 1977 mean is significant at 5 percent level.
* Indicates significant differences from 1985 mean. N is sample size.
This table is also reported in Ippolito and Mathios (1994).

9 percent of the fat for women. Other substantial sources of fat are Poultry, Eggs, Fats and Oils, Dressings/Sauces/Gravies, Milk, Cheese/Cream/Yogurt, and Desserts, though together these sources contribute less fat than the Meat category. Thus, in 1977, foods in the Meat category were the most identifiable, concentrated source of fat. Most public education messages from this period reflect this concentration, advising consumers to reduce meat consumption. For instance, in the Dietary Goals for the United States (U.S. Senate, 1977), meat, along with butterfat and eggs, are the only specific foods singled out for reduced consumption for cardiovascular disease reasons.

Perhaps not surprisingly, given the focus of public health messages at the time and the significance of the category as a source of fat in 1977, the most striking change in fat consumption between 1977 and 1985 occurred for the Meat/Mixtures category. Fat consumption from this category declined by more than any other food category in both percentage and absolute terms.

Fat consumed from the Meat category fell 10.3 grams per day for women between 1977 and 1985. Only 5 other food categories experienced declines in fat, and 3 of these are specifically mentioned in public health messages from the period. Fat consumption declined for Eggs, Fats and Oils, and Milk, adding an additional reduction of approximately 1.9 grams fat per day for both sexes. Fat consumption from Poultry and Fruit/Vegetables also declined.

Despite the large declines in fat consumed from the Meat category, and the modest decline in the other identified categories, overall fat consumption from all foods declined by just 3.2 grams per day. Thus, much of the gain achieved in the Meat category was overcome by increased fat consumption in other food categories. Some of this is expected, of course, as individuals who cut meat consumption increased their intake of poultry, fish, and grain-based mixed dishes. This expected shift does not fully explain the overall compensation observed, however. Women increased fat consumption by only 3.3 grams from the Poultry, Fish, and Grain/Mixtures categories combined. Thus, much of the compensating increase in fat came from other changes in diet during this period.

As shown in Table II the most substantial increases in fat consumption came from increases in Desserts and Snacks. Between 1977 and 1985, women consumed an average of 4.6 grams more fat from these two categories combined. More women increased fat consumption in 9 of the 16 food categories listed (6 of the 10 increased by one-half gram or more), indicating that increases in fat came from a broad range of food choices.

Overall, the results indicate that during the 1977–1985 period fat decreases were concentrated primarily in the Meat category. Moreover, the increases in fat consumption from other foods were sufficient to eliminate approximately 70 percent of the fat reductions achieved in the Meat category, resulting in moderate overall fat reductions between 1977 and 1985. We now turn to a discussion of changes between 1985 and 1989/90.

2. *Changes in fat consumption between 1985 and 1989/90.* The decline in average daily fat consumption was considerably larger between 1985 and 1989/90 than the decline between 1977 and 1985; average daily fat consumption fell by 7 grams for women between 1985 and 1989/90. In contrast to the changes that occurred between 1977 and 1985, no single food category was responsible for the decline in fat consumption in the post-1985 period. Instead, the overall decline in fat consumption occurred across a larger number of food categories. For example, the largest decline in fat consumption again occurred in the Meat category. Compared with the earlier period, the rate of improvement was slower; fat from Meat fell by 4 grams for women between 1985 and 1989/90. If Meat, Poultry, Fish, and Grain/Mixtures are considered together, fat from the categories fell by 1.1 grams for women.

In evaluating changes during this period, note that the fat reduction in the Dessert category is the second largest for women, exceeding 3 grams in each case. More broadly, fat fell in 10 of the 16 categories and by more than one-half gram per day in 7 of the 16 categories. In contrast, only two food categories increased by more than 0.5 grams for women (Poultry and Grains/Mixtures). Thus, in contrast to the 1977–85 period, fat reductions occurred across a broad range of food categories, once the regulations on producer claims were relaxed, including food categories that were a substantial but not the primary source of fat, leading to a more rapid rate of decrease in fat consumption overall. The results are essentially the same if the 1987/88 data is used as the basis for comparison in the post-1985 period.

3. *Summary of results on fat sources.* The nature of the declines in fat consumption across food categories changed during the post-1985

period. Between 1975 and 1985, reductions in fat were highly concentrated in the Meat category and to a modest extent in Eggs, Fats & Oils, and Milk for both women and men. These results are consistent with the hypothesis that government and other sources of information would have their greatest effect in those food categories where general statements could be made. Also, as expected, these sources of information did not do well in the less concentrated food categories or in those that include a broad mix of foods; fat actually increased in 10 of the 16 food categories during this period, so that approximately 70 percent of the reduction in fat from Meat was lost due to this compensation.

In contrast, the reductions in fat during the post-1985 period occurred across a broader range of food categories. Fat continued to fall in the Meat category, but fat also fell in other categories that are not as easily characterized, such as Desserts or Cheese/Cream/Yogurt. In fact, fat consumption fell in 10 of the 16 food categories. This evidence is consistent with the hypothesis that the relaxation of the ban on producer health claims, in the presence of public and other sources of information, would lead to fat reductions across a wider range of food categories. The results are inconsistent with the alternative hypothesis that such claims are inherently deceptive or sufficiently incomplete to lead to a deterioration of food choices across a broad range of food categories, other things equal.

The data examined above indicate that the average consumption of fat, saturated fat, and cholesterol declined during the years 1977 to 1989/90. The evidence also indicates that the rate of decline was generally greater during the period when diet-disease claims were permitted. Thus, at this aggregate level, the data on average fat, saturated fat, and cholesterol consumption are consistent with the hypothesis that the change in policy that allowed producers to mention diet-disease issues in advertising and labeling added information to the market led to a faster rate of improvement in consumers' diets. The evidence is inconsistent with the hypothesis that these producer claims led to a deterioration in the rate of improvement in consumers' diets.

Obviously, this type of data cannot establish that advertising and labeling claims were responsible for the increased rate of dietary improvement, since, for instance, government and other public and private organizations continued their efforts to inform the public during this period and could have found more effective ways to accomplish their goals. Nonetheless, these data provide no support for the view that the introduction of producer health claims adversely affected consumer food choices overall or led consumers to reverse dietary improvements that were underway. Moreover, the data are consistent with the hypothesis that these claims, and the competition they spurred among producers, helped consumers to improve their diets more rapidly during the period when producers were freer to explain why these nutritional characteristics were important.

Producer Health Claims and Food Production Trends

Food production data, or disappearance data, measure basic food supplies moving through trade channels towards domestic consumption, and thus, measure the amount of food available for human consumption. For most commodity categories, this available food supply is measured as the sum of annual production, beginning inventories, and imports, minus exports, industrial non-food uses, farm uses (seed and feed), and end-of-year inventories. Whenever, possible, we use data reflecting retail level production, which reduces the amount of waste reflected in the data. Throughout this section, we will refer to the disappearance data for a food category as simply *production* data, though it is important to emphasize that these data include only food destined for U.S. consumption.

These data are collected by USDA directly from producers and distributors using techniques that vary by commodity (see Putnam & Allshouse, 1993, for a detailed description). These data are not collected from individual consumers, and thus they provide an independent basis for examining food consumption changes that does not have the potential problems implicit in consumer survey data. If waste and other losses in the system are relatively constant over time, these data provide an independent measure of changes in food consumption patterns. Thus, trends in these production data can also be used to test the hypotheses that government and general sources of diet-health information were affecting consumers' food choices prior to 1985 and that the change in the regulations governing health claims in 1985 provided an additional source of this information, with a corresponding incremental effect on consumption patterns.

This production or disappearance data exist for various types of food. Ippolito and Mathios (1993) examine trends for 22 different food categories. This section summarizes the results for six of these 22 trends. These six trends discussed here are typical of the results found for the others. First, trends in food categories that comprise a sizable portion of fat in the overall diet are examined to determine if consumption shifted away from higher-fat and cholesterol food categories towards lower-fat and cholesterol categories. In particular, trends in red meat, eggs, and animal fat are examined, as are those in lower-fat categories, such as flour and cereals, fruits, and vegetables. The data are obtained from the USDA's *Food Consumption, Prices, and Expenditures, 1970–92* (Putnam & Allshouse, 1993).

For each food group the changes in per capita production were analyzed between the years 1977–1985 (the pre-health-claims period) and 1985–1990 (the health-claims period). Ippolito and Mathios (1993) examine these trends from a statistical perspective, using spline regression techniques. However, these regression analyses are not presented in this chapter.

First, we review three high-fat or high-cholesterol product categories: meat, eggs, animal fat. Per capita production of red meat, which is a major source of fat and saturated fat in the U.S. diet, declined during both the 1977-1985 period and the 1985–1990 period. In 1977 production of red meat per capita was 132.2 pounds per year and by 1985 production per capita had fallen 7.3 pounds to 124.9 pounds per year. In the next five years, however, the decline accelerated. Per-capita production of red mean fell from 124.9 to 112.4 , a decline of 12.5 pounds per year. Regression results indicate that the underlying negative trend was significant and that it fell further during the health claim period, though this increment to the rate of decline is not significant.

Production of eggs per capita also declined during both periods. In 1977 production of eggs per capita was 34.3 and by 1985 this had declined by 1.4 pounds to 32.9. During the next 5 years per-capita product fell an additional 2.8 pounds, again showing an acceleration in the decline of production. Regression results show that the underlying trend was negative and significant, with a significant incremental reduction during the health claims period.

Surprisingly, the production of animal fat rose dramatically from 1977 to 1985 and then fell as dramatically during the 1985–1990 period. Production of animal fat per capita was 10.6 pounds per year in 1977 and 13.3 pounds in 1985, a rise of 2.7 pounds during this period. By 1990, per production fell to 9.7 pounds per year, a remarkable reduction of 3.6 pounds (a 27 percent reduction) in just 5 years.

Regression results indicate that during the pre-health claims period (1977–1985) there was a significant upward trend in the production of animal fat. In contrast, during the health claims period, there was a significant reversal of the trend.

In each of three cases analyzed, the production of high-fat, high cholesterol foods declined during the health claims period. Moreover, this decline was more accelerated compared to the period when producers were prohibited from using health claims. Of the 10 high fat food trends considered in Ippolito and Mathios (1993), all ten showed similar accelerated declines (relative to the underlying trend during the 1977–1985 period) during the health claims periods. Perversely, six of the ten production trends has positive underlying trends during the pre-health claims period.

We now analyze three low fat and low cholesterol product categories: flour/cereal, vegetables and fruits, which are major categories of foods recommended for increased consumption as a replacement of fats in the U.S. diet. The data show that per capita production increased during both periods under analysis for all three food categories, and that the rate of increase accelerated during the health claims period. For example, between 1977–1985 per capita production of flour/cereal products increased by 15.4 pounds (rose from 140.7 to 156.1). In the next 5 years, production rose another 27.4 pounds to 183.5 pounds per capita. For vegetables, production rose 10.3 pounds (from 200.5 to 210.8) between 1977 and 1985, and rose 17.6 pounds between 1985 and 1990. For fruits, production rose 3.4 pounds (from 96.1 to 99.5) between 1977 and 1985 and rose 7.1 pounds between 1985 and 1990. Regression results confirm that the upward trend in production was statistically significant for all three categories and that this trend accelerated significantly during the health claim period for flour and cereal products and for vegetables. The trend also increased for fruit but not significantly.

Summary. Examination of per capita production data for major food categories during the years 1977–1985, when health claims were restricted, provides some substantial evidence that information about fats, cholesterol and disease was spreading to consumers, leading to improvements in important aspects of the diet. The evidence for this period is not entirely consistent, however. Per capita production of some higher fat products increased during the pre-health claims period.

The production evidence for major food groups provides a more consistent picture during the 1985–1990 period, when the policy was changed to allow producers to make explicit diet-disease claims. During this period, per capita production from food categories with high fat and cholesterol levels either stayed on the trend that existed during the government period or experienced a decline in the trend relative to the earlier period. None of the increments to the trends were significantly contrary to expectations under the information hypotheses, and thus, the production data for broad food categories provide no evidence consistent with the view that the addition of health claim advertising had adverse effects on the rate of improvement in dietary patterns. In fact, during the advertising period, movements away from the high fat food categories increased for all major categories. Similarly, the production data show a consistent pattern during the advertising period of additional movements towards lower fat food categories.

APPLICATION TO THE DISSEMINATION OF BIOTECHNOLOGY INFORMATION AND CONTROLLING DECEPTION IN ADVERTISING AND MARKETING

The U.S. experience with respect to the impact of science-based claims in advertising and labeling has implications for issues relating to the regulation of the dissemination of biotechnology information. First, advertising and labeling are likely to be important sources of information to many consumers. Just as was the case with knowledge about the relationship between nutrients and disease, knowledge about the relationship between biotechnology and health is likely to increase if advertising and labeling are permitted to add to the flow of information already being disseminated by government and the news media. This appears to be especially true if producers of biotechnology are permitted to discuss truthfully the health implications of biotechnology products. As indicated in Table I in the report *Understanding the Consumer Interest in the New Biotechnology Industry* (Optima Consultants, 1994), 84% of the surveyed consumers reported an interest in health issues with only 4% indicating no interest. Similar types of responses are found with respect to attitudes towards food safety. Consequently, the potential role of the food industry to promote the benefits of biotechnology (or the lack of harm) are significant and may be similar to the role that producers played in educating consumers in the U.S. about the science linking diet to disease.

In the vast majority of surveys, Canadian consumers express a desire for the labeling of important features of products, including those involving biotechnology. For instance, in the Canadian survey referred to above (Optima Consultants, 1994), 83% of consumers surveyed expressed a desire that product labels should describe the biotechnology method used. Given the relatively low levels of knowledge about the relationship between biotechnology and health risks or benefits, labeling of the method of biotechnology alone, is unlikely to significantly alter consumer welfare. To be effective labeling must be used in combination with other promotion activities. In the study of the cereal market, firms were always able to disclose fiber content but it was not until they could explain to consumers why they should care about increasing their fiber content that observable changes in knowledge and behavior were seen. Moreover, the introduction of healthier products accelerated once firms were permitted to promote the health benefits of their products. Labeling alone was not sufficient to generate the innovation in products. In recognition of the importance of combining a health claim with a content disclosure, the U.S. Food and Drug Administration recently amended its regulation of health claims to permit not only the relationship between certain nutrients and disease, but the consumption of certain foods and disease as well. Thus, in 1997 oatmeal and whole-oat flour products will be permitted to promote the relationship between consumption of these foods and reduced risk of coronary heart diseases. Despite the large amount of media attention devoted to the disease prevention attributes of oats, the FDA chose to permit firms to promote these disease prevention attributes. This reflects the expectation that combining disease prevention messages with content messages is an effective way of educating the public about risk reduction.

Controlling Deception in Advertising

From a public policy perspective it is important to balance the potential benefits of allowing manufacturers greater leeway in making science-based biotechnology claims with the risks of increasing deception in advertising. Clearly, claims that are blatantly false should be stopped. However, once we move beyond the issue of explicitly false statements, questions about which policies are best to control misleading or deceptive claims become much less straightforward, and as might be expected, generate much more controversy and debate.

One type of proposal that is often made when regulating claims linking scientific results to consumer health is that government regulators should pre-approve all claims before firms are permitted to make them, that is, that some form of *ex ante* regulation should be adopted to discipline claims. Approval of labeling as well as specific language that must be used for labeling are examples of this type of regulation. The attraction of ex-ante regulation is, of course, that claims are reviewed before they are used, and thus, deceptive claims are stopped before they can do any consumer harm. The problems with this approach arise when one considers the other goals of public policy, that is, encouraging the dissemination of truthful biotechnology information that keeps pace with scientific discovery. Beyond the standard problems that face large bureaucracies, the potential for interested parties to intervene in the decision making process is an added cause of delay in areas where firms' profits can be affected by the decisions. For example, in the areas of drug regulation in the U.S. there is substantial evidence to suggest that we should be concerned about the ability of the regulatory agencies to keep pace. The relatively straightforward decision to add cholesterol and fat composition information to the nutrition label had been under consideration for over a decade. The over-the-counter (OTC) monograph program at FDA, which was designed to define approved language for claims for OTC drug labels, began in the early 1970s and is was still far from complete in the early 1990s: Only 13 of the 48 drug categories have final monographs. The FTC's experiences with rulemaking in the 1970s to define overall rules for OTC drug advertising and usage of certain terms in advertising were also long processes that were terminated in the 1980s. If a similar process is used to regulate the flow of biotechnology information many truthful and informative claims will not reach consumers until years after consumers could have benefited from the information.

The evolving nature of the biotechnology industry is an essential feature that should be considered in evaluating alternative regulatory approaches for controlling deception in biotechnology promotion. Static regulatory approaches that require lengthy procedures each time a change is needed raise the concerns discussed above. In this case, the spread of some truthful claims will be unnecessarily delayed, and some claims will be allowed to continue after the scientific support for them has weakened. More dynamic regulatory approaches, which automatically allow claims to be made if they meet some basic principles that reflect the scientific support available when the claim is made, are usually more suited to this type of situation. However, such dynamic approaches contain more uncertainty for firms, who must decide whether scientific support is adequate to justify a particular claim.

Should such a dynamic approach be adopted it is essential that firms that overstate or otherwise misrepresent the state of the science be punished. The U.S. government has a variety of mechanisms at its disposal for preventing false and/or misleading advertising and labeling. In the next section, I review the variety of methods the U.S. uses to address these issues. The description of each of the institutions that deter false advertising is based on Mathios and Plummer (1988).

The Institutions That Deter False Advertising

The Federal Trade Commission (FTC). The Federal Trade Commission is the agency that has direct responsibility for ensuring the truthfulness of advertising, the predominant form of information dissemination in the U.S. The FTC advertising substantiation doctrine requires firms to substantiate all claims that are made. Rather than issuing orders that ban a risky product from the marketplace or alter its design, the FTC requires that firms disseminate truthful information and disclose the danger of a risky product if omission of this fact creates a false or misleading impression of the product.

In order for the information approach to regulation to be effective firms that disseminate false information must suffer in the marketplace, either through consumer reaction in the private market or through public institutions that punish firms who violate truthfulness claims. Firms have incentives to meet these requirements because the FTC can penalize firms who engage in false or misleading information. The FTC derives its regulatory power from the Federal Trade Commission Act of 1914, and subsequent amendments. The most important amendment occurred in 1938, when the commission was enabled to prevent "unfair or deceptive acts or practices," as well as "unfair method of competition."

An FTC advertising case begins with an investigation and can result in a complaint which is voted on by the five person commission on a majority basis. If a firm elects to contest a complaint, the case goes before an Administrative Law Judge (ALJ). If overturned the case goes back before the commission. The commission reviews the ALJ's decision and can either dismiss the case or vote out a final order which can be stricter or more lenient than the original ALJ case. While it may seem odd that the Commission gets to review its own initial complaint at the appeal level, usually enough time has elapsed that the composition of the Commission has changed. A firm may appeal an FTC final order to a federal appellate court. This court may uphold the Commission decision, modify it, or overturn the decision. Reversals of FTC final orders are very rare, however.

During any point during this process a firm may opt to negotiate a settlement or consent agreement with the FTC. Most of the cases are resolved through negotiated consent. The consents often involve the firm agreeing to cease and desist from engaging in the contested advertising now and in the future. Once an agreement is reached, violation of the agreement comes with severe monetary penalties. The FTC may seek an injunction to halt an ongoing advertising campaign though this remedy has rarely been used in general advertising cases. Instead, remedies are imposed after final resolution, either by consent or by order. Before 1975, remedies were limited to the cease and desist orders discussed above. With the FTC Improvements Act of 1975, the Commission was enabled to recover monetary civil penalties for knowing violations of the finding of previously litigated cases and to recover consumer redress. Even after 1975, however, the predominant form of remedy is a cease and desist order.

Many have argued that FTC cease and desist orders have little ability to deter false advertising since a cease and desist order is perceived to give firms a "free bite of the apple" with no monetary penalty for violating the law. The lack of monetary penalties, however, does not imply a lack of deterrence. First, an advertiser may be forced to stop a current advertising campaign and forego its expected benefits. Second, the FTC can impose restrictions on future advertising, sometimes with very broad coverage. Third, and most importantly, a firm's reputation can be damaged by involvement with the FTC. There are several studies that demonstrate that firms involved with the FTC suffer a severe penalty. For example, Mathios and Plummer (1989) and Peltzman (1981) find that firms suffer a significant decline in their market value when it is announced that the FTC is investigating a firm's advertising claims. These declines dwarf any realistic monetary penalty that the FTC would assess. These results suggest that the FTC has significant ability to deter false advertising and that monetary penalties may be unnecessary to significantly punish firms. Mathios and Plummer (1989) also find that firms eventually found innocent still suffer significant market value declines. Therefore, poor case selection or vague legal standards could lead to overdeterrence, reducing the amount of advertising that provides useful information to consumers.

The National Advertising Division (NAD) and the National Advertising Review Board (NARB). In 1971, the Council of Better Business Bureaus in conjunction with three advertising associations created a formal procedure designed to resolve charges of false and misleading advertising. The procedure consists of an initial review agency, the National Advertising Division (NAD), and an appellate board, the National Advertising Review Board (NARB), the latter used for appeals from the former. The review process is limited to national advertising and complaints to the system are generated by consumers or competitors. If the NAD determines that an advertisement is questionable, they can request substantiation. After evaluating this substantiation, the NAD reaches a decision. Any of the parties can appeal this decision to the NARB. Remedies dictated by the NAD or NARB are not legally binding and participation in the process is voluntary. Despite the absence of statutory power to punish firms, the NAD has a record of substantial compliance. In over 95% of the cases where the NAD has found an active advertising campaign unsubstantiated, the advertiser has discontinued or modified that campaign (Armstrong & Ozanne, 1983). As with the FTC, NAD/NARB decisions receive publicity and involvement with the agency may harm a firms reputation.

The Lanham Act. The Lanham Act was passed in 1946. Section 43(a) of the Act allows a firm to sue another firm for false or misleading advertising. A plaintiff may bring suit if it is a business enterprise and if it is likely to lose sales as a result of the false advertising in question. The Act does not grant standing to consumers harmed by false advertising nor to firms whose business is unrelated to the defendants. Immediately following the passage of the Lanham Act, judges interpreted the law very narrowly, providing relief only when an advertiser infringed upon a firms trademark or "palmed off" his goods as the competitor's goods. After the landmark case of *L'Aiglon Apparel v. Lana Lobell*,[5] firms harmed by false advertising could pursue relief under Section 43(a). It was not until the mid 1970s, however, that one could observe a rapid increase in the use of this law. This coincided with the advent of comparative advertising making explicit references to competitors' products.

A plaintiff may obtain a preliminary injunction against an advertiser to immediately halt an ongoing campaign. An advantage of the Lanham Act is that it is quick – it is not unusual to find a preliminary injunction granted within a few weeks of the suit being filed. While this decision can be appealed, the injunction is binding during the appeal process. A plaintiff may also seek monetary damages, but the evidentiary burden is much greater for damages. In some cases, treble damages are awarded.

CONCLUSION

In the report, *Understanding The Consumer Interest in the New Biotechnology* (Optima Consultants, 1994), consumers are found to be twice as likely to anticipate beneficial than harmful effects from biotechnology in regard to food safety and to the nutritional value of food (p. 28). Thus the potential for firms to communicate this information to consumers is similar to the diet-disease links discussed at length in this paper. Moreover, consumer sentiment towards industry and their role in the dissemination of information is strikingly similar to consumer sentiment about producers role in the dissemination of diet-disease claims. In Office of Consumer Affairs Quarterly (1996), it is reported that consumers are skeptical of the credibility of profit-oriented organizations while governments and universities were considered the most likely sources of credible information. It was precisely these beliefs that, in part, led the government to ban the use of health claims in advertising and labeling following the innovative science linking diet to disease that occurred after the 1950s. Additionally, government and consumer advocate groups believed that science-based claims in advertising and labeling are too complex for most consumers to understand and put in perspective. This complexity would then present opportunities for deception and misleading claims so that, in aggregate, market competition would lead to less informed choices. The empirical evidence with respect to science-based claims in the nutrition area is inconsistent with this view.

NOTES

[1] See 39 Federal Register, November 11, 1974, 39862, Rosch (1975), or Weitzmann (1975).

[2] See 48 Federal Register, May 24, 1983, 23270–71.

[3] See 52 Federal Register, August 4, 1987, 28843.

[4] Use of only one person per household allows us to avoid the potential statistical problems introduced when multiple individuals from the same household are used in this type of analysis. Also since part of this study will explore changes in the types of food consumed over time, the calorie restriction is added to remove individuals who essentially ate nothing on the interview day. Our tests of this criterion indicate that the results are not sensitive to this requirement.

[5] See 214, F.2nd. 649 (1954).

REFERENCES

Armstrong G., & Ozanne, J. (1983). An evaluation of NAD/NARB purpose and performance. *Journal of Advertising*, *15*(3), 15–28.

Calfee, J. E., & Pappalardo, J. K. (1989). How should health claims for foods be regulated: An economic perspective. *Economic Issues Paper*. Washington, DC.: Federal Trade Commission.

Feick, L. F., Herrman, R. O., & Warland, R. H. (1986). Search for nutrition information: A probit analysis of the use of different information sources. *Journal of Consumer Affairs*, *20*, 173–192.

Grossman, S. J. (1981). The informational role of warranties and private disclosure about product quality. *Journal of Law and Economics*, *24*, 461–483.

Hutt, P. (1986). Government regulation of health claims in food labeling and advertising. *Food Drug Cosmetic Law Journal*, *41*, 3–73.

Ippolito, P. M. (1986). Consumer protection economics: A selective survey. In: P. M. Ippolito & D. T. Scheffman (Eds.), *Empirical approaches to consumer protection economics*, pp. 1–33. Washington, D.C.: Federal Trade Commission, Bureau of Economics.

Ippolito, P. M. (1988). The economics of information in consumer markets: What do we know? What do we need to know? In: E. S. Maynes & The ACCI Research Committee (Eds.), *The frontier of research in the consumer interest*, pp. 235–263. Columbia, MI: American Council on Consumer Interests.

Ippolito, P., & Mathios, A. D. (1989). Health claims in advertising and labeling: A study of the cereal market. *Bureau of Economics Staff Report, Federal Trade Commission*. Washington, DC.: FTC.

Ippolito, P., & Mathios, A. D. (1990a). Information, advertising and health choices. *Rand Journal of Economics*, *21*(3), 459–480.

Ippolito, P., & Mathios, A. D. (1990b). The regulation of science-based claims in advertising. *Journal of Consumer Policy*, *13*, 413–445.

Ippolito, P., & Mathios, A. D. (1991). Health claims in food marketing: Evidence on knowledge and behavior in the cereal market. *Journal of Public Policy and Marketing, 10*(1), 15–32.

Ippolito, P., & Mathios, A. D. (1993). New food labeling regulations and the flow of nutrition information to consumers. *Journal of Public Policy and Marketing, 12*(2), 188–205.

Ippolito, P., & Mathios, A. D. (1994). Information, policy, and the sources of fat and cholesterol in the U.S. diet. *Journal of Public Policy and Marketing, 13*(2), 200–217.

Ippolito, P., & Mathios, A. D. (1996). *Information and advertising policy: A study of fat and cholesterol consumption in the United States, 1977–1990*. Washington, D.C.: Government Printing Office.

Levy, A., & Heimbach, J. T. (1990). Recent public education efforts about health and diet. In: *Proceedings of the 36th Annual Conference of the American Council on Consumer Interests*. Columbia, MO: American Council on Consumer Interests.

Levy, A., & Stephenson, M. (1990). *Nutrition knowledge levels about dietary fats and cholesterol: 1983–1988*. Draft. Washington, D.C.: Food and Drug Administration, Division of Consumer Studies.

Mathios, A., & Plummer, M. (1988). Regulation of advertising: Capital market effects. *Bureau of Economics Staff Report to the Federal Trade Commission*. Washington, D.C.

Mathios, A., & Plummer, M. (1989). The regulation of advertising by the Federal Trade Commission: Capital market effects. *Research in Law and Economics, 12*, 77–93.

National Research Council (1989). *Diet and health, implications for reducing chronic disease risks*. Washington, D.C.: National Academy Press.

Niskanen, W. A. (1971). *Bureaucracy and representative government*. Chicago: Aldine Press.

Office of Consumer Affairs (1996). Biotechnology: What is it all about? *Consumer Quarterly, 1*(3), 1–4.

Optima Consultants (1994). *Understanding the consumer interest in the new biotechnology industry: Study findings*. Ottawa: Office of Consumer Affairs.

Peltzman, S. (1976). Toward a more general theory of regulation. *Journal of Law and Economics, 19*, 211–248.

Peltzman, S. (1981). The effects of FTC advertising regulation. *Journal of Law and Economics, 24*, 403–448.

Putnam, J. J., & Allshouse, J. E. (1993). Food consumption, prices and expenditures, 1970-1992. *Statistical Bulletin No. 867*. Washington, D.C.: U.S. Department of Agriculture, Economic Research Service.

Rosch, J. T. (1975). Trade regulation rule on food advertising – Remarks. *Food Drug Cosmetic Law Journal, 30*, 172–182.

Russo, J. E., Staelin, R., Nolan, C. A., Russell, G. J., & Metcalk, B. L. (1986). Nutrition in the supermarket. *Journal of Consumer Research, 13*, 48–70.

Stigler, G. J. (1971). The theory of economic regulation. *Bell Journal of Economics and Management Science, 2*, 3–21.

U.S. Senate (1974). *National nutrition policy study hearings*. Select Committee on Nutrition and Human Needs. Washington, D.C.: Government Printing Office.

U.S. Senate (1977). *Dietary goals for the United States*. Washington, D.C.: Government Printing Office.

U.S. Senate (1981). *Dietary guidelines for Americans*. Washington, D.C.: Government Printing Office.

U.S. Surgeon General (1979). *Healthy people: The Surgeon General's report on health promotion and disease prevention*. Washington, D.C.: Government Printing Office. DHEW(PHS) Publication No. 79-55071.

U.S. Surgeon General (1988). *The Surgeon General's report on nutrition and health.* Washington, D.C.: Government Printing Office. DHEW (PHS) Publication No. 88-50210.

Weitzman, S. A. (1975). Trade regulation rule on food advertising – An analysis. *Food Drug Cosmetic Law Journal, 30,* 140–171.

THE AUTHOR

Alan D. Mathios is Associate Professor, Department of Policy Analysis and Management, Cornell University, 120E MVR Hall, Ithaca, NY, 14853, USA. E-mail: adm5@cornell.edu.

Some parts of this article are taken directly from Ippolito and Mathios (1989, 1990a, 1990b, 1991, 1993, 1994, 1996) and Mathios and Plummer (1988).

Andrew N. Kleit

Using Advertising to Generate Information and Signals for Product Quality: Lessons for Biotechnology Markets in Canada from Pharmaceutical Markets in the United States

ABSTRACT. Recent innovations in biotechnology have made a number of products available to the consumer, and may provide more products in the future. Yet innovation is not enough to guarantee a product's success. A successful product requires appropriate marketing for commercial success. A crucial part of that marketing involves advertising. Yet advertising is one of the most controversial aspects in microeconomics. This paper addresses some of the relevant issues for biotechnology by looking at the debate over pharmaceutical advertising in the United States. The experience there teaches that for biotechnology advertising to serve consumers, there needs to be appropriate laws and government regulation in place. This legal framework should include proper regulatory and litigation structures in order to seek both to discourage deception and to foster advertisements that convey valuable information.

Recent innovations in biotechnology have made a number of products available to the consumer, and may provide more products in the future. Yet innovation is not enough to guarantee a product's success. A successful product requires appropriate marketing for commercial success. Further, a successful product must be able to deal with competitors as they arise. From the consumer's point of view, both the conveyance in the market of important information and market competition are crucial to economic well-being. Information allows the consumer to buy the product that best fits his or her need. On the other hand, consumer well-being would be reduced if advertising were used in the biotechnology field to create deception or to deter entry by a competitor.

Advertising is one of the most controversial aspects in microeconomics. As Shughart (1990, p. 122) puts it, "[o]n few other subjects do the opinions of economists differ as sharply as they do on the subject of advertising." Some economic theories of advertising (e.g., Nelson, 1974; Klein & Leffler, 1981), imply that advertising serves

Journal of Consumer Policy **21**, 1998.
Bartha M. Knoppers & Alan D. Mathios (eds), Biotechnology and the Consumer.

an important role in the marketplace by informing consumers, both explicitly and implicitly, about the characteristics of certain products. Other theories (e.g., Comanor & Wilson, 1979, and Schmalensee, 1983), imply that the role of advertising is to deceive customers and to increase barriers to entry and therefore advertising acts to harm consumers. Issues related to advertising may well become crucial to the economic success for biotechnology products.

This paper proposes to address some of these issues in both economic and regulatory contexts by looking at the debate over pharmaceutical advertising in the United States. Most biotechnology products currently on the market are designed for sophisticated consumers who are also producers (i.e., farmers). Some products that are intended for widespread consumer use are not widely advertised for either economic or legal reasons. Thus, at this point in time, pharmaceuticals in the United States are the best example of the mass advertising of biotechnology products, and the most useful place to examine the importance advertising can be expected to have in biotechnology markets.

Section II will address the marketing problem facing biotechnology firms. Section III of this paper will review the relevant theories of advertising. Section IV will apply those theories to the debate in the U.S. on pharmaceutical advertising. Section V will examine the available evidence on the effects of pharmaceutical advertising, while Section VI will analyze that evidence. Section VII will contain the paper's conclusions for both biotechnology firms and government regulators.

THE CHALLENGE FOR BIOTECHNOLOGY

Firms that create biotechnology products must often spend millions and millions of dollars on their manufacture. But once these goods have been created, these firms face a second, equally difficult problem: How to convince consumers to buy these products.

This problem may be especially difficult in the case of biotechnology products. As other chapters have noted, consumers may have significant resistance to biotechnology. Einsiedel (1998) points out that there is good reason to believe that the Canadian consumer will be highly skeptical of biotechnology products. In general, biotechnology products are credence goods, that is, their attributes cannot be determined even after they are consumed. Thus, consumers may well find it difficult to properly evaluate the risk associated with the consumption of such goods. In such circumstances, consumers can be expected to fall back upon their prior expectations about the risks of biotechnology products. Einsiedel also points to a series of surveys which indicate that Canadian consumers believe that these products may well pose substantial risks. Given this, consumers may seek out credible sources of information before purchasing and consuming biotechnology products. However, survey results also show that the major distributors of information to the consumer, government and industry, are distrusted by the public.

Thus, government certification of the safety of certain products may not be credible. The public appears to view the governmental process as beset by various political factors that affect the credibility of government positions. Not necessarily any more credible are simple statements by the firms selling the product. In this case, the public realizes that the firms have a profit motive for their statements. Something more is needed to credibly convey the relevant information.

One possible method by which credible information about biotechnology products could be conveyed to consumers is through the media. Here "the media" refers to news articles presented in newspaper, radio, and television, rather than advertisements presented in these forums. Unfortunately, as Strauss (1998) points out, there are a number of reasons to believe that the media will not present an accurate picture of the attributes of these products. In particular, media reporting is often done under intense time pressure. This deters the authors of such reports from gaining substantial information about the matter at hand. With a deadline looming, reporters may simply not have the time to research a project thoroughly. Further, given the competition for space within a newspaper (or time on a broadcast medium) the authors of a particular story are often not the ones at the institution who have the relevant (and relative) expertise on a particular topic. This may be especially important in a complex area like biotechnology.

In addition, the media may often be biased in the type of news that gets reported. Put simply, good news may be no news at all. The media like to present the dramatic. It is dramatic to report that a certain product is harmful. It is not dramatic to report that a product is safe. It is also dramatic to report some good news, in particular that a new product has wonderful attributes that are now available to the consumer. Dramatic good news, however, appears to be less common than dramatic bad news. Given this, and the probability that there will be at least some reported (though perhaps not always accurate) problems with biotechnology products, there is a strong possibility that the overall media reporting of such products will be negative. This in turn will create negative impressions in the minds of potential consumers, not only with respect to the particular product but perhaps to many or all biotechnology products in general (this may be offset somewhat by the positive news on newspapers' financial pages announcing a new product, to the extent that consumers read the financial pages).

This challenge of presenting credible information to final consumers is especially difficult for most biotechnology firms, given their experiences. In the past, biotechnological products have been largely marketed to sophisticated producers of other goods. For example, new herbicides made possible by biotechnological innovations have been marketed to farmers. Because so much of their livelihood depends on the effectiveness of the products they buy, farmers are likely to be sophisticated users of these products who pay a great deal of attention to the characteristics of particular products. Further, because they have extensive experience in the relevant area, they may be more easily able to assess product attributes.

The consumer market, however, has different characteristics. Final consumers will have little experience with the relevant goods, and limited incentive to invest heavily in the characteristics of these goods. It will be a challenge for biotechnology firms to market their goods on a widespread basis to consumers with little knowledge about such products. Accentuating this problem is that biotechnology products in general (and pharmaceutical products in particular) are credence goods. The market for such goods is thus vulnerable to deception, as consumers may be unable to distinguish false from true claims. In such a market, consumers may be economically harmed by buying inferior products, that is, products of lower quality than the consumer expected, and producers of superior products are often harmed because they cannot find a credible way to signal to consumers that their products are indeed superior. A good example of firms producing credence goods is automobile body repair shops. Many or most consumers are unable to evaluate the quality of the products produced by these firms, and high quality firms in the market have difficulty signalling their quality to consumers.

Thus, marketing directly to the consumer is a task that will be largely new for this sector. Further, as discussed above, these consumers, as well as the media, are likely to view the newly produced goods with skepticism. Given these problems, advertising may well be very important to firms in the emerging biotechnology sector.

THEORIES OF ADVERTISING

This section will outline the various theories of advertising and suggest ways to explore how those theories apply in the biotechnology marketplace.

Direct Information

Perhaps the simplest theory of advertising is that it directly conveys product information to the consumer, as Nelson (1974) suggests. Ippolito and Mathios' work (1990) on health claims for food products presents one example of how information on a product attribute is conveyed to the marketplace, and how the marketplace reacts to that information. Ippolito and Mathios studied the effect of information on consumers and producers during a time period when cereal producers were allowed to advertise the health benefits of their products. They found that during this period consumers moved towards buying healthier products, while producers sought to innovate by providing healthier products to the market. Thus, one important issue will be to explore how pharmaceutical firms directly convey information about the attributes of their products (especially their complicated products) to consumers, and how consumers react to that information.

The direct conveyance of information, however, may have its limits. The attributes of a product may be too complicated to convey through direct advertising. This is especially true in broadcast advertisements, where complicated messages do not fit well into the medium.

Signals of Product Quality

Perhaps even more important, however, is the fact that advertising may serve as a performance bond for a high quality product. Klein and Leffler (1981) suggest that advertising may serve to signal to consumers that products are of high quality. If firms renege on their implicit promise of quality, consumers punish them by refusing to buy their products again. As the vice-president of a major pharmaceutical firm has written, "[i]f a product does not do what it says it will, there is no chance for a repeat sale" (Ingram, 1991, p. 70). Such a bond may be of special importance when consumers cannot determine the attributes of a product upon purchase.

A necessary condition of signaling would be that either product attributes are too complicated to convey through advertising, or attributes cannot be determined upon purchase. This condition appears to apply quite well to the pharmaceutical market in particular (and to credence goods in general).

Does the Signal of Product Quality Depreciate?

The relevant theoretical literature seems to imply that once put in place, performance bonds do not depreciate. Yet the limited relevant research on this issue (Thomas, 1989) implies that advertising has a relatively high rate of depreciation. It is therefore important to observe how quickly the signal quality of advertising depreciates. Put another way, it would be informative to determine how product sales respond when a manufacturer ends the advertising that acts as a signal of product quality.

Advertising as a Compliment to Other Marketing Strategies

Advertising is not the only marketing strategy available to producers. Producers can choose the marketing channels through which they sell their products, as well as the particular conditions under which their products can be sold. Kleit (1993) suggests that advertising is an important complimentary tool to other marketing strategies. For example, a firm could invest in point-of-sale services in order to explain to the customer the attributes of their products. Such investment, however, might be of little use without advertising to encourage consumers to shop for such products. This paper will therefore investigate the ways in which advertising and other marketing strategies complement each other in the pharmaceutical market.

Advertising as a Barrier to Entry

Kaserman and Mayo (1995, p. 110) define a barrier to entry as "something that keeps firms from entering a market that is generating economic profits." As they note, however, the definition of barrier to entry varies among economists. An important strand of the economic literature (for example, Schmalensee, 1983) implies that advertising serves as a barrier to entry for protecting a dominant incumbent firm from other competitors. This may be especially important in the new area of biotechnology, where one firm innovates, enters the market, and then may have to defend itself against entry by other firms.

This hypothesis may prove difficult to discern against other hypotheses. For example, consider an incumbent firm that is able to credibly convey a commitment of product quality through advertising. Rivals may be reluctant to enter the marketplace against such a firm. However, what makes that firm so difficult to compete against is not that it has created a barrier to entry, but that it has done a good job of serving the needs of consumers.

A corollary to this theory is that advertising serves to deceive consumers into buying products that are not useful for them. As we will see, this claim has arisen in the debate over pharmaceutical advertising.

THE LEGAL REGIME FOR ADVERTISING IN THE UNITED STATES

Historically, pharmaceutical advertising was done largely through "detailing" – promotion directly from the manufacturer to the physician, either through visits by representatives or through advertisements in professional journals. Since the mid-1980s, however, drug companies in the U.S. have turned increasingly to direct to the consumer advertising, in particular after the end of a three year (1982–1985) FDA "voluntary" moratorium on direct-to-consumer pharmaceutical advertising. This advertising largely takes place through advertisements on television and in newspapers. This change in advertising approach has its share of both critics and advocates, and has placed new stress on the regulatory system.

The Legal Regime for Pharmaceutical Advertising

Pharmaceutical advertising in the U.S. is highly regulated, in fact if not in law. The Food and Drug Administration (FDA) "asks" that firms "voluntarily" submit their advertisements for prior approval from the FDA. It appears that all major pharmaceutical firms have indeed submitted their advertisements for approval, and all have followed any conditions the FDA placed on presenting any advertisements.

The reasons the pharmaceutical firms have followed the FDA lead is clear. The FDA has complete authority to approve or deny the ability to sell new drug products. This approval is very valuable to pharmaceutical firms. It therefore appears that no firm is willing to risk its goodwill with the FDA by not complying with the FDA's "voluntary" advertising screening process. As Kessler and Pines (1990, p. 2410) put it, "[c]ompanies interested in maintaining positive relationships with the FDA usually agree to the FDA's remedy" (see also Rubin, 1991, p. 21). Given this peculiar bureaucratic arrangement, where the FDA regulates both drug approvals and drug advertising, the FDA appears to have almost unlimited power to determine the regulatory regime for pharmaceutical advertising, regardless of the merits of its position.

The FDA has not issued a formal list of rules for advertisements. In addition, because no pharmaceutical firm has challenged the FDA's authority, there are no court decisions on the FDA's rules. One clear FDA requirement for advertisements, however, was that if a firm wished to have an advertisement state both the drug and the relevant illness, it must also issue a description of the appropriate usage of the drug. Such a listing can be done in the context of a newspaper advertisement, where space is available for such information. No such space, is available, however, in broadcast advertisements.

The net result of this restriction was that television and radio advertisements could name a product, or name a disease, but could not name both. Instead, most advertisements named one or the other, and suggested that a potential patient "ask their doctor" for more details. Despite this handicap, there was a good amount of broadcast advertising for pharmaceuticals under these rules. In August 1997 the FDA relaxed its regulation, allowing pharmaceutical firms to state in television advertising both the names of their products and the diseases that those products treat.

Regulations for Other Industries

In contrast to pharmaceuticals, the legal responsibility for acting against deception in most industries in the U.S. rests with the Federal Trade Commission (FTC). Unlike the FDA, the FTC does not require prior approval of any advertisements. Rather, it seeks injunctions and penalties against those firms that it believes have acted to deceive consumers. These are civil legal actions, with decisions based on "the preponderance of the evidence." This approach is in contrast to the criminal law standard used in the corresponding Canadian law enforcement program. While the criminal approach metes out stronger penalties for those found in violation of a law, it has several practical difficulties. Specifically, the criminal approach requires a stronger standard of proof ("beyond reasonable doubt") and can be inflexible with respect to changing circumstance (see Addy, 1991, pp. 410–412, for a discussion of these issues).

The FTC actively applies an "advertising substantiation program," in which firms, if they are being investigated for deception, are required to present documentation of the basis for their claims. The documentation, however, must have been generated prior to the advertising campaign at issue. In this way, the FTC seeks to make sure that firms have the basis for their advertising claims before they make those claims, while avoiding the "chilling effect" and inherent arbitrariness of prior approval. Using such approaches, the FTC has shown itself to be able to actively enforce the laws against deception in court (Beales, 1994, p. 1391).

Government agencies are not the only enforcers of the law in the U.S. against deceptive advertising. Under the Lanham Act, competitors of a firm which is engaged in advertising can sue in Federal Court. If the suit is successful, the plaintiffs gain an injunction against further such advertising, and defendants are required to engage in corrective advertising. Court cases (which have dealt with products ranging from dog food to prescription drugs) have focused on whether the claims made in the advertisements were true, based on the available evidence (Walsh & Klein, 1992, p. 415). Beales (1994, p. 1370) asserts that Lanham Act decisions have established a clear and effective legal framework for the litigation of deception cases.

It is important to note that economic incentives imply that private firms will have strong incentives to resort to Lanham Act procedures if their rivals engage in deception. If a firm increases its market share through deception, both its customers and its rivals will be harmed. The rivals, who likely have the financial ability to reach the court system, will also have important incentives to do so. Thus, private action under the Lanham Act appears to complement public enforcement of deception laws. It may be especially important if bureaucratic processes and financial limitations preclude government agencies from acting vigorously against deception. Unfortunately, it is possible that the Lanham Act could also have anticompetitive effects. A successful, non-deceptive advertisement by a firm would also harm its rival and give that rival incentives to sue under the Lanham Act. If such action deterred truthful advertising, it would harm, not aid, consumers.

THE DEBATE IN THE UNITED STATES

Many of the theories discussed above have been brought out in the debate over pharmaceutical advertising in the U.S. This section will review that debate and attempt to place it in the context of the possibilities for using advertising to market products in the biotechnology area.

Advocates of Advertising

The beginnings of the current policy debate may date from Masson and Rubin (1985). In a brief communication in a major medical journal, they make several points about the merits of consumer advertising for pharmaceuticals. First, advertising may cause consumers to realize that they do indeed have a disease. Advertisements by Pfizer, for example, apparently caused many people to realize that the thirst they were suffering from was a symptom of diabetes. As the leading producer of diabetes-treating drugs, Pfizer had important incentives to run this advertising, even though FDA rules prevented it from identifying its product by name. Similarly, advertising may reveal to consumers that a new treatment may exist for a condition currently being treated by another drug. The consumer may then respond by asking their physician for a prescription for the new drug. Masson and Rubin (1985) also point out that if manufacturers were able to discuss drug side effects in the advertising, this could lead to a reduction in such effects. The FDA, however, does not allow such advertising.

Two lessons for biotechnology appear out of the Masson and Rubin piece. First, advertising may be an effective means of informing consumers about the existence of new products. However, it may be necessary to advertise in conjunction with point-of-sale expertise (like a physician). Biotechnology products may well be complicated to use, or it may be difficult to determine when particular usage is optimal. Thus, biotechnology products may be best marketed with a combination of advertising and retail expertise made available to those consumers who are attracted by the advertising.

Opponents of Advertising

The basis of the opposition to pharmaceutical advertising is that it enhances the possibility that consumers may be victims of deception and fraud when buying pharmaceuticals. As discussed above, pharmaceuticals are credence goods whose attributes often cannot be determined even after they are taken by the consumer.

Kessler and Pines (1990, p. 2414) suggest that the time that consumers are most susceptible to harm is when a drug has first been introduced. At that point in time some consumers may be eager for treatment for a previously untreatable condition. Further, the adverse effects of the relevant drugs may not be well known. While all drugs are subjected to extensive clinical trials, no finite set of trials can test for every possible effect of a new treatment. Given this, Kessler and Pines believe that it is important in the early stages of the marketing of a drug for that drug's effects not to be overstated.

Cohen (1988) suggests that advertisements may interfere in the doctor-patient relationship. Under this theory, patients eager for advertised remedies may pressure physicians into prescribing inappropriate drugs, reducing health care benefits and increasing health care costs. Note that this is the opposite of the Masson-Rubin theory, which implies that advertising will help physicians make more informative choices.

Cohen also points out that advertisements, by their very nature, are going to stress the positive benefits of a drug and de-emphasize the negative. It is possible, as the FDA attempts to do, to require a more "fair" presentation of the facts. Using cigarette smoking as an example, however, Cohen points out that clever advertisers are often able to minimize the effect any "fair content" rule has on advertisements.

Along these lines, Pearson (1995) asserts that pharmaceutical advertisements, like all other advertisements, are one-sided by nature. That is, a company will accentuate the positive in its advertisements and play down any negative aspects. This may be especially harmful in the case of pharmaceuticals, which may have dangerous side effects.

EVIDENCE FROM THE PHARMACEUTICAL MARKETPLACE

As the section above implies, there has been a great deal of controversy over the economics of pharmaceutical advertising. The purpose of this section is to review the available evidence on the issue.

In an early study, Telser, Best, Egan, and Higinbotham (1975) analyze entry decisions in the pharmaceutical industry as a function of a variety of factors, including advertising and promotion. By "promotion," the authors refer to both promotion to physician and promotion direct to consumers. Their database did not differentiate between the two. Telser et al. find that promotion intensity is at its highest for firms just entering a particular product line. It is at its lowest in concentrated industries with little or no entry. This evidence appears to support the hypothesis that promotion and advertising are important tools for selling a new product to consumers. The lack of promotion intensity in concentrated industries indicates that those industries are not using advertising to create barriers to entry. Both results are consistent with the pro-competitive theories of pharmaceutical advertising.

Leffler (1981) conducts a study of advertising and promotion in the pharmaceutical industry similar to that of Telser et al. He finds that promotion takes two major forms. First, promotion is used heavily by both entrant firms and the firms that those firms are entering against. This is consistent with advertising being important to both facilitating entry and competing against new entrants. Second, advertising often takes the form of "reminders" to consumers and physicians to use established drugs. This result is consistent with both the idea that advertising depreciates, and that advertising is used to generate a barrier to entry. (Leffler did not attempt to differentiate between these two hypotheses.) Reminder ads are also consistent with a theory of firms placing new performance bonds on the sale of their goods as the old ones depreciate.

Hurwicz and Caves (1988) conduct a similar study to that of Leffler. Once again, they find that advertising is an effective means for new products to enter a market. They also find, however, that advertising is a useful method for incumbent firms to defend their market shares. In addition, Hurwicz and Caves find that promotion is an effective substitute for the "goodwill" built up by a monopoly producer during the period of that producer's patent monopoly on the relevant drug. While the patent for all drugs is 14 years, some drugs gain regulatory approval earlier in their patent life then other drugs. This allows these drugs to be sold longer as monopolies. The longer a manufacturer is able to sell a drug under patent, the stronger its goodwill appeal once that patent expires. Firms are able to replace that goodwill, however, through effective advertising. In effect, this goodwill serves as a performance bond in the context of Klein and Leffler (1981).

Keith (1995) explores the physician-patient interaction in the consumption of aspirin. Aspirin is useful in deterring a variety of heart-related diseases, and various advertisements have pointed this out. Keith found that physicians do not generally suggest that patients with a high risk of heart disease use aspirin unless those patients suggest it, and that pharmaceutical advertisements may encourage such suggestions. She also found that there was widespread under-usage of aspirin for high-risk patients. Given the FDA's discouragement of advertising, Keith suggests that a liberal regulatory regime would lead to greater advertising of aspirin, and a greater and more appropriate usage of aspirin by high-risk patients.

Berndt, Bui, Lucking-Reiley, and Urban (1997) conduct a study of the impact of marketing in the U.S. anti-ulcer drug industry during the 1970s and 1980s. They find that initial drugs in a particular area have a "pioneering advantage" due to the fact they are the first into the market. They also find that marketing in the form of journal advertising and direct contact with physicians, is crucial for a new firm in this market. Thus, prohibiting advertising would serve to create a barrier to entry in this market, rather than the reverse. Berndt et al. also find that the depreciation of the effect of marketing effort is statistically zero. This effect may be a function of the fact that antiulcer manufacturers target a highly sophisticated group, physicians, rather than consumers.

Roth (1996) conducts a marketing study of the content of pharmaceutical advertising in general interest newspapers and magazines. He makes several findings relevant to the debate over drug advertising.

As in the articles cited above, Roth finds that new drug treatments are often advertised, implying that advertising is an important mechanism for teaching consumers about the existence of new drugs. Roth also finds that market leaders are more likely to use advertisements. While theories of advertising predict this will occur, it is unclear how to interpret this finding. This result could be generated by firms creating entry barriers. Or, this could be the result of successful brands renewing the already depreciated advertising, and increasing the performance bond attached to their products.

Roth finds that pharmaceutical advertising is employed for drugs with relatively few side effects and relatively well known conditions. As Roth points out, this refutes the idea that pharmaceutical advertising is focused on deceiving consumers into using drugs with dangerous effects. Rather, drug advertising appears to be concentrated in the areas where it can do the least consumer injury.

Roth notes that many of the advertisements imply an increased role for both physicians and pharmacists. Thus, advertising in pharmaceuticals does appear to be a complement to receiving information from other sources.

The most commonly cited example of danger from pharmaceutical advertising directed at consumers is the experience with the drug Oraflex. In 1982 Eli Lily launched a major publicity campaign for its then-new anti-arthritis drug Oraflex. Lily, however, was forced to pull Oraflex off the market after only a few months because of adverse publicity regarding the drug's side effects. Wolfe (1995) charges that the problems associated with Oraflex were exacerbated by the massive media campaign conducted by Lily. Wolfe asserts that this campaign caused a great deal of damage as it induced many patients to use a drug that was not helpful for them.

Sheffet and Kopp (1990, p. 43) point out, however, that Oraflex may not be a good example of the dangers of direct-to-consumer advertising. While Lily did mount a significant publicity effort for Oraflex, it in fact used no direct-to-consumer advertising. Rather, it relied on the more traditional methods of appealing for "free media." That is, Lily made a good deal of effort to induce media outlets to write and broadcast stories about Oraflex, generating positive publicity for itself. This appears to be an example of how the media overemphasize the dramatic, as Strauss (1990) suggests. In any event, Oraflex does not appear to be a good example of the dangers of pharmaceutical advertising. Further, when the bad news about Oraflex became known, it was prominently presented in the media, helping to generate the recall of the drug (and likely acting to punish Lily for its poor performance).

A recent study by Labor, Schommer, and Pathak (1995) reviews the ability of consumers to process complex and complicated information. This study reviews the response of human subjects to information presented to them in the laboratory. Labor et al. find that people who receive complex information about drugs (like the materials the FDA requires to be presented in most newspaper advertisements) often become quite confused about what they are reading. In effect, they become "overloaded" with information. Such individuals were about as likely to become confused as those individuals only given a small amount of information. In contrast, people who receive intermediate-level information on pharmaceuticals often have a much clearer picture of the issues at hand.

Williams and Hensel (1995) surveyed residents at several retirement communities to determine their responses to direct-to-consumer pharmaceutical information. They found that such advertising often induced consumers to seek out more information about the relevant drugs from physicians, pharmacists, or friends. This is consistent with the idea that advertising is a complement to other ways of capturing information, as indicated by Kleit (1993). This study also found that consumers with lower levels of education were more likely to be responsive to pharmaceutical advertising. This result is consistent with two contrary hypotheses about advertising. First, it is consistent with the idea that advertising is useful as a supplier of information to consumers who do not have easy access to other sources. On the other hand, it is also consistent with the theory that advertising is used to generate deception among less educated, and therefore, more vulnerable, members of society.

There is at present only a limited amount of information available about how skeptically consumers view drug advertisements. A privately-conducted survey, reported by Brandstatter (1995), indicates that only about 25 percent of consumers consider advertising by pharmaceutical firms to be objective, and less than half believe it to be reliable. Despite this, more than 70 percent of consumers surveyed found the advertisements to be valuable and worthwhile. Of course, since this was a privately conducted study whose methodology and data are not publicly available, its results may not be as credible as an academic study.

Economic theories of advertising as performance bonds indicate that firms will be punished when they break their promises. There is no direct examination of this issue in the pharmaceutical industry. In a more general study, Mathios and Plummer (1989) find that firms suffer severe reductions in their stock prices when they are cited for deception by the FTC. These losses are far in excess of any fines levied by the FTC. This result is consistent with the stock market reacting adversely to information that these firms have sacrificed their performance bonds with consumers by not selling the quality of products they had promised.

ANALYSIS OF THE DEBATE AND THE ECONOMIC EVIDENCE

The primary lesson that comes out of the experience with pharmaceutical advertisements in the U.S. is that advertising is a crucial component of the competitive strategy of a new entrant into the market. Consumers need to know what new products are available, and to be credibly convinced of the quality of these products. Pharmaceutical firms have responded by using advertising intensively when they introduce a new product. These advertisements appear to act as complements to other sources of information (here physicians and pharmacists), along the lines discussed by Kleit (1993). Biotechnology products may well require similar advertising campaigns when they enter the market.

There is less conclusive evidence dealing with advertising as a barrier to entry. The work by Telser et al. (1975) suggests that it is not a barrier, though Hurwicz and Caves (1988) imply the opposite. Berndt et al. (1997) find that advertising is conducive to entry, at least in the anti-ulcer industry. There is also little evidence concerning the depreciation rate of advertising. Leffler (1981) implies that such depreciation is significant, while Berndt et al. do not find it to be so.

There is, however, very little reason at this point to suggest that pharmaceutical advertising has led to increases in the level of deception in that market. Perhaps this is because consumers do understand that advertising in general only presents one side of an issue, and that therefore advertising should be viewed skeptically. Indeed, some evidence indicates that if manufacturers are required to generate more information for consumers, consumers may not be able to consume it. In addition, it is reasonable to believe that consumers of biotechnology products in Canada will regard such advertising with even more skepticism. As the pieces by Einsiedel (1998) and Strauss (1998) point out, Canadian consumers are already reluctant to consume biotechnology products, and the Canadian media is prone to highlight the dangers in these products.

This context may generate large costs for firms who heavily advertise with poor products. As the saying goes, "nothing kills a bad product faster than good advertising." If adverse news comes out about a product that has been heavily advertised, such news will naturally be more "newsworthy" than similar news about a more obscure product. Such news will therefore gain larger attention in the media, and cause greater harm to the firm with the poor product.

CONCLUSION – IMPLICATIONS FOR BIOTECHNOLOGY FIRMS AND FOR REGULATORS

Firms attempting to sell new biotechnological products to consumers face significant challenges. These firms have little experience of marketing directly to consumers, and consumers may be reluctant to buy such products. Advertising may well represent the method by which biotechnology firms mass market their products to consumers. In particular, biotechnology firms may use advertising as a complement to a network of knowledgeable retailers who can instruct consumers on when it is appropriate to use particular products.

For such advertising to serve consumers, there need to be appropriate laws and government regulation in place. This legal framework should seek both to discourage deception and to foster advertisements that convey valuable information. The U.S. experience suggests that the government agency regulating advertisements for biotechnological firms should not be the same agency that grants regulatory approval to sell such products in the market. Such a legal regime gives firms strong incentives not to fight the agency when that agency has made an improper decision. Rather, such regulation should be carried out by a regulatory agency with authority over advertising in general. Legal cases should be settled based on the civil code, using a legal definition of deception. It also may be appropriate to allow other firms to sue to stop deceptive advertising, if such advertising is harmful to both the economic interests of such firms and the interests of consumers in general.

REFERENCES

Addy, G. N. (1991). Deceptive marketing practices. In: R. S. Khemani & W. T. Stanbury (Eds.), *Canadian competition law and policy at the centenary*, pp. 389–414. Halifax: The Institute for Research on Public Policy.

Beales, H. P. (1994). FDA regulation of pharmaceutical advertising. *Seton Hall Law Review, 24*, 1370–1397.

Berndt, E. R., Bui, L. T., Lucking-Reiley, D. H., & Urban, G. L. (1997). The roles of marketing, product quality, and price competition in the growth and composition of the U.S. antiulcer drug industry. In: T. F. Bresnahan & R. S. Gordon (Eds.), *The economics of new goods*, pp. 277–328. Chicago: University of Chicago.

Brandstatter, L. (1995) representing Ehrenthal and Associates, *Direct-to-consumer promotion, Food and Drug Administration Public Hearing* (October 18–19, 1995). http://www2evolvingtech.com/etc/industry/hearings/95N227contents.html

Cohen, E. P. (1988). Direct-to-the-public advertising of prescription drugs. *The New England Journal of Medicine, 316*, 373–375.

Comanor, W., & Wilson, T. (1979). The effect of advertising on competition: A survey. *Journal of Economic Literature, 17*, 453–476.

Einsiedel, E. F. (1998). The market for credible information in biotechnology. *This volume.* Also in *Journal of Consumer Policy, 21.*

Hurwicz, M. A., & Caves, R. E. (1988). Persuasion or information: Promotion and the shares of brand name and generic pharmaceuticals. *Journal of Law and Economics, 31*, 299–320.

Ingram, R. A. (1991). Some comments on direct-to-consumer advertising. In: D. S. Pathak, A. Escovitz, & S. Kucukarslan (Eds.), *Promotion of pharmaceuticals: Issues, trends, options*, pp. 67–74. New York: Pharmaceutical Products Press.

Ippolito, P. M., & Mathios, A. D. (1990). Information, advertising, and health choices: A study of the cereal market. *Rand Journal of Economics, 21*, 459–480.

Kaserman, D. L., &. Mayo, J. W. (1995). *Government and business: The economics of antitrust and regulation*. New York: Dryden.

Keith, A. (1995). Regulating information about aspirin and the prevention of heart attack. *American Economic Review, 85*(2), 96–99.

Kessler, D. A., & Pines, W. L. (1990). The federal regulation of prescription drug advertising and promotion. *Journal of the American Medical Association, 264*, 2409–2415.

Klein, B., & Leffler, K. (1981). The role of market forces in assuring contractual performance. *Journal of Political Economy, 89*, 615–642.

Kleit, A. N. (1993). Efficiencies without economists: The early years of resale price maintenance. *Southern Journal of Economics, 59*, 597–619.

Leffler, K. B. (1981). Persuasion or information? The economics of prescription drug advertising. *Journal of Law and Economics, 24*, 45–74.

Labor, S. L., Schommer, J. C., & Pathak, D. S. (1995). Information overload with written prescription drug information. *Drug Information Journal, 29*, 1317–1328.

Masson, A., & Rubin, P. H. (1985). Matching prescription drugs and consumers: The benefits of direct advertising. *New England Journal of Medicine, 313*, 513–515.

Mathios, A. D., & Plummer, M. (1989). The regulation of advertising by the Federal Trade Commission: Capital market effects. *Research in Law and Economics, 12*, 77–93.

Nelson, P. (1974). Advertising as information. *Journal of Political Economy, 82*, 729–754.

Pearson, C. (1995) representing the National Women's Health Network, *Direct-to-consumer promotion, Food and Drug Administration Public Hearing* (October 18–19, 1995). http://www2evolvingtech.com/etc/industry/hearings/95N227contents.html.

Roth, M. S. (1996). Patterns in direct-to-consumer prescription drug print advertising and their public policy implications. *Journal of Public Policy and Marketing, 15*, 63–75.

Rubin, P. H. (1991). The FDA's prescription for consumer ignorance. *Journal of Regulation and Social Costs, 1*(3), 6–24.

Schmalensee, R. (1983). Advertising and entry deterrence: An exploratory model. *Journal of Political Economy, 91*, 636–653.

Sheffet, M. J., & Kopp, S. W. (1990). Advertising prescription drugs to the public: Headache or relief? *Journal of Public Policy and Marketing, 9*, 42–61.

Shughart, W. E. II (1990). *The organization of industry*. Homewood, IL: Irwin.

Strauss, S. (1998). Biotechnology and the media. *This volume*.

Telser, L. G., Best, W., Egan, J. W., & Higinbotham, H. W. (1975). The theory of supply with applications to the ethical pharmaceutical industry. *Journal of Law and Economics, 18*, 449–478.

Thomas, L. G. (1989). Advertising in consumer goods industries: Durability, economies of scale, and heterogeneity. *Journal of Law and Economics, 32*, 163–193.

Walsh, C. J., & Klein, M. S. (1992). From dog food to prescription drug advertising: Litigating false scientific establishment claims under the Lanham Act. *Seton Hall Law Review, 22*, 389–465.

Williams, J. R., & Hensel, P. J. (1995). Direct-to-consumer advertising of prescription drugs. *Journal of Health Care Marketing, 15*, 35–41.

Wolfe, S. (1995) representing Public Citizen's Health Research Group, *Direct-to-consumer promotion, Food and Drug Administration Public Hearing* (October 18–19, 1995). http://www2evolvingtech.com/etc/industry/hearings/95N227contents.html.

THE AUTHOR

Andrew N. Kleit is Associate Professor in the Department of Energy, Environmental, and Mineral Economics, The Pennsylvania State University, State College, PA 16802-5030, USA. E-mail: ank1@psu.edu. The views expressed herein are solely those of the author and not necessarily those of Industry Canada.

Stephen Strauss
Biotechnology and the Media

ABSTRACT. This contribution begins by describing the semi-chaos inside of daily or weekly journalism into which reporting on biotechnology must fit. Within this framework it is not intrinsically obvious that even the most profound technological change easily lends itself to anything like systematic news coverage. A model to keep in mind is imagining how nineteenth century journalists would have floundered trying to report on the emergence of electricity. A citation analysis indicates that despite its significant social implications newspapers and magazines generally view biotechnology as a business story. A way around this economic focusing may be discussion groups and web sites on the Internet. Finally, it is not clear how influential the news media is in forming people's opinions about biotechnology. It is suggested that fiction, and in particular movies and television, may have a greater impact on public consciousness than news organizations.

PART I: ON JOURNALISM IN GENERAL

There are two truths in this world: one of the laboratory and the other of the media. What people perceive of as the truth is truer in a democracy than the same little grubby experience in a laboratory notebook (Daniel Koshland Jr., former editor-in-chief of Science magazine, 1990).

It is impossible to argue about what the media's role is, or should be, in the portrayal of biotechnology to the general public without taking a giant step backward. Everything basic about journalism – sources, biases, economic self-interest, its decisions as to what constitutes news, and the power of those decisions to change people's minds – has been subject to immense and generally inconclusive debate. In a general sense, suspicious outsiders sometimes see the media, and particularly daily journalism, as a sort of never-ending plot on behalf of special interest groups against the common weal (Gitlin, 1980). On the other hand journalism's own practitioners quite often see it as a kind of daily struggle to make tentative sense out of the chaos of experience. This view is often summed up in the old definition of journalism as the "first rough draft of history."

Therefore, it does not seem possible to sensibly ask how the media function in relationship to biotechnology without first understanding how the media operates in general. With that in hand I will then analyze what, if anything, is unique about journalism's relationship to the warts, gnarls, and twists of modern biotechnology. This two-headed approach is taken in the hopes of having conference attendees appreciate both just how much biotechnology is exactly like every other area which its advocates wish to see reports on, and how, in many ways, it is also quite unique. Finally, the paper will make some suggestions about what the special case of biotechnology will mean in terms of future reportage.

Before going on, a limitation must be noted in what I will present to you. The basic source for my discussion will be daily newspapers, even though studies of how Canadians get their information reveal a tremendous reliance on television. A 1995 survey of Canadians (Conquest Research, 1995, pp. 4–5) said that 81 per cent of them got most of their information about science and technology from television, 55 per cent from newspapers, 46 per cent from magazines, 33 per cent from radio. However, television coverage is extremely difficult to retrospectively get a handle on. Television is both feature documentaries on specialty channels, 90 second news clips on network news programs, and round table discussions on cable channels. But equally important, television is also situation comedies, movies, soap operas, and dramatically presented advertisements, material which can have a host of contrary messages and information. These can come at a viewer with the swiftness of a channel clicker (Dubois, 1996). How consumers would ever influence the latter in a democracy is not something apparent to the author of this paper. But more importantly, methodologically the problem is that once the television program has been broadcast it is extremely difficult to track down what and how things have been portrayed. A recent health message study of a bare three weeks of television programming on only two channels in Quebec required looking at 756 hours of television (Dubois, 1996). All one can say to anyone who would undertake studying all that television presents to the consumer which might relate to their understanding of biotechnology is: Wow.

With this in mind, and considering that an overwhelming amount of other media's news stories will be mirrored in the newspapers, I have concentrated on newspapers with an occasional television and magazine aside. However, for those wanting more about television, and more Canadian information in general, help is on the horizon. A joint University of Calgary and University of Quebec at Montreal study to be released next year is systematically surveying, in both a quantitative and a qualitative way, how the print and electronic media is covering biotechnology (Einsiedel & Schiele, forthcoming).

The Inherent Chaos of the Exercise

The power to determine each day what shall seem important and what shall be neglected is a power unlike any that has been exercised since the Pope lost his hold on the secular mind (Walter Lippman, quoted in Nelkin, 1995).

One of the great illusions of journalism to those outside it, is the illusion of planning. Since the paper appears every day, looking more or less the same, the public's feeling is that there must be an overriding mind which decides what goes in, and what does not. Whereas to those inside journalism the operative word is controlled chaos, or as DeFleur and Dennis (1996, p. 446) have put it: "News is the imperfect result of hurried decisions made under pressure."

The chaotic elements come from almost everywhere imaginable. To begin with there is the need to fill up the paper with something. This seems self-evident, but can actually be inherently contradictory. How much space there is to fill up is a function of what day or time of year it is. "On Saturday, newspapers must have enough information on hand to fill the news columns of ad-crammed Sunday editions. On days when there are fewer ads (especially weekdays between Christmas and February 1) editors may have to discard information that might have been printed on a Sunday," says Gaye Tuchman (1978), who has studied the newsmaking process extensively.

The size of the "news hole" – that is the amount of space devoted to news coverage – is directly influenced by advertising but not in a way many people think. In a large city newspaper it is unlikely that a given advertiser can wield much authority over what gets into the paper.

However, the advertising collective certainly can. Publishers generally allot a fixed ratio of the total volume of their paper to the news hole. This generally ranges from 30 to 40 per cent, depending on the paper. Thus, if advertising overall is down, people will get less news, or news in a more abbreviated format. And conversely, during times of economic growth there is also a growth in the amount of information people are given.

Deciding how the news hole is to be allotted is an even more arcane task. Territoriality is everywhere. Space in the general news sections can be reserved for geographical areas – Ottawa-Parliament, Europe

– or areas of interest – sports, entertainment, business, science. While space in generalized areas may be allotted in varying ways, wholesale cutting and chopping is not done. That is, there is never a day in which the sports pages are entirely eliminated, because there was too much news elsewhere.

Nonetheless, there is still intense internal competition for space, particularly prime space. Another of the ironies of journalism is that, in the popular mind, journalists are tremendously scoop driven in order to "beat" their external competition. The view from inside is that your true competition is often your fellow journalists and section editors who, in their zeal, may simply fill up so much space that there is nothing left for you. As Thompson (1992, p. 117) comments, "A newsroom is characterized by competition. It is a place filled with bright, ambitious people. They all want to have the best and sexiest stories to dazzle their editor. There is a competition for the front page. It's the only place to be for a journalist. The best journalists have an instinct for what makes a page-one story."

This competition means that a person who knows more about a subject area – say a science writer in the case of biotechnology – will not necessarily be consulted when a reporter in another area is writing a piece pertaining to that subject. The Globe and Mail's initial coverage of the Biochem AIDS announcement, which will be discussed subsequently, is a good example of this.

Over the course of the day the nature of this competition keeps changing. Possible stories are discussed at various meetings among editors, each pushing hard for the works of his or her reporters. The reporters are also pushing, coming up to editors with various degrees of excitement, trying to describe the news-worthiness of what they are trying to do, as well as its timeliness. In this mix it should be pointed out that some stories – science related stories in particular – often do not fare that well. As Perlman (in Friedman, 1986, p. 21)) points out, when editors have to balance "the latest news on the plasmid front" against the latest news about property tax cuts, tax cuts are going to win every time.

There is often a different kind of spatial pressure. This might be considered to be time as a space. Things often have to get into the newspaper the next day and, as such, have to be written by a certain time in the day. A science reporter can learn about a complex new development in biotechnology at 4:30 and be expected to have something written by 6:30. A survey of 92 Canadian science writers found that 73 per cent said time pressure was a problem and 67 per cent considered it a serious problem (Dubas & Martel, 1975). Time effects not only how long pieces can be, but how much effort can go into them. Time pressure is also a most likely source of errors in stories. Three quarters of the same group of Canadian science writers said making stories both scientifically accurate and understandable was a significant problem (Dubas & Martel, 1975).

Many of the same issues apply to television news. On television the news hole is the time that the program runs. The length of each item and its relative placement in the program, e.g., the most important first and longest, is the equivalent of what is on the front page and what isn't. However, the great temporal virtue of TV is that it, unlike a newspaper, can simply empty out everything it previously scheduled and bring you live, extended coverage of what it deems are major news events. It doesn't have to "go to press," and this is one reason why it has become such a popular vehicle for breaking events.

What Is News?

When dog bites man, that is not news. When man bites dog, that is news (attributed to John Bogart of the New York Sun, 1880).

All this bargaining and maneuvering presupposes that people who work for newspapers know what the news is. In point of fact, the definitions of news and news-worthiness are subject to constant and often contradictory debate. There are some general features, as defined by a famous remark of U.S. broadcaster, David Brinkley. He stated that "news is the unusual, the unexpected. Placidity is not news. If an airplane departs on time, it isn't news. If it crashes, regrettably, it is" (quoted in Dennis & Merrill, 1996). A more inclusive standard criteria states that news is about conflict, or progress, or disaster, or novelty, or human interest, and that things are made more newsworthy by their timeliness, their proximity to readers, and the prominence of those involved (Dennis & Merrill, 1996).

However, it is also true that who decides what is news can vary from newspaper to newspaper and day to day as different opinions about what is news, and how to cover it, are presented. When asked who decided what should be covered in her paper, one Canadian science writer jadedly replied: "Me; the managing editor; the publisher; the city editor; the assignment editor; the feature editor; the national editor; the political editor; the foreign news editor; three assistant managing editors; a couple of news editors; the editor-in-chief; and the publisher's doctor" (Dubas & Martel, 1975, p. 149). More quantitatively, a survey of 53 Canadian daily science writers indicated that 33 decided themselves what was an important story; 21 mentioned the city editor, and eight the managing editor (Dubas & Martel, 1975).

What they do not decide is what is often the most contentious conveyor of information: Headlines. These are both written by some one else and have their own space shortages. Everyone complains – nearly half of the Canadian daily science writers polled felt that erroneous headlines were a serious problem for them (Dubas & Martel, 1975). However, given the different time frames in which things are done at the paper and the inability to predict exactly what size space a headline must fit into, it is hard to see how this problem could be solved.

The Dramatic Imperative

The press, the media in general, are much more interested in the story, the news, than in the facts (Arnold Relman, former editor of The New England Journal of Medicine, quoted in Dubas & Martel, 1975, p. 146).

Finally, it is true that news is not simply information, but also has a dramatic context. Reporters will describe their most factual matters they report upon as "stories." And given the chance they will often rely on dramatic tools to frame these facts. They do this because they believe narrative is how people actually view the world. The dramatic approach means focusing on individuals, on conflicts, on all kind of dramatic devices to "tell" the story. "We have whole new stories to choose from . . . stories of discovery, of dashed hopes of geniuses and fools. Stories that are never dull, and that never get printed around tire ads," is how Baltimore Sun reporter Jon Franklin counsels science writers to approach their craft (Friedman, 1986, p. 145). In terms of biotechnology this means that information, which various stakeholders think is of paramount importance, may sit around for ages as it awaits a dramatic framework – what is called in the business a "news peg." In practice this translates into the general public only receiving an intense lesson in DNA testing when the

O.J. Simpson trial takes place, and human cloning – with all its ethical, legal, commercial implications – remaining largely in the domain of fictional reporting until the birth of Dolly, the cloned lamb.

The dramatic imperative also means that journalism is much more poetic than either the science, the business, or the public policy it reports on. That is, metaphor is intrinsic to what is being communicated. For example, typically DNA is not a scientific explanation in newspapers, but "the blueprint of life," or the "photocopying machine of life," or, as one poetical reader once suggested to me "the web which spins the spider." There is a real tendency to shy away from telling people what something actually is, and to tell them, instead, what it is like. Or as Nelkin (1995, p. 5) tartly remarks, "imagery often replaces content." In television this imagery is often actual visual images. And this is the reason why Greenpeace, one of the world's most media-smart organizations, had its protesters dress up in rabbit suits when it recently protested against Nestle's use of genetically engineered foodstuffs. No matter what the merits of their case, a bunch of people in bunny suits carrying a banner on which a half-bird, half-octopus is emblazoned is very arresting TV (Greenpeace, 1997).

Less dramatically, it must simply be pointed out that if the information does not fit into any recognizable "story" format it may never make it into the newspaper. Victor Cohn, a former science editor with the Washington Post, once classically defined what the need to fit information into a story structure means for medical journalism. "There are only two kinds of medical reporting: New Hope and No Hope," he said (Nelkin, 1995, p. 119). What this means in a general sense is that the journalist is the captive of events. Processes which have not been encapsulated into a dramatic form, that is have no news pegs on which to hang a story, often simply don't get reported. This is a point which should be underscored over and over. Process, slow change, of the kind which clearly takes place in biotechnology, doesn't get reported until something happens. As a recent book on the media and risk analysis puts it: "Even a scoop, every journalist's dream, only anticipates the occurrence of an event – or more likely other journalists coverage of an event – by a little time" (Singer & Endreny, 1993, p. 21).

Sources

There is a tendency on our part to believe anyone who has a Ph.D. or M.D. after his name. We automatically assume he knows what he's talking about, and we rarely think he's either wrong or self-serving (former Philadelphia Inquirer science reporter Joel Shurkin, quoted in Friedman, 1986, p. 28).

A clear issue for people outside the media is who gets access to the media, both in terms of a voice to be listened to and a voice which speaks out in news stories. As a general rule, in Tuchman's words (1978, p. 4), "news is the ally of legitimated institutions. The Secretary of State can float an idea on the news media. The 'average' man or woman does not have the same access to the media." And to say that the general public is not consulted in most stories in which there are technical issues is a breath-taking understatement. The Canadian study of 92 science writers mentioned previously demonstrated a clear, and not surprising, predilection to consult experts. At the top of the list were university scientists and engineers who were deemed essential sources of information by 53 per cent of those surveyed. This was followed by attendance at seminars and conventions (46 per cent), and doctors, government scientists, and medical personnel (37 per cent). Interestingly, industry spokesmen were in sixth place at 30 per cent, and industry scientists were only deemed essential sources by 18 per cent of science writers (Dubas & Martel, 1975). The list reflects a general level of trust. Another survey of the science writers indicated that 76 respondents found university scientists and engineers and academic conferences generally reliable, 73 approved of university reports, 52 of industry research and development scientists. And while 58 found industry spokesmen or PR people reliable, 25 found them either unreliable or with varying degrees of reliability. Interestingly there was no indication that non-scientists of any sort were even on the table (Dubas & Martel, 1975). And it should be pointed out that, while comprehensive at the time, the study is more than 20 years old and thus probably reflects an era in which Canadian university researchers had many fewer connections with industry than today. This would particularly be true for industrial biotechnology, an area which would not have existed in any meaningful way when the survey was conducted.

This reliance on "experts" can create conscious or unconscious biases. There is a continual fear among all journalists that they have become defenders and champions of what they legitimately should only report upon. As David Perlman, science writer of the San Francisco Chronicle has put it: "So many scientists are personal friends; so many others have shared their feelings so intimately . . . that one often becomes over-protective – like a police reporter when the cops misbehave" (quoted in Friedman, 1986, p. 28). However, the line-up of officialdom in newspapers makes it seem as if all you had to be is some institutional pooh-bah to get your name and your opinion into the newspaper. Not so. For example, a recent study of a year's worth of food, health, and medical stories in four media outlets between October 1, 1993 and December 31, 1994 quantified where stories come from. Roughly 90 per cent all of the stories were in response to news, that is new scientific publications, court decisions, government announcements, conferences, a disease outbreak, an anniversary, or a book publication. Neither companies nor activists had much success in breaking through this externally set agenda. Of the 4,029 stories looked at, company PR was the source for roughly five percent in the Associated Press, New York Times, and the Globe and Mail. It was 8.4 per cent in the Kitchener-Waterloo Record. In contrast, press releases from activist environmental or consumer groups generated around 3 per cent of the science stories surveyed in AP and the KW Record; this fell to 1.6 per cent of the Globe and only 0.1 per cent of the New York Times (Powell, 1996).

But the larger reality, and one often missed, is that indications are that the majority of what comes in from all sources is ignored. A British medical journalist counted four days of his mail in October of 1995 and was able to quantify the information tide which threatened to drown him. In that time he received three books, a 30 page pamphlet, three articles sent for publication, 30 invitations for press conferences (two of which he attended), 15 newsletters and bulletins of which eight were seen as useful, 47 releases from the computer industry, 56 general science/medical releases, of which 17 were deemed worth reading, and 24 journals and magazines, of which 13 were seen as useless. This accounted for what was estimated to be 500,000 words and completely ignored the phone calls, faxes, e-mail, and wire service stories which are increasingly a part of science reporting (Wilkie, 1996). In this regard, Wallace Immen – a Canadian medical reporter and one of the standard conduits for biotechnology stories – estimated he could easily in one week see 30 faxes and 20 e-mails (personal communication).

Given the immense amount of material which comes their way,

sometimes it seems as if the journalist's most pressing daily task is filling up the wastepaper basket. "[Journalists] can also detect hype, and once they start getting rubbish from a particular source will soon stop opening the envelopes," remarks a British observer of science journalism (Wilkie, 1996, p. 1309). A survey of Canadian science writers indicated that 60 per cent of them found discovering the small amount of news in the large amount of press releases to be a major problem in their work (Dubas & Martel, 1975). The result is a situation in which, said one business biotechnology reporter (Marion Stinson, personal communication), most press releases are born dead. "Companies, for whatever reason, generate huge amounts of press releases, which, as far as I can tell, never are reported anywhere."

All of which should make clear that, in a very real sense, journalists see much of the efforts to communicate to them to be in neither their institutional nor their personal interest. They are suspicious of people sending things, or holding press conferences. Or as Jerry Bishop, Wall Street Journal science and technology writer, has put it: "I have never been at a press conference that was called for the benefit of the press, whether it was called by the White House, Merck, Harvard, UCLA, or the volunteer fire department" (quoted in Burkett, 1986, p. 128).

What Effect Does Journalism Have on People?

Do we know for certain the media are in themselves powerful? Or are they simply secondary factors, triggering mechanisms if you will, which have various impacts on society under certain conditions? (John Merrill, journalism professor at the University of Missouri, quoted in Dennis & Merrill, 1996, p. 85).

If you want to know how much effect the media has on people making decisions, pick a theory. The "bullet theory" suggests there is a direct impact; the "limited effects theory" suggests media is just one of several things which effect motivation; the "cultivation theory" says the more one watches, reads, or listens to a given medium the more importance it has; the "McLuhan theory" says that the nature of different media changes the nature of thought; the "spiral of silence theory" says the ubiquity, cumulative effect, and consonance of the media can effect public opinion; various Marxist theories say the media is a tool of the ruling class (Willis, 1996).

Without subjecting any reader to 75 year's worth of communications theory and research, suffice it to say that there is no general agreement as to exactly how much or indeed how the media affects the way in which people think and act. At best there is a sense that all the media working in concert has the power to focus people's minds on a given issue – "telling people not what to think but what to think about" (Dennis & Merrill, 1996, p. 84). However, it is true too that people also demonstrate the power to ignore what the media is telling them to think about, and to pose very different questions. In essence, the public shows a strong ability to tune media out.

All of which leads to a great contradiction. Generally, reporters do not have a sense of either their audience or the importance of what they do. This is particularly the case because the amount of general reader feed-back to any given piece of reportage which a reporter sees generally hovers around the number zero – except when some very simple error is involved. As an example, in my capacity as science writer at the Globe and Mail over 17 years, the largest reader response I ever received was when 15 people wrote to tell me I had mistaken the number of electrons and protons in an isotope of oxygen. It should also be pointed out this mistake made absolutely no difference to the story.

The Economics of Journalism

We aren't a candy bar and we aren't a bar of soap. But damn, we are a product (Chris Anderson, editor of the Orange County Register, quoted in Underwood, 1993, p. XI).

Newspapers are a business, and a very profitable business at that. In 1989 the top eight newspaper publishers in Forbes' annual report on American industry ranked second only to soft drink makers in long term return on equity (Underwood, 1993). The same is true in Canada. However, the nature of what makes for profitability in papers is often misperceived by those outside. They assume that it is paper sales which generate profit and that therefore papers are financially rewarded for generating scoops that increase newsstand sales. In fact the reverse is true. Generally, the cost of the paper's production is not covered by the price of buying it. That is, the costs of paper, ink, salaries, rental space, distribution, etc. which go into producing an individual newspaper is not recovered by its sale price. This leads to the counter-intuitive situation where a large individual day's sales – a potentially scoop driven event – actually loses a newspaper company money. What doesn't lose a company money is advertising. In North America it typically generates 80 per cent of revenue, and advertising takes up 65 per cent of newspaper space (American Newspaper Publishers Association, 1982). The price to advertisers, in turn, is based on circulation figures. What this means in practice is that it is a paper's ability to sustain circulation, not its ability to generate a single day's interest by a scoop, which matters to it economically.

What the relationship between advertising and circulation also means is that in small towns with relatively few advertisers, and only a few large business which employ people, papers must be very guarded. Negative stories can lose both readers and advertisers, or as one U.S. reporter (Nelkin, 1995, p. 116) described it: "I cannot even call a company, a company, without comment from the company, because we're so afraid of losing ads." However, in larger communities there generally are so many advertisers that no individual company exercises any overwhelming influence on the paper's livelihood. And, if there is only one paper in town, it is difficult for advertisers who want to reach people to go elsewhere.

Bias

We talk of objective reporting. There is no such thing. I have been a reporter for many years, and I can tell you that no reporter worth a hoot ever wrote a purely objective story. You get a point of view in it (H.L. Mencken, quoted in Dennis & Merrill, 1996, p. 108).

Journalists can have all sorts of prejudices – race, religion, sex, economic and social status – but as the section on the chaos in production should make clear, it is hard for any one person's point of view to get into the paper unalloyed. This would apply less to a columnist or editorial writer, but they too would face constrictions in terms of time and space. What is not difficult to get into print are presumptions so wide ranging that they simply are not seen to be there.

In terms of what happens to newspapers in general, and what might influence biotechnology in particular, let me simply mention what some years of experience have taught me is the most important. To do this I will do no more than quote Robert Fulford (1965), who has spent better than 40 years in Canadian journalism: "In a country dominated by its middle class, where all our important institutions reflect a middle class point of view, the newspaperman is now preeminently a middle-class citizen. His attitude is not far from the dead-center suburbanite: an attitude that looks to organizations to sustain it" (p. 153). And although people outside the media may find it difficult to appreciate it, the person the journalists generally write for is some image of themselves, or their friends, or their family. In practice this means that the hardest stories in which to detect your own bias are those featuring people most like you.

As an example, I once had a rather noisy argument in the Globe newsroom with a reporter who had written that the incidence of AIDS among women was rising dramatically. His source was a social worker who counseled such women, and who made the claim in a talk she had given. What my colleague had difficulty appreciating was the self-interest of a professional, middle-class person who he perceived as doing good. That is, even when I battered him with a barrage of statistics affirming that the rate of AIDS among women remained low, he still was unconvinced. What he was all but blind to was that a good woman, and her good organization, had an agenda. He didn't want to confront the sad truth that the worse the problem of AIDS among women was, the more money and the more attention the social worker's group would receive. Because of his unseen prejudices, it was almost impossible for my colleague to understand that he had to hold a good, middle-class person like the social worker to as high a factual standard as he would a company spokesperson, a public policy advocate, or a politician.

Added to all the journalists' biases is the complementary truth that readers of newspapers – the consumers of biotechnology – have some contrary and biased natures of their own. One interesting English study suggests that readers are significantly more likely to believe information about genetic engineering which first appeared in a newspaper, if they don't know that this was its place of origin (Frewer & Sheppard, 1994).

PART II: BIOTECHNOLOGY IN THE EYES OF THE MEDIA

Journalists are the second hand [of the clock] of history. This hand, however, is usually not only of inferior metal to the other two hands, it also seldom works properly (Arthur Schopenhauer, 1970, p. 395)

At the present time there is a great paucity of materials pertaining to how the media – both print and electronic – covers biotechnology in general, and what happens in Canada in particular, although the study referred to in the opening section (Einsiedel & Schiele, forthcoming) should address some of that in the near future. In its absence, much of what follows will take on the character of arguments and anecdotes. What has been added as a piece of original research is a quantitative look at how several different newspapers cover biotechnology.

Biotechnology Is Too New to Be Understood

It is generally accepted that biotechnology is not a "unitary entity" but an "enabling process" (Plein, 1991). It is new medicines, different foods, changed insurance rules, tests for genetic diseases, custom-designed babies, pesticide resistance, animal organ transplants, waste treatments, energy generation. It is new companies not wanting too much government regulation so they can expand rapidly. It is new ethical considerations and new fears. Given this, there is no one way to report on biotechnology, because it is not one thing.

To appreciate the journalists' dilemma imagine that this conference was being called in the 19th century and the new phenomena which government and the public were worried about was electricity. What hope in hell would journalists have in reporting sensibly and consistently about electricity in 1820? Do you report about it as a physical process; do you describe the rise of electrified industrial practices; do you sing the praises of some possible future events in which all homes would be electrified; do you worry about what environmental effects huge hydroelectric dams might cause? Or are you – as it seems to me is the case in biotechnology – buffeted along by many individual things, assuming all the while that later generations will make sense of it all.

And with that as a presupposition, let me give you an example of the real confusion which exists when the press tries to cover a story which stretches over many areas and largely involves process. In November of 1994 the Montreal biotechnology company, Biochem Pharma, reported that scientists at a conference in Scotland had found that in combination with other drugs, its new AIDS drug, 3TC, dramatically cut down the virus. The story was flashed across the front page of the Globe and Mail with the headline: "Aids Drug Shows Dramatic Results, Montreal Firm's Discovery Reduces Virus in Combination Therapy." The story described, in a general sense, that over a one year period the combination drug therapy had both reduced the virus and led to an increase of the body's immune cells which naturally attack the virus. Quoted in the article were Biochem's CEO, their head of research, a spokesman for the British drug company Glaxo, which had the license to develop 3TC, as well as two doctors who work with the drug Glaxo had sponsored. There was a note of caution in everyone's voice. "I cannot guarantee what we will find in the future," said Biochem's CEO.

Some negative side effects of the drug were discussed. However, the article concluded with a description of how the announcement had buoyed the company's stock and a discussion with two stock analysts about their view of what the announcement meant for the company.

The following day a story about the matter appeared on page seven of the front section of the Globe; it was written by the paper's two medical reporters. Its headline was: "Reports on New AIDS Therapy Assailed, Not Enough Data to Say Whether Drug Will Extend Life, Researcher Says." The researcher in question came from the University of Toronto. His skepticism was mirrored in spades by two members of an AIDS support group and a doctor specialized in dealing with the disease. The latter commented caustically: "It is irresponsible for drug companies . . . to promote their drug in this high profile fashion which gives people a hope which doesn't exist in reality. And it is equally irresponsible for the media to co-operate in that type of promotion. . . . It is reckless and it causes a great deal of pain."

No mention of stock prices were made, though undoubtedly this cold water couldn't be seen as a boost for Biochem's promoters. Why such a different approach? In part it was a question of timing. The original story had appeared over the week-end and the business reporter had to rely on wire services and his contacts in the business community to write the story. When he wrote it, he wrote it to appear in the business section, but when editors looked at it they decided it should be catapulted to the front page. Part of the medical reporters' story had to do with differences in journalistic experiences. The medical reporters had experience covering AIDS breakthrough stories that hadn't panned out. They had also written and reported on the agonies and desperation of people with the disease. To them AIDS wasn't a business story, it was a health Greek tragedy. They were also greatly troubled by the release of non-peer reviewed, industry sponsored research at a conference – research which, without much ado, became the stuff of a potential instant newspaper cure.

Who was right? How should the story have been covered? There was much internal debate in the newspaper about that question. At least four other reporters were asked at one time or another to take a look at the data, at how it had been released, at how the findings had been reported, to see if there wasn't a story to be written about what was really going on. None of them – one of whom was the author of this paper – could clarify the situation any more than the two strikingly different articles.

Nearly three years later, it now appears that the initial excited report of multi-drug therapy, plucked out of the business section and thrust onto the front page, heralded a revolution in the treatment of AIDS which made its initial proponent Time Magazine's 1996 Man of the Year. Were the medical reporters reactions wrong? No. Supposed AIDS breakthroughs were a dime a dozen before this; findings released at conferences have not been properly vetted by the scientific community. It just so happens this was an exception which disproves the general rule. What happens in process stories like biotechnology is that journalists have to wait upon time to sort out their preconceptions. It takes time to figure out the meaning of the new.

Lack of Dramatic Context

Dramatically speaking, how does one characterize biotechnology? Is it a risk? Is it a benefit? If both, how much of both? As we have stated previously, at present much of the risk part about biotechnology is theoretical. Altered genes in a Flavr Savr tomato do bring about a more natural ripening, but despite all assurance, who knows, they might ultimately affect the humans who eat them. Only so far they haven't. Genetically engineered crops seem to be better able to resist pesticides and so require farmers to use less chemicals on their field. Who knows, they might pass on their pesticide resistant genes to other plants, but so far they haven't. Transgenic laboratory animals have been created. Who knows, true monsters might be made out of genetic engineering experiments, but so far they haven't. The dramatic problem with newness is: How do you report on dangers whose riskiness cannot be quantified and which indeed have not happened yet?

The answer is that, without a news peg, you don't do it very much. There has been a paucity of defining framing events in the recent past which allow journalists to truly portray biotechnology's hopes and fears. One can compare this to something like Chernobyl and the dangers of atomic power. Without a defining accident, the media is left to report on the process of nuclear safety. And process is, as we have been saying over and over, something the media hardly ever reports on without a triggering event.

What is interesting, from an historical point of view, is that biotechnological dramas have in fact existed and also have been the source of considerable public debate. Rae Goodell, a professor of science writing at the Massachusetts Institute of Technology, has described the rise and fall of the debate over the safety of recombinant technology (Friedman, 1986). Initially, scientists wanted to involve the press in alerting the public to what many viewed as a potential danger in the release of bacteria whose DNA had been manipulated by man. It created a dramatic new framework and ready made headlines. An event – a 1974 press conference by scientists about the potential dangers – was reported in one paper as "Genetic Scientists Seek Ban – World Health Peril Feared." Drama of the highest order. The scientists concerned then tried to prevent reporters from attending a recombinant DNA safety conference in Asilomar, California later that year, but were threatened with a press revolt and clear bad publicity. Drama of another order. They, read the Scientists, don't want Us, read the General Public, to know about this incredible possible danger. Later that year opponents of the way in which the Asilomar scientists defined the problem – "a narrow question of health risk – the scientists' approach to a solution: self-regulation by the researchers" – arose and were duly noted by the press (Friedman, 1986). Again a dramatic framework. Proponents and opponents. For and against. Another event was U.S. Senatorial hearings on the matter in which public input was sought. In 1977 bills were introduced into the U.S. Congress which again served as a focusing event, and again allowed the media to write about the matter. Dramatic situation after dramatic situation wherein journalists could find a news peg and on which they could hang discussions of recombinant DNA, its values and its dangers.

Finally, as the scientific fears of the dangers of biotechnology subsided, a biotechnology industry arose and with it came very different news events – press conferences from new companies, and massively successful stock offerings. Headlines now read: "The Miracle of Spliced Genes," "The Miraculous Prospects of Gene Splicing," or "The Industry of Life" (Friedman, 1986, p. 176). The drama was: How much money will be made; what products will be produced? Goodell (1977) is ultimately critical of the press's failure to have reservations about what she typifies as a public relations campaign by the industry. Others suggest that the newspapers were attune, in ways that they weren't conscious of, with the emerging economic importance of biotechnology (Pfund & Hofstadter, 1991). I would suggest neither critic realizes how much a prisoner of events journalists are. The focus of the coverage changed because, as journalists could best determine, the scientific community believed the dangers it once feared weren't going to happen. And – and this is critical – nothing did happen. There were no recombinant DNA Chernobyls. Fears about the technology became much less mainstream, less dramatic, and ultimately less covered.

Similar dramatic debates have raged over the injection of bovine somatotropin (BST) to increase milk production and the announcement earlier this year that Scottish scientists had cloned a lamb from tissue taken from its mother's udder. In terms of BST, the dramatic elements were: big U.S. farmers using the hormone to drive little farmers out of business; animal rights activists defending public health against greedy farmers; "natural" farmers fighting hormone-injecting farmers to preserve public health; environmental activists fighting against the giant chemical company, Monsanto; governments in Canada and the U.S. taking different positions on whether the substance was absolutely safe. Even the fact that Montsanto had launched a press campaign supporting the substance was portrayed as suspicious. As one article argued, in essence: Why would they have to do this, if BST is so benign (Milk's raging hormone, 1995)? Powell (1996) has a long description of how these dramatic issues have so far played themselves out in media coverage in both the U.S. and Canada. His most salient point is that the company's and government's attempts to define how the drama should be resolved, namely, by "attempts to reassure citizens on the basis of safety," have been utterly futile (p. 155). I would argue this is largely the case because the various dramatic elements of the situation – those issues discussed above, – have been more compelling both to the public and the media than what one would describe as arguments in a rational debate. But I would also argue that, if over the long run nothing bad happens with BST augmented milk, the ethical issues will cease to interest journalists. The critical debate will die, like the recombinant DNA debate, from a lack of a focusing event.

A similar dramatic scenario occurred with the cloning of the lamb Dolly. While this will be discussed in greater detail later, I will just point out here that the presentation of the cloning was, in almost all cases, immediately given a dramatic twist. It wasn't sheep we cared about, but humans. Nothing more needs to be said than quoting the opening paragraph of the first Vancouver Sun story on the matter: "Researchers have cloned an adult mammal for the first time, a controversial feat that could presage the ability to clone human beings" (Sheep cloning opens ethics debate, 1997).

Even if There Is a General Interest in the Area Among Readers, in the Newspapers Biotechnology Is at Present Largely a Business Story

I will attempt to give the reader a flavor of the sort of play which biotechnology gets in the Globe and Mail, La Presse, Maclean's, and the Winnipeg Free Press. As the Globe has been on-line since October of 1977 there is 20 years worth of coverage, which can be easily scanned by computer search. No other paper in Canada has a similarly long and easily analyzable set of data. What I have come up with isn't definitive, but it is suggestive of how non-intuitive one's uninformed impressions can be.

Near the end of 1996 there had been 1,927 articles in which the term biotechnology appeared and 676 in which the term "genetic engineering" was present. More pointedly, there had been 558 articles which classifying librarians decided were really about biotechnology, and 112 which were really about genetic engineering. By way of comparison, over the same time frame something more than 1.52 million articles appeared in the paper. Obviously, the flow of information about biotechnology in the paper is more a rivulet than an Amazon.

But more interesting is where the stories appear. If one is just looking for what section the word biotechnology appeared in, roughly two thirds (69 per cent) of the time it was in the business section of the newspaper. The figure was a little higher (72 per cent) when stories were judged to be about biotechnology. Interestingly, a different skew appeared when the key words were "genetic engineering." About 30 per cent of the stories were in the business section, either as key words or as stories judged to be about biotechnology. By way of comparison, the most common place for genetic engineering stories to appear is in the paper's science page – the spot where 56 per cent of the references occurred.

Finally, there were 59 stories on the front page paper where the word "biotechnology" appeared, but only 8 before 1989. There were only 13 stories focusing on biotechnology on the front page, the first of which appeared late in 1987. There were 25 references to genetic engineering and 6 specifically about it. The earliest reference was in 1987.

While the Globe is a heavily business oriented newspaper, and as such is skewed in ways other Canadian media aren't, a look at other publications reveals similar patterns. In the period July 1990 to March 1997 – the period of time over which computerized searches of La Presse are possible – the word "biotechnology" has appeared in 329 stories, and "genetic engineering" in 62. However, when one looks for stories which the newspaper judged to be specifically about those two subjects, "biotechnology" appeared only once and genetic engineering not at all. Where did they appear? Again the business impact is dramatic. One hundred and fifty-one stories with the word "biotechnology" in it appeared in the business section; two of the genetic engineering stories appeared there. Only five stories mentioning the word "biotechnology" appeared on the front page and three that talked about genetic engineering. There were three mentions of biotechnology on the editorial page and none about genetic engineering.

Both the Winnipeg Free Press and Maclean's magazine reveal similar patterns when looked at in early 1997. There have been 14 major stories about biotechnology in Maclean's since 1988, six of which were seen to have a basically business slant. There were three major stories about genetic engineering, none of which were judged to have a major business slant. Roughly 29 per cent of the 52 biotechnology stories appearing in the Winnipeg Free Press since 1993 were found in its business section, but only 2 of 20 dealing with genetic engineering.

Without an in-depth look at all the stories involved it is impossible to say, for certain, what the differences means; however, the dramatic difference of where genetic engineering stories and biotechnology stories appear suggests that a kind of verbal division has appeared in the media's mind. Genetic engineering is about science; biotechnology is about business. One is experimental; the other is applied.

The other point is that, unless one is reading the entire newspaper, front to back, one is misreading the sweep of the coverage. One might well ask how many consumers – what the paper would call general readers – pore over the business section to find answers to their questions and doubts about biotechnology. Conversely, how many of a business oriented readership turn to the newspaper's science pages to find out what is experimental and could be potentially scary about the enterprise? The truest truth, however, is that nobody is going to be much moved if all they read is the headlines on the front page. In Canada neither biotechnology nor genetic engineering is a front page subject.

Nonetheless there is a sense in some quarters that, no matter where it appears, the media has been particularly pernicious in its reporting of biotechnological matters. Of particular note should be the recent suggestion by the French National Bioethics Committee that some kind of regulator – perhaps an ombudsman – was required to rein in the excesses of medical and biological journalism in that country. The Committee claimed that an "ethical threshold" had been crossed, a Rubicon marked by an increased occurrence of "rash statements, retention of information, self-interested connivances, attempts at manipulation of decision makers, and unrepentant spreading of false ideas." The group suggests much of the blame lies in the media's "obsession with audience figures," its preoccupation with "scoops," and its love of "the sensational and the emotional" (French ethics panel, 1995). In this regard, I would point to the earlier discussion of how little a scoop effects any given paper. One also wonders whether the French clamor would have any effect in this country, given the little claim biotechnology has to being central to Canadian news agendas. To give this lack of centrality another numerical context let me simply point out that over the last four years the Globe and Mail, for all its reputed stodginess, had 31 stories about O.J. Simpson and his various murder trials on its front page. This is nearly two-and-half times more front page stories than the Globe has published about biotechnology over nearly the last 20 years.

Who to Believe?

Because biotechnology is at present playing itself out as a business story, it creates huge "source" problem for journalists. Who is an unbiased source? When a scientist is both funded by a company and the part owner of that company, who is he or she speaking for? The BioChem affair highlights this dramatically. When industry-university research is actively encouraged by the government through such programs as its centers of excellence efforts, is a journalist supposed to take "job creation" into effect when discussing its significance? What effect on their reporting of their findings comes into play when it is not possible for reporters (as with the 1975 study of Canadian science writers discussed earlier) to generally have confidence in the impartiality of research scientists?

One of the things that business reporters do to get around this problem is call financial analysts, as was done in the BioChem story. For non-business reporters, finding impartial commentators in biotechnology is probably becoming more and more difficult all the time. A recent study by Tufts University science policy expert Sheldon Krimsky found that 34 per cent of 1,000 Massachusetts scientists who had published in major science journals had financial stake in the work they presented. This included shares in biotechnology companies (Conference notes, 1997). Moreover, another U.S. study suggests that scientists with significant commercial backing are more than twice as likely to pick research topics with an eye toward their commercial applicability as non-industry related scientists. Industry-sponsored scientists were three times more likely (14.5 per cent versus 4.7 per cent) to keep aspects of their work secret (Commercial backing, 1996).

It is not a mistake that Industry Canada surveys indicate that the general public trusts no information source with the exception of David Suzuki (Creative Research, 1996, p. 11). Journalists don't know who to trust when it comes to biotechnology, and that includes David Suzuki.

Is It Hopeless Trying to Influence the Media to Have It Say What You Want? No. Just Play the Game

Despite all that has been said about how hard it is to get inside the newspaper if you are not an official source, outside groups can break into the closed circle when it comes to biotechnology. Generally, the way that they do it is by appealing to those dramatic, story-telling tendencies that journalists love, and doing it with "expert" labels which they tend to trust. Powell (1996) describes how the Chicago-based Cancer Prevention Coalition, the Vermont-based environmental group, Food and Water, and the Council of Canadians got considerable media attention by holding a press conference and releasing a new study, which concluded that "milk from cows injected with rBGH increases risks of breast and colon cancers in humans."

The long running arguments about BST's safety in some environmentalist circles becomes a "new" news story in that a "new study" showed new dangers as revealed by Samuel Epstein, a professor of environmental medicine at the University of Illinois School of Public Health. According to Powell (1996), in addition to CBS This Morning, USA Today, and CNN, 350 radio and TV outlets in the U.S. carried the story. A large number of Canadian newspapers and radio stations picked up the story as written either by a CP reporter or a Montreal Gazette reporter. While Powell describes problems with the study and with Professor Epstein, I would suggest that the most significant feature to what went on is how absolutely predictable it was. Put information in a dramatic framework, wrap it up with an expert bow, and the media will respond.

Another classic example of creating news occurred in 1996, when a biochemist from the Maharishi University of Management (MUM) in Fairfield, Iowa, in conjunction with a controversial retired University of Western Ontario genetics professor, toured parts of Canada to argue against biotechnology in general. They sent out a press release with the provocative headline "Fish Genes in Tomatoes??" In this they claimed that a genetically engineered food supplement had killed 37 people in North America and permanently disabled 37 more. At the bottom of the release there appeared in large italic type: "We are all guinea pigs in this dangerous global experiment on our food supply." Included with the release was the curriculum vitae of MUM professor John Fagan, which highlighted his mainstream background – Ph.D. Cornell University, seven years research at the prestigious U.S. National Institutes of Health (NIH). Professor Cummins had taught at Western for 20 years. It was also pointed out that Professor Fagan had turned down a grant of more than $600,000 (U.S.) from the NIH because of his disagreement with the legal status given biotechnology. What we had was a drama and a story. An "honorable" and apparently reputable scientist was so appalled by the government's portrayal of biotechnology that he was willing to sacrifice his scientific career to rectify the situation. And contrarily, the apparent efforts on the part of officialdom to ignore him suggested something was amiss.

Dr. Fagan and Dr. Cummins got the press's attention. A story appeared in the Toronto Star's food section headlined "Altering Food Called 'Dangerous Experiment'." A headline on the next page said "Genetically Engineered Food Facing Tough Foes." The opening two paragraphs said "Critics call it Franken-food and issue dire warnings against it. Genetically engineered food that's currently on your grocer's shelves poses an unexpected threat to humanity, they say. And unless the situation is controlled, we face a future rife with genetic mutations, the spread of dangerous diseases between species, and super viruses resulting from genetic recombinations." In it Fagan's claims that genetically engineered foods place entire populations at risk, "have a reduced nutritional value," and may produce hazardous heath and allergy problems, were highlighted. The only rebuttal was a single paragraph at the end of the piece quoting Margaret Kenny, associate director of the biotechnology office of Agriculture and Agri-food Canada (Altering food called dangerous experiment, 1996). A story appeared in the Calgary Herald headlined "Altered Food Labeled Dangerous." It created a more classic newspaper debate as it juxtaposed Fagan's remarks with denials from University of Calgary biochemist Maurice Maloney. It began: "Genetically engineered food is dangerous and should be banned until properly tested and labeled, a scientist who quit a successful career in genetic research said Monday. Foods contaminated by "unnatural" genes – such as frost resistant fish genes inserted into a tomato – are being pushed by companies wanting to recoup billions of dollars spent on biotechnology research, John Fagan . . . told reporters." However, Professor Fagan's bona fides were undermined by the noting that his institution had some relationship to "yogic flying" (Altered food labeled dangerous, 1996).

A Canadian Press wire story – that is one not written by any staff reporter – appeared in the Globe and Mail. Its headline was "Scientists Fear Humans Used as Guinea Pigs." The subheadline read "Genetically Engineered Food Approved in Canada Is Being Questioned in Europe." The press conference was used as a vehicle to tie in dramatic confrontations in Europe. The first paragraph said: "Shipments of genetically engineered soybeans prompted protesters in Europe to occupy offices and chain themselves to gates this week. In Canada, the food was approved months ago." It went on to describe Professor Fagan's (who it wrongly said taught at Cornell) and Dr. Cummins' protests. Fagan is described as being funded by "a fringe political group that became the butt of jokes by advocating better government through meditation." Much space was devoted to protests of Greenpeace and the Green Parties in Europe. Again the only voice supporting biotechnology belonged to Margaret Kenny, and that only in a single paragraph (Scientists fear humans used as guinea pigs, 1996). Again, we have the issue of biotechnology and food safety cast, not as process or debate, but as political drama. The scientists speaking up. The Greens protesting. The Canadian government responding with only words.

I should point out in passing that while Fagan appears to have understood where the media's hot buttons are, the daily press had no similar sense of what questions to ask Dr. Fagan in response. None of the articles actually described the death-dealing biotechnology he claimed had killed people. But even more to the point, none of the reporters asked Professor Fagan about his own food beliefs. Had they, they might have heard some interesting discourses on the Hindu ayurvedic theories of the relationship between various kinds of foods and health – theories which readers might have found clashed rather dramatically with their own. They might have also learned that MUM is a university set up to expound the theories of Maharishi Mahesh Yogi – the Maharishi who achieved fame as the Beatle's guru and the founder of Transcendental Meditation. His theories try to connect ancient Indian teachings with modern scientific practices and then produce a new natural law. (This, by the way is the philosophic underpinning of the Maharishi's political party – the Natural Law Party.) Attempts to combine science with 4,000 year old Indian teachings have led MUM researchers to produce studies claiming that people in a certain meditative state can fly, and that if a small number of people meditate they can reduce traffic accidents and bring peace to the Middle East. And it is likely that Professor Fagan believes in these apparent violations of the basic laws of physics – points of view which might have diminished his scientific expert status in some readers' eyes.

Why weren't these questions asked? Several reasons probably. With his Ph.D. and national grants, Professor Fagan looks like a bona fide expert. Also daily reporters probably didn't know much about the Maharishi University of Management. But I will also suggest most strongly that the dramatic framework of how the issue was presented – lone scientist fights for truth against the world, large protests against potential danger in Europe, nothing in Canada – simply swept away more critical thinking. Professor Fagan's little-man-fights-uncaring-government story was simply too compelling not to tell. Additional evidence for this dramatic imperative is how journalists elsewhere reported Dr. Fagan's statements. When he first announced that he was returning his grant money in a press conference in the U.S., representatives from 40 media outlets, including such well respected organizations as Science Magazine, USA Today, Newsweek, and U.S. News and World Report, as well as three Asian television stations attended (John Fagan aligns with Mother Nature, 1994).

Maybe It Doesn't Matter What Appears in the Daily Press, Maybe People Already Are Involved in the Debate

In some sense one of the greatest confounding factors about biotechnology is that it is an area of scientific advance about which consumers care. It is not like research in astrophysics, or geology, or the sex life of polar bears. People care because it affects the food they eat, the medical treatment they receive, the environment they live in. They want to know more. This may mean that, in ways which don't happen in other areas, they ignore what the media says and carry on an inner and outer debate amongst themselves.

A classic example of this is Priest's (1994) study of 608 people who looked at nine representative newspaper articles. One thesis was that the framing of newspaper coverage in biotechnology – that is what was talked about and what was ignored – would strongly influence what readers talked about after having been shown the articles. Not so. While it was judged by external reviewers that 76 per cent of the time the articles focused on the benefits and economic importance of biotechnology, when given a free range of discussion, focus groups had other things on their minds. They discussed benefits only 12.2 per cent of the time, while biotechnology "dangers," which only accounted for 1.5 per cent of the articles' coverage, were discussed 18.4 per cent of the time. The newspaper reports spent about 30 per cent of their time discussing economic issues, while Priest's subjects spent less than ten per cent of their time on the same discussions. Some topics, such as the environment, ethics, and regulatory issues which were either completely ignored or only minimally discussed in the articles, were discussed collectively more than 20 per cent of the time.

Part of the problem may simply be, as stated earlier, that the media is not a Newtonian force which effects behavior in a simple way, but is more like a restaurant where you get to decide what should be on the menu. "It is essential to realize that 'the media' do not form a monolithic group. . . . In a modern society, swamped with information, individuals tend to specialize in what they are interested in – science, economics, foreign policy, interior decorating. People get information from many different sources, and use no one source consistently," admits Priest about a result which contradicted her thesis (Thompson, 1992, p. 114).

Potentials for Change – Are Papers Listening More?

If people truly care about biotechnology, is there a way to have the media write or broadcast about it? Maybe. As I have described in the first section of the paper, the old ways of simply trying to convince newspapers to write about what you want them to write about generally do not work. There are too many other agendas at work. However, in ways in which they haven't been in the past, newspapers are now more interested in what their readers might care about and care to hear about. There has been an erosion in the relative power of newspapers in recent years and consequently in their future potential affluence. While the numbers of readers have remained the same on an absolute level, the market penetration of the press has waned. The number of readers as a per cent of the population in North America fell from 69 per cent in 1973 to 53 per cent in 1990 (Thompson, 1992). This has made newspaper owners and publishers extremely nervous about the future of their news product.

One of the things which follows from this is that there is a newly-found strength on the part of readers to influence both the content and direction of papers. The old "the reporter or editor knows best" approach to news has been challenged as publishers worrying about falling circulation figures, have instituted focus groups and other reader interest polling techniques. The idea: Give the readers the kind of paper they want. While this is opposed by old school journalists, it has very much become a war cry of many managers. "Newspapers are to be of service to readers and are not to be staffed by a Brahmin class that was chosen to lecture the population. People who refuse to be service-oriented will leave in disgust and say we're pandering and will call us bad names – but they will leave," one U.S. newspaper executive has said (Conquest Research, 1995, pp. 4–5).

However, this demand for a more "service oriented" paper assumes there is an easy way for the outside world to tell the papers what should be reported on. And there may be a way for that to happen. Even if no easy pathway exists, people may soon garrulously start giving one another information outside of official media.

The Internet, or, Gossip as Media

Powell (1996) reports that eight per cent of those polled in a study said that the Internet was a prime information source. This is an unregulated media where information is passed on without the fact-checking, error-monitoring, self-corrective features of daily journalism. He describes a variety of cases in which incorrect information about bovine somatotropin (BST) was passed along via Internet discussion groups and newsletters. What is interesting about this is what one English study showed: that as regards the trust that 139 test subjects had in the information they got from a particular source, there was no statistical difference between information from a quality newspaper and information from friends (Frewer & Shepard, 1994). Since Internet discussion groups often feel like friends telling things to friends, it may grow in importance as an information media.

Indeed some companies are already hiring people to monitor Internet discussion groups so that they can respond to what they believe is false information being spread about them. Conversely, the Internet is already opening up the traditional news organizations to other voices and other concerns. A recent survey says that 69 per cent of Canadian media are hooked up to the Internet (Ernst & Young, 1996). They are using it both to receive press releases – there were 44 biotechnology-related releases on the Internet for the week of June 16–21 alone – to listen in on discussion groups, and to look at information posted on web sites. One can, for example, go beyond the press release described earlier and get John Fagan's entire 10,000 word explanation of why he is against genetically engineered food from his personal web page. The same instant entree is also available for the general public, which if it wishes can use the Internet to access anything from a Greenpeace video to a high school student's term paper and a U.S. government report on public perceptions of genetic engineering. In a more organized fashion, the University of Guelph (1997) is posting material related to food safety (the Food Safety Network or Fisnet) and agriculture (the Agriculture Network or Agnet) on the Internet. Food biotechnology is one of the issues dealt with. It is interesting to note however, that a search of the Internet in July of 1997 failed to turn up a general discussion group on the Internet exclusively devoted to biotechnology or genetic engineering issues.

But there may also be a more direct way in which the Internet influences news coverage. An example of this occurred recently when the Globe and Mail wanted to write a piece on the implications of the cloning of the lamb Dolly. The questions were: What do people want to know, and what do they already think? In preparation for the piece, and as a way to circumvent expert opinion, a note was put into the paper soliciting e-mail on just such matters. Approximately 20 "ordinary" people responded with a variety of concerns, and two were quoted in the paper.

But let me continue to repeat my caveat to the Internet's information boom. The question is not what is available, but what its quality is.

Finally, Fiction

It is often expressed, although sometimes in an oblique fashion, that when people try to study the relationship between the media and biotechnology, they unnecessarily restrict themselves to non-fiction. Over and over again commentators on the field argue that movies and novels are the general public's true educators in biotechnology (Thompson, 1992).

For example, "The public is left to pick up what they can from fictional accounts (the popular book Jurassic Park and its film sequel)," argues Priest (1995, p. 45). She adds: "A steady stream of good news about biotechnology – in the absence of ethical, social, economic, political, and policy issues – is unlikely to quell public concerns, whereas science fiction that presents pessimistic visions of a technological future serves a constructive (not necessarily an anti-scientific) purpose if it stimulates meaningful public debate." To some scientists, the fiction is too interesting and too hard to counter. As an editorialist in a scientific journal suggests, Jurassic Park "will have an affect on public attitudes, no matter how much people are willing to enjoy films such as this as straight entertainment" (Wilkins, 1993).

And the newspapers themselves seem to accept the fact that what they are responding to when biotechnology stories come out, is not just the science, but the ideas about science which science fiction put in people's heads. When Jurassic Park came out it was a common ploy for journalists to get scientists to comment on how real a piece of genetic engineering Crighton's feat of imagination was. When the lamb Dolly was cloned, there was not only an immediate leap to the issue of human cloning, but a discussion of where truth and fiction intersect (Genetic marvel, 1997; With cloning of a sheep, 1997). The Globe and Mail felt obliged in its week-end wrap-up piece on the cloning to couch, at least part of it, as a rejoinder to The Boys From Brazil, Brave New World, and In His Image (Hello Dolly, 1997). Perhaps, it is because they already bear the imprint of highly believable fiction. Hollywood understood immediately, too. After Dolly hit the headlines, it was reported that any old scripts having clones in them was suddenly hot property. No matter what the dramatic conclusion of the official debate about Dolly in newspapers, the world's public was going to shaped by (a) a four-hour mini-series on cloning; (b) discussions in Alien IV of the resuscitation of a main character as a clone, and (c) perhaps another movie about a nurse who meets a character in the emergency room who is a clone of herself (Send in the clones, 1997).

In Conclusion

Interested parties would no doubt like some kind of magic formula which would allow them to communicate what they think are correct and socially important messages about biotechnology to the consumers of that technology. As I have tried to show the media doesn't work that way. The coverage of biotechnology in newspapers – and undoubtedly television as well – has been diffuse and will likely remain so until a focusing event appears which suddenly turns general information into specific drama. Process, what is actually happening with biotechnology on a day-to-day or week-to-week basis, is not dramatic enough to be reported upon very regularly or very prominently. Everyone's – consumers, business, scientists, government, public interest groups – efforts to enliven or shape the debate await these dramas. Otherwise the media does things in the media's own time.

What other issues should be considered?. The question of bias on the part of biotechnology experts quoted in the media seems likely to grow as the line between scientist and exploiter of information blurs. Ultimately, it is impossible to report on the true history of that which is new until sometime later in its evolution. Journalists really don't know what to write and what the important issues in biotechnology are. The Internet holds open some possibilities to influence both media coverage and to circumvent it. Finally, if various interest groups are truly interested in influencing the public mind, they might consider hiring a good science fiction writer and tell him or her the philosophical points they would like to appear in a novel or movie.

REFERENCES

Altered food labeled dangerous (1996). *Calgary Herald*, November 19, p. B2.

Altering food called dangerous experiment (1996). *Toronto Star*, November 20, p. E1.

American Newspaper Publishers Association (1982). *Facts about newspapers.* Washington, D.C.: ANPA.

Burkett, W. (1986). *News reporting: Science, medicine and high technology*. Ames, IA: Iowa State University Press.

Commercial backing could impair academic influence (1996). *Nature*, December 13, p. 801.

Conference notes (1997). *Globe and Mail*, February 22, p. D6.

Conquest Research (1995). *The 1995 national science literature survey*. Prepared for the Royal Society of Canada. Ontario: Ontario Science Centre and Science Network Ontario.

Creative Research International (1996). *Environmental applications of biotechnology: Focus groups*. Ottawa: Office of Consumer Affairs, Industry Canada.

DeFleur, M., & Dennis, E. (1996). *Understanding mass communication* (5th ed.) Boston: Houghton Mifflin.

Dennis E., & Merrill, J. (1996). *Media debates*. White Plains, NY: Longman.

Dubas, O., & Martel, L. (1975). *A research study on science communication.* Vol. 2. Ottawa: Minister of State Science and Technology.

Dubois, L. (1996). Le discourse sur la santé à la television: differentes types de locateurs pour differentes types de representations. *Canadian Journal of Public Health, 87*(1), 56–61.

Einsiedel E., & Schiele, B. (forthcoming). *Biotechnology, media coverage, public perceptions and public policy*. A study for the Social Sciences and Humanities Research Council and the Canadian Institute of Biotechnology.

Ernst & Young (1996). *The media and the internet.* Toronto: Ernst & Young.

French ethics panel warns of crisis in science reporting (1995). *Nature*, July 13, p. 108.

Frewer, L., & Shepherd, R. (1994). Attributing information to different sources: Effects on the perceived qualities of information, on the perceived relevance of information, and on attitude formation. *Public Understanding of Science, 3*, 385–401.

Friedman, S. (1986). *Scientists and journalists, reporting science as news*. New York: Free Press.

Fulford, R. (1965). The built-in bias of the press. In: L. Lyons (Ed.), *Reporting the news*, pp. 153–159. Cambridge, MA: Belknap Press of Harvard University Press.

Genetic marvel spawns potential ethical nightmares (1997). *New York Times*, February 24, p. A1.

Gitlin, T. (1980). *The whole world is watching*. Berkeley, CA: University of California Press.

Goodell, R. (1977). *The visible scientist.* Boston: Little Brown.

Greenpeace appeals to Nestle shareholders (1997). www.greenpeace.org/-comms/97/geneg/press/june05.html.

Hello Dolly, it's so easy to see you (1997). *Globe and Mail*, March 1, p. A8.

John Fagan aligns with Mother Nature (1994). *Fairfield Weekly Reader*, December 8, p. 1.

Koshland, D., Jr. (1990). Two plus two equals five. *Science, 247*, 1381.

Milk's raging hormone (1995). *Globe and Mail*, July 13, p. A13.

Nelkin, D. (1995). *Selling science, how the press covers science and technology*. New York: Freeman.

Pfund, N., & Hofstadter, L. (1991). Biomedical innovation and the press. *Journal of Communication, 41*(2), 138–153.

Plein, L. C. (1991). Popularizing biotechnology, the influence of issue definition. *Science, Technology & Human Values, 14*, 474–490.

Powell, D. (1996). *Applying risk communication theory to the Canadian agri-food sector*. Guelph: University of Guelph. Unpublished Ph.D. thesis.

Priest, S. H. (1994). Structuring public debate on biotechnology. *Science Communication, 16*, 166–179.

Priest, S. H. (1995). Information equity, public understanding of science, and the biotechnology debate. *Journal of Communication, 45*(1), 39–54.

Schopenhauer, A. (1970). *Essays and aphorisms, selected and translated by R. J. Hillindate*. Harmondsworth: Penguin Books.

Scientists fear humans used as guinea pigs (1996). *Globe and Mail*, November 15, p. A8.

Send in the clones: Hollywood sees Dolly potential (1997). *Vancouver Sun*, March 1, p. D7.

Sheep cloning opens ethics debate (1997). *Vancouver Sun*, February 24, p. A1.

Singer E., & Endreny, P. (1993). *Reporting on risk – How the mass media portrays accidents, diseases, disasters, and other hazards*. New York: Russell Sage Foundation.

Thompson, L. (1992). *Genes and human self-knowledge*. Iowa City, IA: University of Iowa Press.

Tuchman, G. (1978). *Making news*. New York: Free Press.

Underwood, D. (1993). *When MBAs rule the newsroom*. New York: Columbia University Press.

University of Guelph (1997). *Listservs look at global food safety and agricultural risks*. Press release, April 3 (listserv@listserv.uoguelph.ca).

Wilkie, T. (1996). Sources in science: Who can we trust. *Lancet, 347*, 1308–1311.

Wilkins, A. (1993). Jurassic Park and the "Gay Gene": The new genetics seen as through the distorting lens of the media. *FASEB Journal, 7*, 1203–1204.

Willis, J. (1996). *Journalism, state of the art*. New York: Praeger.

With cloning of a sheep, the ethical grounds shift (1997). *New York Times*, February 24, p. A1.

THE AUTHOR

Stephen Strauss is science reporter at Globe and Mail. His address is 365 Sackville Street, Toronto, Ontario, M5V 3G5, Canada. E-mail: sstrauss@globeandmail.ca.

G. Bruce Doern and Heather Sheehy

The Federal Biotechnology Regulatory System: A Commentary on an Institutional Work in Progress

ABSTRACT. The authors describe and assess key institutional aspects of the Canadian federal biotechnology regulatory framework. Noting that Canada's system for regulating biotechnology relies on five federal departments playing a substantive role, the authors explore how Canada's system addresses central features of normally functioning regulatory systems and how it interacts with the macro regulatory environment. From these discussions it is shown that each of the departments involved in regulating biotechnology applications in Canada have very different approaches which are, in many respects, sensible and supportable. However, the paper concludes, further work needs to be done in order to secure the confidence of a wide-range of interests, noting especially, that the current system has not established an arena in which ethical issues can be addressed.

The purpose of this paper is to describe and assess key institutional aspects of the Canadian federal biotechnology regulatory framework. This is done both through a look at selected features of the emerging federal biotechnology regulatory system and through a commentary which draws on basic insights from broader regulatory institutional analysis. The attention to regulatory theory as such is very basic and limited. Such an assessment also implies an awareness of the particular features of science-based regulation where the overall legitimacy and efficacy of the regulator must meet both a test of scientific-technical capacity and objectivity and a political test of reasonably transparent public and interest group involvement (Doern, 1981). Such a regulatory system can be technically competent but fail the larger political litmus test. Or it can be open and reasonably democratic but falter in its underlying scientific and technical capacity.

The paper is necessarily cast as a commentary on an institutional "work in progress" because the Canadian biotechnology regulatory system is itself relatively new, emerging, and institutionally dispersed. Its key players will understandably say that the system, having approved many biotechnology products on the market, is more than "in progress." It is established.

Journal of Consumer Policy **21**, 1998.
Bartha M. Knoppers & Alan D. Mathios (eds), Biotechnology and the Consumer.

As authors, however, we will use the more tentative "work in progress" designation for several analytical and cautionary reasons. The first is that in most areas of regulation, one can readily identify the core regulatory institutions centred around one main regulatory body and surrounded by a set of secondary or non-core players. This is true for the regulation of competition, telecommunications, transportation, and atomic energy to name only a few regulatory realms (Doern, 1995; Schultz & Doern, 1998). As we will see below, the biotechnology regulatory system does not have a core in the same sense. This does not mean that it must have a core body to be successful but if it is too dispersed and difficult to understand, it may well have difficulty meeting the test of political legitimacy, which along with scientific-technical objectivity and capacity is crucial for science-based regulators. This is because Canadians will simply not be aware of the key features of the system.

The core of the biotechnology regulatory system, by deliberate design, is itself a secondary set of arrangements and processes agreed to by several regulatory bodies whose primary regulatory tasks are far broader than biotechnology.

This central fact induces some caution because it is difficult to assess the biotechnology regulatory system against any fully agreed criteria. The paper will comment on the twin issues of political legitimacy (including accountability, transparency, participation) and scientific-technical capacity but these of course have many degrees of meaning and thus their adequacy is the stuff of political and technical disagreement. For example, it will be evident that some interests see the current system as being seriously flawed precisely because it is *not* more centrally controlled by Environment Canada through CEPA legislation. Others would view it as profoundly illegitimate if it was so centralized.

In addition, the paper's cautionary status as a commentary also arises because it must deal with an industrial and technological activity which is not only a relatively new industrial sector itself but also an enabling technology for other traditional sectors of the economy (Chemical and Bio-Industries Branch, 1996, p. 2). The authors are also aware of the fact that the paper does not and could not deal fully or evenly with all of the elements of the present system. The limits of time and resources thus affect what can be covered in a relatively short exploratory paper. More than the normal premium is placed on simply getting an accurate sense of what the regulatory system looks like.[1]

But even this can only be partly gleaned from public documentation. Some of the broader dynamics have been derived by the authors through interviews with governmental and nongovernmental practitioners involved in or with the several regulatory and policy bodies involved. Alas, this picture too must be considered a preliminary one. Nonetheless, these parts of the developing regulatory story can be instructive, especially if related to some of the mainstream literature and analysis of regulatory institutions in general and in other health and safety spheres (Ayres & Braithwaite, 1992; Dewees, 1983, Ch. 1; Doern & Wilks, 1998).

The structure of the paper reflects the key practical tasks at hand. We first describe the current stated federal framework for the regulation of biotechnology, including the six principles which govern it. In the second section, we highlight briefly and in a general way several key issues involved in assessing regulatory institutions, and why these go beyond the stated federal framework's six principles. In the absence of a single institutional core body, we look first and more generically at three central features of any normally functioning regulatory system: (a) the statutory base and the overall regulation making process; (b) the case handling and approval process; and (c) the compliance and enforcement process. When applied in the third part of the paper these aspects are our surrogate for the regulatory core since there is no single regulatory body. We also highlight generally the existence and influence of four even broader regimes of regulation whose role is secondary or non-core in the context of this paper but still important in helping to understand why the biotechnology regulatory system may or may not work and what kind of relationships it has with a broader regulatory world.

In the third and longest part of the paper, we look concretely at the dispersed core functions of the biotechnology regulatory system itself (regulation making, case handling, and compliance and enforcement). Conclusions and final commentary then follow.

THE FEDERAL FRAMEWORK FOR THE REGULATION OF BIOTECHNOLOGY

In 1993, the federal framework for the regulation of biotechnology was announced (Office of Consumer Affairs, 1996, p. 1). But biotechnology as a concerted policy or strategic initiative can be traced to

1983 and received a renewed emphasis, a significant relaunch, in 1991. At time of writing, a new mandate for the strategy is being prepared. Biotechnology can be defined in statutory terms (as in CEPA) or in more general ways but for our overall purposes it refers to "the applied use of living organisms or their components to make or modify products, to improve plants or animals or to develop microorganisms for specific uses" (Chemical and Bio-Industries Branch, 1996, p. 2). Cast in the context of genetic engineering and advanced applications based on recombinant DNA, the biotechnology field raises many concerns about the public interest, commercial regulation and the nature of risk, and ethical issues (Breyer, 1993; Mausberg, Winfield, et al., 1993; Standing Committee on Agriculture and Agri-Food, 1994). The federal framework was accordingly developed in the light of both interdepartmental and stakeholder consultation processes with diverse interests. But, as we will see, there are diverse views as to the adequacy, openness, and legitimacy of the processes used and of the regulatory system that is presently in place.

The framework is centered on a set of six principles with respect to the regulation of biotechnology. The regulatory system is intended to:

- Maintain Canada's high standards for the protection of human health and the environment.
- Build on existing legislation and institutions, clarifying jurisdictional responsibilities and avoiding duplication.
- Develop guidelines, standards, codes of practice, and monitoring capabilities for pre-release assessment of the risks associated with release to the environment.
- Develop a sound scientific database, upon which risk assessments and evaluations of products can be made.
- Promote development and enforcement of Canadian regulations in an open and consultative manner, in harmony with national priorities and international approaches.
- Foster a favourable climate for development, accelerating innovation and adoption of sustainable Canadian biotechnology products and process.[2]

In various ways the six principles are intended to reflect a reasonable and balanced trade-off between securing the practical benefits of biotechnology products/processes and Canadian competitiveness in this sector, and ensuring the protection of individuals and society in relation to environmental, human, and animal health and safety. There are also vital ethical concerns about some aspects of biotechnology but these are not directly reflected in the six principles.

As we see below, there are four federal departments with direct regulatory roles – hence the absence of a normal institutional core – but the notion of the biotechnology framework actually being a "framework" also owes its existence to the role of Industry Canada, a fifth departmental player, which itself has only a limited regulatory role (albeit an important one in intellectual property). The Industry Canada mandate necessitates the search for a balance as well between business and consumer interests in the marketplace because the Minister has a consumer protection mandate. The consumer interest would arise even if Canada produced no biotechnology products itself and instead imported them. In crucial terms, consumer concerns are also related to the price and quality of biotechnology products and to whether the nature of biotechnology products are generally understood and accepted by consumers (Sheehy, 1996). Industry Canada's minister was caught in the midst of some of these linked roles and political controversies when the "rbST case" (see more below) came on the direct regulatory agenda in the early 1990s (Standing Committee on Agriculture and Agri-Food, 1994).

However, of more immediate importance in this initial account is the fact that the biotechnology regulatory framework emerged out of the 1991 National Biotechnology Strategy (NBS) developed by Industry Canada, and that several institutional bodies are partially coordinated by Industry Canada. Thus, Industry Canada chairs the Interdepartmental Committee on Biotechnology and the Biotechnology Coordinating Group. It also supplies the secretariat for the National Biotechnology Advisory Committee (NBAC), a broad industry and academic advisory group, and an Industry Coordinating Group.

The newness of the biotechnology regulatory framework and of the biotechnology regulatory system means that there is only a scattered and eclectic analysis of its form, workings, or efficacy. One study prepared for the National Biotechnology Advisory Committee compared the Canadian system with the U.S. but its focus was on the latter (National Biotechnology Advisory Committee, 1995). It broadly saw the U.S. system as being more advanced and established, but also more rigid and rule-driven. The U.S. was also seen to have had a longer and more sustained debate about the underlying values and ethical issues in biotechnology. The study saw virtue in the inherently greater flexibility of the nascent Canadian biotechnology regulatory system. It was argued that this flexibility could be capitalized on, especially in developing a system based as much as possible on "equivalency and reciprocity" rather than detailed duplicative regulation. But, the study also saw Canada as lagging behind the U.S. in the degree to which the basic issues of biotechnology were understood by the public.

Another source of major criticism has come from the non-governmental organizations, particularly environmental groups. These critical voices go back to the mid-1980s when biotechnology strategies were first unveiled, but more recently they have been channelled through, and reflected in, the positions of two Parliamentary Committees in reports on rbST (Standing Committee on Agriculture and Agri-Food, 1994) and on CEPA (Biotechnology regulation in Canada, 1996; Standing Committee on Environment and Sustainable Development, 1996).

These criticisms are wide ranging and frequently go well beyond the biotechnology regulatory system but they focus on several related points. The first is that the biotechnology strategies are being led by Industry Canada which is seen by environmentalists as the "lobby from within" for the biotechnology firms and as a department seeking to adopt strategies which give insufficient weight to environmental and human health issues and protection. Second, NGOs and the public more generally have not, according to this view, been allowed into the debate in a serious way except by going to Parliamentary committees such as those noted above. Third, Agriculture and Agri-Food Canada is often singled out as having a conflict of interest in regulatory matters in biotechnology because it is both a strong promoter of the industry and its regulator.

More will be said about the federal regulatory framework, its core underlying principles, and criticisms of its functioning. However, before doing so we need to have a complementary reference point for the understanding of regulatory institutions, in this case, drawn from a literature which deals with both long established areas of health and safety regulation as well as regulation (as in the biotechnology field) where the regulatory realm is emerging or rapidly changing (Francis, 1993; Harris & Milkis, 1989; Hill, 1994).

ISSUES IN ANALYZING AN EMERGING REGULATORY SYSTEM: CORE FUNCTIONS AND SURROUNDING REGIMES

The first point about the biotechnology regulatory system is that it must be properly related to its core functions and also situated in the world of broader regulatory regimes in which it operates. Thus, we need to draw initial attention to two levels of analysis.

The Core Regulatory Functions

In a regulatory system without a core regulatory body, it is even more crucial than normal to examine issues related to the three key functions of any operating regulatory system: regulation making and how core policies about it are made; case handling of ongoing product or substance applications; and compliance and enforcement. Since these are the core of the analytical sections in the following portion of the paper, their main features are initially only highlighted here (Doern, 1978; Meier, 1985; Ogus, 1994; Schultz & Alexandroff, 1985).

The first aspect is the statutory base of the regulatory system and the basic overall regulation making process. The concerns here are not only the actual statutory terms and what regulatory compromises are enshrined therein, but also how open and consultative the regulation making process is or seems to be. Statutes and regulations are made and changed only every few years and thus these processes are less frequent than the processes for case handling and for compliance, the other two processes examined below. The development of guidelines and codes may also be considered to be a part of this process. In other words there is a hierarchy of norm or goal-setting modes that starts with framework principles (such as those for biotechnology), extends to law or statutory expressions and policies, continues into regulations (delegated legislation passed through the RIAS-Canada Gazette-centered review processes) and includes as well guidelines and codes (Baldwin, 1995; Doern, forthcoming; Stanbury, 1992).

The second element of an operational regulatory system is the case handling and approval process. This is more than simply the paper flow of day-to-day regulation. It is the process through which real micro technical, economic, and political decisions are made through the exercise of discretion by officials. Working within the confines of the enunciated statutes, regulations, and guidelines, this process is one which, in the case of biotechnology, enables a potential and actual new biotechnology product or substance to be assessed scientifically and technically by the staff of the regulatory body or bodies. If the product meets the standards or risk assessment approaches, it is approved and is allowed on to the market or some other approved action is specified. If the product does not comply, then the biotechnology product or process cannot proceed or is sent back to the drawing board.

The case handling process must in some respects be seen to start within the firm or research institution long before it reaches a formal regulatory stage as such. At these stages, the firm or entity is in many respects "policing itself," both because it knows there is a regulatory system it must eventually satisfy, but also because there will be concerns within the firm about actual product efficacy and concerns about private legal product or damage liabilities (Dewees, 1983). One of the hidden aspects of any regulatory system or case-handling process is how one accounts for the cases that never reach the starting line because of these early and anticipatory stages. These are "non-events" that do not get counted in regulatory statistics (Sparrow, 1994).

The early case handling process will also involve preliminary discussions and perhaps even negotiations between the proponent or applicant and the regulator or regulators. The regulator encourages such discussions both to avoid future problems and to potentially save time. Concerns about regulatory lag and lead times are central in the case handling process. Global comparisons are here increasingly crucial since larger multinational firms in particular are aware of the relative efficiency of different countries' regulatory systems and are applying pressure to speed up the process.

The third element of any regulatory system is the compliance and enforcement process. In most basic conceptions of regulatory systems, the compliance and enforcement process would include all the aforementioned aspects of the case handling process (Baldwin, 1995; Grabosky, 1995; Sparrow, 1994). After all, such case handling is at the centre of implementing regulations, in our case the various biotechnology regulations and guidelines. However, if ultimate implementation embraces, as indeed it does, both softer compliance activities and approaches and harder enforcement actions (Hood, 1986; Hutter, 1989) then it goes beyond case handling per se. This third element of regulation deserves separate mention also because of the philosophical and political changes that are driving governments in an age of reduced funding and staffing and in an era of real or alleged "reinvented" government (Grabosky, 1995; Pildes & Sunstein, 1995).

Compliance activity can be seen to include aspects of education and training at several levels (among regulators, and within and among firms and consumers). Compliance in science-based regulatory realms turns also on actual scientific and technical capacity (and its demonstrated objectivity) and whether that knowledge base of the regulator is up-to-date in a fast changing economy. Enforcement activity will depend not only on the size and consequences of penalties in law but also on whether the Justice Department will pursue enforcement, especially regarding expensive cases. Legal action to ensure enforcement may also come from, or simply be threatened by various players: competitor firms, consumer and public interest groups, as well as the regulators themselves (Ayres & Braithwaite, 1992).

The Surrounding Macro Regulatory Regimes

While issues related to the three core functions of a regulatory system are crucial to regulatory analysis, it is increasingly evident that no regulatory system functions in isolation. Hence it is important to see in a general way how any particular regulatory system is "nested" within a larger interacting set of four macro regulatory regimes (Doern, 1998). Each regime is understood for analytical purposes to be an interacting set of organizations, statutes, ideas, interests, and processes. In other words, the first test of there being a regime is that there is some inner core of features and characteristics that warrant such a designation for analytical purposes. Anyone using the concept of regimes must do so knowing that there are varying usages of the term (Harris & Milkis, 1989; Hoberg, 1993). In domestic regulatory areas, it is often used as suggested in Regime I below where quite formal organizations are present. In international relations literature it is often used to convey softer arrangements and agreements where there often is no real organization at its centre. Economists will also often refer to regimes in regulation by referring to "rate of return" or "price-capping" as different regimes for public utility regulation. Despite these difficulties, the term has analytical value for the reasons set out further below.

In a recent comparative analysis these four macro regimes have been identified as follows: Regime I, the immediate sectoral regime of the regulatory body; Regime II, the horizontal framework regulatory regime; Regime III, the overall governmental executive regime for making and managing regulation; and Regime IV, the international regime, embracing global, regional, and bilateral rules-based approaches or de facto co-decision making by national or international bodies (Doern, 1998).

Regime I draws attention to the regulatory body itself (single statutory regulator, commission, ministerially headed department). A basic sectoral or public utility regulator such as the CRTC or NEB comes quickly to mind as examples of established Regime I regulators. In the biotechnology sector, as we have stressed, there is no one central regulator in the sense implied by Regime I. Industry Canada and the various biotechnology committees may well be the nominal central catalyst but it remains the case that Regime I is weak in the biotechnology case.

Regime II refers to a set of regulators such as environmental, competition, intellectual property, and consumer regulators (and many others) whose mandates are horizontal or framework oriented. Therefore, its core logic, in principle at least, is that such regulatory mandates are intended to cut across economic sectors and not discriminate between sectors. Such realms are rarely as pure as this description suggests, but there is little doubt that Regime II is of growing importance and certainly is a crucial part of most countries' regulatory institutions. The federal biotechnology regulatory system is in many ways a part of Regime II in that multiple intersecting departmental and product-based health, safety, and environmental regulatory processes characterize it. As we have seen above, however, the biotechnology regulatory system is dispersed across parts of this regime.

Regime III embraces the players and processes that exist in an even broader cross-governmental sense, particularly within the executive levels of government. This is often the least appreciated regime in conventional regulatory literature perhaps because it is so easy, at first glance, to think of regulation only in terms of well-known regulatory bodies. But the core logic of Regime III is simple and compelling. Regime III forces an understanding of how the regulatory function is managed within the executive or even within the state as a whole including the courts (Jacobs, 1992; Martin, 1995; McGarity, 1991; Stanbury, 1992). In the first instance, this leads to a focus on so-called delegated legislation or the "regs" that can be made under the authority granted by hundreds of parent statutes. Every government has some kind of process for "managing" regulation and this process can exert independent influence. In the biotechnology regulatory system, Regime III may exert its influence through pressures to regulate in certain ways (incentive-based versus command-and-control), to cut back staff (or occasionally expand staff), and to respond to government-wide concerns about regulation (e.g., the paper burden effects on small business, or requiring regulators to charge for their services through fees).

Regime IV, the international regulatory regime, can, in one sense, be seen as a virtual commonplace 1990s' reflection of contemporary globalization and of the internationalization of most areas of public policy, including regulation (Doern, Pal, & Tomlin, 1996; Trebilcock & Howse, 1995). Regulators are increasingly constrained by, or must interact with, a potential array of (a) international agencies (regulatory and otherwise) and other countries' national regulatory counterpart bodies; (b) international rather than just national coalitions of interest groups; and (c) international rule-based dispute settlement processes or prescribed cross-boundary consultation processes (Trebilcock & Howse, 1995). Pressures to harmonize or to mutually recognize rules and to reduce the costs and/or increase the speed of regulation can arise as large international firms "shop" globally for the best entry points for regulatory approval.

The biotechnology regulatory framework was deliberately based on a decision not to create a separate new regulatory agency and hence in a very real sense its creators knew that it was embedding the biotechnology framework in a set of existing regulators. In part this decision was perfectly understandable in that it would probably be possible only with hindsight to know the precise moment when a food product went from being conventionally to biotechnologically produced. But the above mentioned macro four-regime model suggests that the biotechnology framework needs to be seen as being located in an even more complex system than that. Some of these deeper four-regime influences and characteristics will be drawn out as we proceed.

The brief discussion above of the three core functions and the secondary set of four macro regimes suggests the obvious existence of complexity and overlap. This will be evident in the sections to follow but two examples are indicative of the subtle kinds of interplay involved. First, in ordinary circumstances, it would be expected that overall statutes, regulations, and guidelines (the first process set out above) would govern and drive decisions in individual cases. However, a particular biotechnology case such as the rbST case may, in fact, drive regulatory policy in that it creates a type of political-technical controversy which sets "precedents" or which simply drives decision-making out of the normal regulatory cycle, in this case into the arms of a Parliamentary Committee (Standing Committee on Agriculture and Agri-Food, 1994).

Synthetic bovine somatotropin (rbST) is a biotechnologically developed product with "equivalence to a naturally occurring bovine growth hormone" (Standing Committee on Agriculture and Agri-Food, 1994, p. 1). It stimulates milk production in adult cows. It had proceeded along the normal biotechnology regulatory cycle but became the subject of hearings and a report by the House of Commons Standing Committee on Agriculture and Agri-Food which recommended in 1994 a one year moratorium on the approval of the product, ending July 1, 1995. As mentioned above, this is also one of the committees where the wider NGO community mounted a significant criticism of what they saw as an industry dominated biotechnology regulatory system.

Or, consider a second practical example. In ordinary circumstances cases might proceed in a regular manner with the regulator applying its normal technical capacities to the case at hand. But if the regulator genuinely lacks the needed up-to-date technical capacity or has cut staff, it may literally be the case that it cannot, in this instance, act with competence. As a result the system simultaneously fails both the firm's commercial interests and the public interest.

In the sections below, the relations among the three core functions of the biotechnology regulatory system must also be seen in the context of the different operating cultures of the different departments. The political-economic and organizational cultures of environment, agriculture-food, health, and industry ministries are part of the story. And such cultures are partly a product of (a) positive pride and belief in their tried and tested approaches and technical expertise and (b) bureaucratic turf wars in an era of shrinking governments.

The brief discussion above of key analytical features of regulatory systems suggests why the six principles of the federal biotechnology framework are important, but also why they are an insufficient base for analysis. The six principles undoubtedly convey important ideas and goals that lie behind the biotechnology framework, but they do not themselves show how an actual regulatory system can work or fail to work. It is in this context that we can now look more closely, but still only illustratively, at the actual federal biotechnology regulatory system.

THE THREE CORE FUNCTIONS OF THE BIOTECHNOLOGY REGULATORY SYSTEM

Legislative Mandates, Policy, and Regulation Making

This section is largely descriptive with only periodic analytical observations. It is obviously essential to have an initial sense of the basics of the system emanating from each department's role. Analytical commentary increases as we move into the case handling and compliance and enforcement sections for the simple reason that by then we will have more and more of the full regulatory system "digested" and in view.

In Canada, legislative responsibility for biotechnology is divided primarily among four departments: Environment Canada, Agriculture and Agri-Food Canada, Health Canada, and the Department of Fisheries and Oceans. In addition, as we have seen, Industry Canada partially coordinates policy development activities related to biotechnology strategy across government departments and has been a strong force within the government in seeking to establish and strengthen an internationally competitive biotechnology industry.

The current Canadian regulation system for biotechnology uses a two-step process to ensure health and safety. Section 26(3)(a) of the *Canadian Environmental Protection Act* (CEPA) outlines an "equivalency" clause, whereby all pertinent federal legislation must comply with the notification and environmental and human health assessment requirements outlined in CEPA as it applies to an assessment of a toxic substance. In other words, meeting CEPA notification and assessment requirements is the condition for products not being subject to notification under CEPA. The notification requirements in subsections 1 and 2 of Section 26 "do not apply in respect of a substance that is manufactured or imported for a use that is regulated under any other Act of Parliament that provides for notice to be given prior to the manufacture, import or sale of the substance and for an assessment of whether it is toxic." It is worth noting that "assessment" does not mean an environmental assessment as would be understood for the purposes of the *Canadian Environmental Assessment Act* (CEAA). CEPA, however, is the ultimate back-up of the biotechnology regulatory system in that if the legislation of other departments fails to meet these provisions, CEPA automatically "kicks in" as a type of safety-net legislation thereby ensuring a minimum environmental, health, and notification standard across government. However, as the Standing Committee on Environmant and Sustainable Development (Biotechnology regulation in Canada, 1996, Recommendation 1) put it, "the concept of equivalence or harmonization does not mean that the same precautionary tests and data are required for all products, but that appropriately comprehensive and stringent standards shall apply to all products, whichever legislation those products fall under."

In this way, each of the four regulatory departments are intrinsically connected but there is considerable debate among agencies and the interest groups that scrutinize them about what substantive equivalent standards are since this goes beyond the "equivalence" provisions of CEPA. The following paragraphs briefly describe the regulatory mandates and requirements associated with each of the four departments and reveal some of the policy and regulatory controversies. Other key aspects of interdepartmental politics and operating cultures are brought out later.

Environment Canada. Environment Canada (and Health Canada) assesses substances rather than products under CEPA. Changes to CEPA, introduced into the House of Commons on December 10, 1996, identify biotechnology substances separately but, as of late 1997, these changes had not been passed. Specifically, Part VI requires that biotechnology substances and/or products pertaining to "living organisms" undergo a notification and assessment of toxic process.[3] An exemption mechanism is also established in sections 106–109 of the legislation. The new biotechnology provisions also require that all living organisms not on the Domestic Substances List be the subject of notification.

CEPA does not find a substance to be toxic or non toxic. Instead a substance is found to be suspected or not of being toxic. Assessment of toxicity in this sense is comprised of two parts: potential exposure and potential hazard. A substance is suspected as being toxic if it

> is entering or may enter the environment in a quantity or concentration or under conditions:
> (a) having or that may have an immediate or long-term harmful effect in the environment;
> (b) constituting or that may constitute a danger to the environment on which human life depends; or
> (c) constituting or that may constitute a danger in Canada to human life or health.[4]

If a product is suspected of being toxic then additional measures are taken to mitigate or confirm any concerns. If a product is suspected of being toxic, then it cannot be placed on the Domestic Substances List but can be manufactured or imported subject to conditions imposed under Section 29. Domestic Substances List refers to a list of substances (although mostly chemicals to date) that do not require notification under CEPA. If a product is assessed as not being suspected of being toxic then it has the potential of being listed on the Domestic Substances List.

Department of Fisheries and Oceans. Currently biotechnology applications related to fish and marine animals are quite minimal and are limited to work done in registered laboratories with appropriate containment procedures and facilities and adequate security, to prevent accidental escape or intentional removal of the transgenic organisms.[5] In anticipation of a future greater role, and as the department responsible under the *Fisheries Act* for the conservation of fish resources and their habitat, Fisheries and Oceans Canada plays an active role in the federal biotechnology regulatory framework.

The Department of Fisheries and Oceans has stated that it expects the use of transgenic fish in aquaculture to appear in the next five to ten years. Specifically, three areas are expected to benefit from genetic engineering technology: indirect application such as drugs, direct applications such as changing the characteristics of fish, and creation of novel organisms or transgenic organisms by the transfer of genes between fish species (Department of Fisheries and Oceans, 1996, p. 2). As such, the Department is prepared to translate policy into regulations as the need arises, perhaps in one or two years. In the interim, it is preparing a policy and guidelines document to begin the process of dealing with new developments regarding CEPA.[6] The Department of Fisheries and Oceans has stated that the intent of its policy is to "ensure wild resources are protected while encouraging the expansion of categories of organisms available for aquaculture. It is not intended to inhibit or control research nor to unnecessarily inhibit use of the transgenic organisms . . . but prevent . . . deleterious impacts of transgenic aquatic organisms" (1996, p. 6).

Health Canada. Health Canada's biotechnology legislative responsibilities include its joint role with Environment Canada regarding CEPA. But beyond this, they fall under the *Food and Drugs Act* and the *Pest Control Products Act.* The former oversees the safety of a wide array of consumer products including food, food additives, drugs, medical devices, and cosmetics. Many of these items currently use biotechnology procedures with many more applications likely to emerge in the future. In order to comply with the notification regulations specified under CEPA, Health Canada is in the process of developing new regulations to guide the required assessment of biotechnology products.

Health Canada is also developing regulations to outline a notification procedure with respect to novel food products.[7] Novel food products is a broader category than biotechnology products but does include the latter. Proposed regulations will require that Health Canada be notified of any novel food product 90 days prior to the sale or advertising of sale of the product. In addition, a novel food can be required to undergo an assessment process if the regulator finds it necessary to do so. In this instance, the producer of the novel food would be required to submit information related to the product's safety for consumption prior to obtaining permission to put the product on the market.

In addition, the Pest Management Regulatory Agency (PMRA) was created in 1995 to provide an umbrella agency to oversee the *Pest Control Products Act.* The PMRA controls the manufacture, sale, and use of pest control products, including those developed through biotechnology. Prior to 1995, responsibilities had been spread across four departments.

Agriculture and Agri-Food Canada. As the government department responsible for agricultural products, Agriculture and Agri-Food Canada (AAFC) has historically been mandated with regulating agricultural products of biotechnology and has been doing so since 1988. As noted in a report by AAFC, agricultural regulations confirm for consumers that "new products are effective and safe to humans, animals and the environment . . . regulations also assist Canadian companies in maintaining the quality and effectiveness of products that are traded internationally."[8]

Agriculture and Agri-Food Canada is also in the process of implementing regulations that will comply with the environmental and human health assessment issues outlined in CEPA. Amended regulations were published in Canada Gazette Part I on August 17, 1996 and final amended regulations went into force on December 19, 1996.

Importantly, since April 1997 the newly formed Canadian Food Inspection Agency (see more below) has taken over major regulatory functions. Historically, however, Agriculture and Agri-Food Canada had identified four principles to guide their regulation:

(i) Regulation based on the characteristics of the products (including a review of the process by which it is made)
(ii) Science-based risk assessments
(iii) Protection of health and the environment
(iv) Building on existing legislation and areas of responsibility.[9]

Generally, this approach recognizes and builds on the knowledge, expertise, and infrastructure already present in the traditional regulatory areas and is based on the concept that new products of biotechnology should be regulated on the same basis as conventionally derived products. In short, they are regulated on their characteristics as products which includes a thorough evaluation of biotechnology products. It also recognizes the principles of risk assessment, and uses the concepts of familiarity, substantial equivalence, and information feedback.[10]

More specifically, Agriculture and Agri-Food Canada was responsible for several federal acts that relate to regulation of biotechnology products: The *Fertilizers Act*, the *Feeds Act*, the *Seeds Act*, the *Health of Animals Act*, and the *Plant Protection Act*. These are now the responsibility of the new Canadian Food Inspection Agency. Each of these statutes with respect to biotechnology are outlined below.

A product-based act, the *Fertilizers Act* requires that manufacturers demonstrate the safety and effectiveness of fertilizers and supplements (including biofertilizers) both in terms of human health and environmental safety. Specifically, the legislation outlines four key activities: standards and labelling, experimental research (safety assessment required), registration (safety assessment required), and post-registration monitoring.

Feeds are defined in the *Feeds Act* as "any substance or mixture of substances manufactured, sold, or represented for use or consumption by livestock, for providing the nutritional requirements of livestock, or for the purpose of preventing or correcting nutritional disorders of livestock."[11] The legislation specifies that all single ingredient feeds be evaluated prior to their use in livestock feeds. This applies to imported or domestically manufactured products. For the experimental use of any novel feed (imported or not) authorization is required. These feeds are also subject to safety assessment for environmental and animal health and safety.[12]

The regulations require the submission of information to permit assessment review focusing on toxicity to livestock, human safety in terms of transfer of harmful residues to human foodstuffs, safety of workers handling feeds, and safety to the environment.

The *Seeds Act* regulates the inspection, testing, quality, and sale of seeds in Canada. Seeds that are developed through the genetic engineering process also undergo the same requirements. The regulations allow for Agriculture and Agri-Food Canada to refuse or cancel the release of a novel plant,[13] that may be damaging to human or animal health and safety of the environment.[14]

Testing involves field testing under both confined and unconfined conditions (larger commercial production). Agriculture and Agri-Food Canada also carry out environmental assessments on plants with novel traits including thorough characterizations of the novel proteins and the modified plant, considerations of weediness, ability to pass genetic information, potential to become a pest, potential to cause unwanted interactions with other organisms in the environment, and potential to cause negative impact on biodiversity. In the end, "the process of developing a plant is examined in relation to the effect on the safety of the final product."[15]

The *Health of Animals Act* and the *Plant Protection Act* are laws which broadly control the importation of "all organisms that could be injurious to animals or plants, regardless of the breeding method or process used to produce them"[16] and animal vaccines and disease diagnostic kits to provide that "veterinary biologics are safe, pure, potent, and efficacious, and are based upon a risk assessment."[17] "The (legislation) and regulations specifically provides authority to regulate the production, importation, field testing, and registration of veterinary biologics in Canada, produced by modern techniques of biotechnology or by traditional methods."[18] The law on plant protection was not amended to include regulations for conducting environmental and health assessments. Only the veterinary biologics portion of the *Health of Animals Act* was so amended in December 1996.

Industry Canada. As we have seen, Industry Canada does not have major legislative responsibilities with respect to biotechnology. However, Industry Canada has been given responsibility for coordination of aspects of policy development across departments through the committees and mechanisms mentioned earlier. In some respects this role for Industry Canada emerged out of its own movement into a political vacuum. The various health and environmental regulators had an interest in policy on their regulatory mandates but no one, with the exception of Agriculture and Agri-Food Canada (see more below) had a focussed concern about fostering the development of a biotechnology industry. It is this area of policy or strategic initiative that Industry Canada made its own at the urging of the biotechnology industry lobby.

In addition, however, Industry Canada's Intellectual Property Policy Directorate does have responsibility for developing the government's policy with respect to patents and patent law. At present, patenting of recombinant DNA *higher* life forms is not allowed in Canada, but in the future, this could be an area where Industry Canada will be involved legislatively with biotechnology.

Joint consultation processes. The above descriptive account of the key departments indicates both longer term biotechnology regulatory development and activity in some departments such as Agriculture and Agri-Food Canada and virtually none in the fisheries sector. Much of the recent regulation making (both in the statute and as regulations and guidelines) was centred around the CEPA amendment process in which biotechnology was a key concern. This de facto regulation making process was carried out in a consultative manner among the key departments and among the key stakeholder interests. But assessments of the genuineness and efficacy of the consultation vary enormously, with the strongest criticism coming, as we have seen, from the environmental lobby which strongly opposed what it saw as the gutting of the CEPA legislation and regulations in particular and the failure to involve the public in any fully legitimate way.

Within the federal government, however, there was a joint process among departments to ensure that their respective changes went through regulatory review and assessment and were published in the Canada Gazette in a coordinated fashion. This process also triggered consultation processes with a range of interests and interest groups, groups which varied among the four sectoral areas of agriculture, health, environment, and fisheries. The federal government's various biotechnology coordinating groups were also important in this regard.

All of this suggests that there was a quite elaborate consultation process but environmental groups in particular saw the process as flawed from the start with the main evidence of a failed process being that CEPA was not the centre-piece of the biotechnology regulatory system and that CEPA was itself weakened in their eyes by a powerful industry lobby and by other economic ministries and ministers in the federal Cabinet.

Space simply does not permit an account of what the interest group constituency of the biotechnology regulatory field totally consists of. Aspects of it are concerted and powerful such as pharmaceutical and chemical interest groups and key companies. They have pressured Health Canada to revise the August 1995 proposed regulations under the *Food and Drug Act* and strongly opposed any biotechnology regime that would have given primacy to CEPA. In this regard they had the support of Industry Canada in the internal government debate and lobbying.

Other actors such as environmental groups are more diffuse but vocal and aligned, to some extent at least, with departments such as Environment Canada. Indeed, this coalition of interests extends well beyond environmentalists to church, animal protection, consumer, and labour groups whose concerns extend to broader ethical issues regarding biotechnology.

The Case Handling and Approval Processes

The case handling and approval process of the federal biotechnology regulatory system is in one sense simple, but in all other senses is complex and can only be described and seen as several processes. The simple notion of their being one overall process derives from the fact that the biotechnology producer goes first to the relevant sectoral regulator in the agricultural, health, or fisheries departments, with Environment Canada and its CEPA legislation held in reserve as the safety-net regulator. The biotechnology producer's entree to the sectoral regulator is through the provisions of the relevant sectoral statute covering the product.

Because it is so central to their commercial or research interests, biotechnology producers know first-hand what the regulatory road map looks like and what the right turns, blind alleys, and cul de sacs are likely to be. In real terms, however, the case handling road map is not objectively very clear or self-evident for those other interests which may be more occasional participants in the regulatory process or whose own incentive systems make it difficult for them to know the regulatory regime as well as business interests do. A particular example of this can be found in a recent effort by the Labour Branch of Human Resources Development Canada to supply a biotechnology regulatory roadmap (Human Resources Development Canada, 1996). Constructed in part from the vantage point of labour and occupational health concerns, the map covers the regulations involved, not the more precise case handling processes, and it involves an even wider array of potential statutes than those highlighted in the previous section.

The brief discussion below of the case handling dynamics of three departments[19] will also show that the case handling process is in fact a series of processes (or roads). However, before looking at this it is useful to ask a prior question. What kind of volume of cases is the biotechnology regulatory system currently handling? In many respects this is a way of asking what the biotechnology regulator's "unit of work" and work load is. This is important because we are dealing with a biotechnology regulatory system that is relatively new but which is also appended to, and embedded in, several existing regulatory systems (and, as noted, even larger macro regulatory regimes).

Complexity arises because the answer to this question about volume is that "it depends" on the particular sector and on its constituent parts and statutes. The biotechnology cases handled in Health Canada are around 20 to date whereas in Agriculture Canada the case count could be in the hundreds or even thousands if one is thinking of approvals of field trials, or as few as 41 if one was thinking of licensed biotechnology-derived veterinary biologic products (Agriculture and Agri-Food Canada, 1996, p. 6). The case load at Environment Canada will eventually have to sort out mainstream CEPA cases (e.g., on toxic substances) from those that are residual biotechnology cases (at present, none).

The case handling dynamics are also complex because of the different regulatory cultures and realities of each of the sectoral regulators. Consider first the example of Agriculture and Agri-Food Canada. Operating under its five different statutes, it is clear that the department sees biotechnology as basically a broadened aspect of the other assessments that it has been doing for decades. Its regulatory focus is on asking whether there is something novel or unique about the organism. Accordingly it focuses on *product* characteristics but it also looks at the method of production. In an overall sense it seeks to establish appropriate safety levels based on the best scientific information with safety seen to embrace a concept of "acceptable risk."[20]

The agri-food department as regulator has seen some demands from business for one-stop or single-window regulation in the biotechnology field. Business thus certainly seeks an efficient process but it also can prefer the strategic option of having more than one point of entry and leverage. The Agriculture and Agri-Food Canada regulator has often offered to firms that it will be the "lead agency" on the firm's behalf. The new Canadian Food Inspection Agency will now act as lead agency on the firm's behalf, to better ensure interdepartmental coordination where multiple regulatory meetings and assessments are needed. Traditionally, companies have gone to Health Canada separately for food safety assessments.

In general, Agriculture and Agri-Food Canada has a higher volume of biotechnology business than other biotechnology regulators. In part this is due to Canada's quite well developed agriculture research community. It could be argued that the political culture of the department is such that it is more sympathetic to pro-agriculture industry needs rather than its "health and safety" clientele interests compared to Health Canada or Environment Canada and their interest group clientele. Agriculture and Agri-Food Canada regulators strongly dispute this charge but it is certainly what is meant by environmental critics who have argued that Agriculture Canada has a regulatory conflict of interest. This may have been closer to the truth when Agriculture Canada was more purely an "agriculture" department (Hill, 1994; Skogstad, 1987). But with its merger with a more explicit food manufacturer's mandate in the 1990s, its health and safety concerns as Agriculture and Agri-Food Canada are undoubtedly increased and more salient relative to the earlier period. The formation of the more arms-length Canadian Food Inspection Agency also reduces the potential for conflicts of interest of this basic kind. The higher volume of biotechnology cases in the agriculture sector may also simply be because it is technically easier to experiment with biotechnology products in the agriculture field. For example, herbicide tolerance has been an easier and definitive screen for the presence or absence of a transformed gene. In addition, plant products are much easier to work with in the scientific sense that you can produce as many as three generations a year. Furthermore, there are no ethics committees to deal with.

Consider next the situation in Health Canada and its regulatory milieu for case handling. Health Canada is similar to its agricultural counterpart in that it has, as we have seen, a multi-statute jurisdiction and hence potential multiple case cycles. It has direct responsibilities for CEPA. But in its own domain it manages the biotechnology case process through two regulatory bodies, its long-standing Health Protection Branch and the recently added (to the department) Pest Management Regulatory Agency. Health Canada also regulates on a product basis using a risk assessment and risk management approach to identify and manage risks to human health and environment.

Health Canada's regulators see themselves as a public health and public interest oriented regulator. Since the early 1980s, it has been able to give institutional focus to aspects of biotechnology, through its biologics program and unit in the Health Protection Branch (Health and Welfare Canada, 1991). Approximately 20 biotechnology products have been approved in recent years and the department has sought to produce single-window regulatory assessment. This has been the goal not only in its own cases where more than one of its statutes might be involved in a given case product, but also interdepartmentally, among other sectoral biotechnology regulators.

It is difficult to disassociate the biotechnology case cycles from the larger health protection case cycles of which they are a part. And in this larger sense, there is no question as to what the direction of pressure has been on Health Canada. In its overall drug regulation role as a whole, the pharmaceutical business lobby has for a decade been pressing for faster and more efficient approval times. Quicker approval processes also come from other particular lobbies such as that on AIDS.

Business pressure, especially from the pharmaceutical industry, has centered on how to set more internationally competitive regulatory approval completion time targets to reduce the average assessment period of 30 months to about 18 months, in line with what some other country's regulators have already achieved. This is seen as vital to a competitive industry and to attracting investment capital since firms have many places where they can locate and/or develop products from a research and development base.

The connecting link here is increasingly centred on how the

regulator is funded. The Health Protection Branch historically has been funded through tax dollars and appropriations. It is now funded up to 60 percent on a cost-recovery basis. Thus industry, through charges and fees, is increasingly paying the bills and wanting to achieve more efficient regulatory throughput.

Business pressure on case loads must be seen, however, in some realistic context. The drug producers and biotechnology producers benefit from the regulator's role in reassuring the public about human health and safety and in weeding out unsafe drugs and products. Thus business pressure in the health industries still seeks effective regulation, not just fast regulation.

Health Canada, nonetheless, has had differences of view on particular biotechnology cases. The rbST case comes to mind here in that Health Canada was institutionally more skeptical and cautious than Industry Canada whose core instinct was to see the approval go ahead as a positive signal to the new biotechnology industry (Standing Committee on Agriculture and Agri-Food, 1994). In the end the Minister of Industry agreed with his Health counterpart in 1994 that any approval of rbST should be postponed for a year pending further analysis. The Parliamentary Committee on Agriculture and Agri-Food itself was influential not only in being cautious about biotechnology but also because the then relatively new Chretien Government had promised that Parliamentary committees would be given a greater independent role in scrutinizing government and in influencing policy.

Finally, we look briefly at Environment Canada. Its role in biotechnology case handling is as the backstop or safety-net regulator.

Environment Canada and other environmental groups had sought a primary role in biotechnology regulation. CEPA, according to this view, would have supplied the umbrella legislation where all products would go *first* to CEPA and then to the sectoral regulators which are the "use specialists."

Environment Canada's role is very much contextualized by the evolution of CEPA as a piece of legislation and the fear of other federal departments that Environment Canada would have too much authority and power and would be applying different standards of assessment and risk (Doern & Conway, 1994). CEPA in its initial form (for assessing chemicals or toxic substances) made no reference to products or to use as did the other sectoral regulators. It assessed substances. Biotechnology regulations are treated as a class of substances for the purposes of CEPA. The biotechnology regulations are technically an amendment to the New Substances Notification Regulations. The basic trigger for notification under the CEPA regulation is the same as for chemicals, whether it is on the Domestic Substances List or not. If it is not, then the substance is notifiable. Nominally this may seem to be a quite different approach to that of Agriculture and Agri-Food Canada which is based on "novelty." To some interests this difference in terminology may be confusing but some officials in Environment Canada see little difference in practice in that the Domestic Substances List is seen simply as a tool to distinguish old from new, i.e., novelty.

Environment Canada knows that it has the safety-net role but this by no means resolves all the case handling dilemmas. This is because the key is what the term "assessment" for suspected toxic effects means in practice. The Department of Justice's interpretation, given as a policy ruling to departments, was that assessment meant a form of "equivalence" but without specifying what that might be regarding the other ministries and statutes. But this is only an opinion. Court cases may eventually rule differently.

There are also some particular bilateral tensions and differences between Environment Canada and Agriculture and Agri-Food Canada (and the Canadian Food Inspection Agency) over styles of regulation that can affect case handling and which are centered in part on the use of guidelines versus regulations (delegated legislation). Some critics argue, mistakenly, that section 26.3 of CEPA requires that some provisions and information requirements of other Acts must be in the form of actual regulations which would then go to Cabinet through the regular regulatory review process.

The reality is that different statutes have different provisions regarding control and flexibility. For example, the New Substances provisions of CEPA do not provide for approval for manufacture or import of a substance. CEPA specifies that a person must provide information prescribed in regulations and wait for an assessment period to expire. There is no requirement to have approval to manufacture or import. The government does, however, have the opportunity to prohibit or control the substance or ask for more information.

Though CEPA requires for its purposes that data requirements are spelled out in regulation, this is not the case in many other statutes regulating biotechnology. Thus, Agriculture and Agri-Food Canada tends to use more flexible regulatory guidelines for some information provisions or a combination of regulations and guidelines. Clearly, this may simply reflect different approaches to regulation regarding how much detail is in regulations versus guidelines.

Agriculture and Agri-Food Canada's approach to its part of biotechnology regulation has also been criticized by some environmentalists regarding the Department's actual legal authority. This criticism centres on a view that existing agriculture-food statutes were established to promote product quality regulation and not a broader concept of environmental and health protection. Hence, it is argued that the Department does not have the legal authority to regulate in this broader fashion. Agriculture and Agri-Food Canada regulators dispute this claim and have proceeded on the basis of its basic statutes. Thus questions are quite naturally raised as to whether Agriculture and Agri-Food Canada's regulations, guidelines, and provisions are "equivalent" to those of CEPA since there are differences in the two sets of laws regarding what is notifiable, and what factors, at what level of detail, are considered in the assessment. The same is true for Health Canada and its various statutes, although regarding CEPA, it must be remembered that Health Canada has a joint responsibility with Environment Canada and its use of regulations in the CEPA aspects indicated above.

While our interest here is in the biotechnology regulatory case handling process, there is little doubt that larger complex pressures are also at work. CEPA's degree of coverage (not in biotechnology but in an overall sense) has also been affected by federal-provincial concerns about jurisdiction and about even larger issues of equivalence where provincial actions are being judged (Harrison, 1994, 1996). Furthermore, in its larger sustainable development mandate, Environment Canada faces genuine dilemmas about how forcefully it can assert its jurisdiction and about how hard other departments will fight back. Having a sustainable development mandate is of course not just an Environment Canada task as such, other departments have such mandates too. But Environment Canada is clearly seen as having led the pressure to extend this on a government-wide basis. Moreover, in more particular ways, there are also concerns in other departments that Environment ministers, as distinct from Environment officials, are too captured by environmental lobbyists who might want to make a *cause célèbre* of a particular biotechnology or other environmental "case" that happens to pop up onto the agenda at any given time.

Finally, there is a sense in other regulatory departments that, even through Environment Canada has a safety-net role in biotechnology, it still has plenty to do in its other CEPA tasks where its case load is considerable and which will only grow in the next few years. In other words, the view in other departments is that there is a lot of overall case work (biotechnology and otherwise) to be regulated by the various regulators and that Environment Canada ought to tend to its own knitting.

The above discussion of biotechnology case handling cycles and processes must be seen as very basic and illustrative. Glimpses of varied cycles can be seen within departments and among them, but always in a way that must be linked to the larger work loads of the broader overall regulatory mandates. The three departments are indeed the biotechnology regulatory case managers but their primary business is still not biotechnology regulation. Moreover, the underlying differences in work units or basic rhythms of regulatory business must be appreciated. As we have seen, regulating *products* is different from regulating *substances* and there are widely varying traditions and needs regarding regulatory flexibility and how it is managed.

Some parts of the biotechnology industry would appear to favour a focus on a broader process approach because by definition it suggests larger clusters of approval, in short, fewer cases and fewer regulatory hoops to go though. Such preferences when suggested informally have been resisted by some federal sectoral regulators.

Compliance and Enforcement Processes

As noted earlier, the third aspect of the biotechnology or any other regulatory system – the compliance and enforcement process – should in fact include the case handling and approval process. However, since compliance and enforcement goes beyond such case handling processes and includes many other kinds of regulatory implementation activity, it is important to draw attention to this wider, denser, and often quite subtle world (Doern, 1994; Law Reform Commission, 1986). In the context of again looking illustratively at the same three federal biotechnology regulators, we draw attention to, and raise issues regarding, three aspects of this fuller spectrum of compliance and enforcement activity: (a) the role of education, information and awareness, (b) actual enforcement activities, and (c) the adequacy of the regulator's scientific personnel in supplying real operational capacity to carry out risk-assessment activity. These aspects must in turn be situated in the practical context of government in the late 1990s where governments are searching for forms of regulatory implementation that are less "command-and-control" oriented and more "incentive-based" (Doern, 1994; Grabosky, 1995; Sparrow, 1994).

All three of the biotechnology regulators examined above are devoting increasing proportions of their resources (money and personnel) to broadly educative and informational activity. This activity can be seen broadly as activity that is, on the one hand, intended to improve the underlying public awareness of the benefits and disadvantages of biotechnology products, and, on the other, seeks to inform people about current laws, regulations, and guidelines.

For example, Agriculture and Agri-Food Canada provides the public with information on approved field testing of genetically engineered material. Labelling of products is another compliance-oriented method of informing consumers about product characteristics. Some of this activity is also based on putting into place more "service oriented" aspects of regulation which are better geared to the needs of quite particular clients or customers and citizens.

Environment Canada in one sense has had less compliance experience to date with biotechnology because the New Substances Notification Regulations only came into force on September 1, 1997 and the renewed CEPA has not been given final passage by Parliament. The department, however, has probably had a more intensive debate internally than other departments about the "compliance versus enforcement" or softer versus hard approaches to implementation regarding its regulatory mandate as a whole. Environment ministers are under pressure from environmentalists to come down hard on polluters while the "Ottawa system," and some key parts of Environment Canada, prefer more varied compliance and flexible incentive based approaches. At present, officials estimate that over 50 percent of Environment Canada's enforcement personnel are devoted to compliance promotion activities. It remains to be seen how the enforcement versus compliance trade-offs will be played out in the biotechnology aspects over which Environment Canada has jurisdiction.

Both Health Canada and Agriculture and Agri-Food Canada also carry out activities which are closer to the enforcement end of the continuum of implementation activity. Quality assurance monitoring and inspection activities are a part of the Agriculture and Agri-Food approach. In the case of Health Canada, these approaches are also important and they are backed up by vital post-use information and reporting from physicians and other health care practitioners.

Another recent institutional change to take note of in the context of compliance is the recent establishment of the Canadian Food Inspection Agency (CFIA). It deals with much more than biotechnology but it was established to enhance food safety systems by integrating the inspection and quarantine services supplied by Agriculture and Agri-Food Canada, Health Canada, and Fisheries and Oceans Canada (Doering, 1996).

In considering the overall compliance and enforcement aspect of the biotechnology regulatory system, it is crucial to reiterate that underlying it is a science-based and informed or objective knowledge-based *capacity* to regulate.

At an analytical and philosophical level, issues of capacity start with debates about what kind of "risk assessment" is to occur both in developing rules and standards and in handling cases, and in deciding how to educate and promote awareness (Breyer, 1993; Harrison & Hoberg, 1994; Mehta, 1996). As we have seen, it seems clear, on the surface at least, that the federal biotechnology system has adopted some version of an "acceptable risk" concept rather than a "zero risk" or even "risk-benefit" approach.

But the translation of acceptable risk as a concept turns fundamentally on whether the scientific personnel and research and information are available to the regulators and whether such personnel are given appropriate training. This issue of availability is a complex question. One aspect which can be raised regarding the three key federal biotechnology regulators is whether the general ageing of their scientific regulatory staff is an indicator of reduced capacity in this regard. An older staff is, on the one hand, more experienced in the full skills of science-based regulation. But on the other hand, a younger age profile may indicate a more up-to-date knowledge of key scientific changes. Another factor regarding technical capacity goes beyond age or being up to date and centers instead on the inherent professional backgrounds of staff. This view suggests that some professional or scientific disciplines are simply too reductionist and thus downplay or lack knowledge in how to assess ecological or synergistic effects. Again such claims and counterclaims are difficult to prove one way or the other in the limited context of this paper.

These issues cannot be easily separated from the impact of budgetary cuts and person-year cuts in the three departments and their regulatory operations. Interviews suggest that cuts have occurred, a fact which would hardly be surprising given federal cutbacks in general (Swimmer, 1996). Health Canada has absorbed heavy cuts in budgets, linked to the aforementioned requirements to obtain more funds from user fees and charges paid for mainly by business firms. Environment Canada has had cuts as well, but in the early 1990s it did quite well because of the availability of Green Plan funding.

CONCLUSIONS

This paper has examined the Canadian biotechnology regulatory system in a three step process: first, an account of the content and origins of the Federal Biotechnology Regulatory Framework; second, the general identification of the core functions and surrounding regimes that exist in most regulatory systems; and third, a closer look at the core functions of the biotechnology regulatory system itself: regulation making, case handling processes, and compliance and enforcement.

The "core functions" approach provides an especially needed focus for the analysis of the biotechnology regulatory system because it does not have, as do many regulatory areas, a core regulatory body. This is one of the reasons why we have cast the Federal Biotechnology Regulatory Framework and the biotechnology regulatory system as a regulatory "work in progress." It is also a work in progress in the sense that it is not clear that it meets the dual general test of any science-base regulator, namely, some reasonable overall political legitimacy and an on-going scientific and technical capacity. The paper has not fully addressed all the factors that might go into such a judgement and hence the authors must offer cautious and qualified conclusions.

The underlying principles of the biotechnology regulatory framework are in many respects sensible and supportable but it is not yet a full-fledged regulatory system in which a wide range of interests have full confidence. For example, the current system has no regular or established arena in which ethical issues can be addressed. Such issues have been raised in Parliamentary committee arenas but they are not a part of the current system or its statement of purposes.

The analysis has shown, however, that a quite wide range of criteria are part of the inherent spectrum of debate. This has been most evident in the main implied polar alternatives for the biotechnology regime. An Environment Canada and CEPA-centered regime is the broadly preferred model for the environmental lobby writ large. In this system, CEPA would be the first stop for biotechnology firms and the sectoral regulators the second stop. One would therefore in theory have more of a focused accountable "centre" to the system. However, such an Environment Canada-headed system would be anathema to key biotechnology business interests who prefer the more decentralized system where CEPA is the safety net.

Neither the current biotechnology regulatory system nor its implied polar opposite model can be fully understood or their actual or potential defects explained without seeing them as being embedded in a much larger array of macro regulatory regimes: sectoral, framework, government-wide, and international. For example, the claimed and actual virtues of the flexibility of the present Canadian biotechnology regulatory system should not be trumpeted unduly unless questions are asked that go beyond the six principles of the biotechnology regulatory framework and embrace the main features of any regulatory process as highlighted in the paper.

The basic statutory foundations and the regulation making process for biotechnology in the early to mid-1990s reflect a reasonably good internal consultative approach among federal departments. The basic two-step sectoral-centered regime with CEPA as the backstop can be defended but is not without controversy and indeed was only arrived at after an intensive intra-Cabinet and interdepartmental battle. Environmental critics, however, see the internal process among departments as having been dominated by industry and Industry Canada and Agriculture and Agri-Food Canada. They see the external consultation process with non-governmental and non-business groups as being very weak and maintain that the real import of biotechnology industries have not yet been seriously debated in Canada, including the ethical issues.

The analysis of the case handling and approval process shows that not all interests have a clear sense of the basic biotechnology regulatory road map, let alone the many side roads. Business interests have a surer sense of the regulatory road than have other interests, a situation that needs further institutional change and resource commitments to enhance elementary accountability.

The compliance aspects of biotechnology regulation do not appear to be beset by extreme contrasts in perceptions of risk. Acceptable risk and the need for good science-based regulation seem to be supported. Other aspects of real compliance capacity such as the adequacy of scientific personnel and resources for educative work are less convincingly present or at least less provable in a persuasive manner.

It is also the case that, whether there is a more centralized biotechnology regulatory system or not, there will always have to be different ways in which regulations versus guidelines may be employed to meet the needs of different product cycles in industries and sectors as diverse as health, agriculture and food, and eventually fisheries as well. These different approaches can often be seen to be the product of interdepartmental turf wars only but in fact they usually have more complex and more valid technical bases and causes than that.

NOTES

[1] In this regard, the authors are especially indebted to several officials and participants (business and NGO) in the overall biotechnology regulatory process for cooperating with interviews and for their careful and constructive comments on earlier drafts of this paper.

[2] Government of Canada, *Backgrounder – A Federal Regulatory Framework for Biotechnology*; January 11, 1993, as quoted in KPMG (1995), p. 40.

[3] Biotechnology substances for the purposes of CEPA as a whole are defined under two categories: (a) Microorganisms (such as bacteria, algae, and fungi) and Organisms other than microorganisms (naturally occurring or genetically modified) and (b) Biochemicals (a biotechnology product, other than an organisms or biopolymer, that is produced by a microorganism) and Biopolymers (polymer produced by a microorganism).

[4] Environment Canada: CEPA Information Day. Held at the Citadel Hotel, September 24, 1996.

[5] Department of Fisheries and Oceans statement to the Standing Committee on Environment and Sustainable Development (1996b, Section 2.24).

[6] The Department of Fisheries and Oceans, Aquaculture and Oceans Science Branch released a draft of their policy guidelines: *Transgenic Aquatic Organisms: Policy and Guidelines for Research with, or for Rearing in Natural Aquatic Ecosystems in Canada*, in June, 1996. These are currently being revised.

[7] Health Canada defines novel foods as "substances that have previously not been used as food in Canada, or result from a process that has not previously been used for food in Canada; existing foods that have been modified by genetic manipulation and exhibit some one or more characteristics that were previously not identified in that food, or food that results from production by a genetically manipulated organism exhibiting such new characteristics; food containing microorganisms that have previously not been used as food or to process food before and; food that is substantially modified from the traditional product, or is manufactured by a process that has been substantially modified from the traditional process" (Canada Gazette, Part I, p. 2987, 1995).

[8] Department of Agriculture and Agri-Food, BioInfo: Regulation of Agricultural Products. Revised, 1996.
[9] Department of Agriculture and Agri-Food, Regulation of Agricultural Products of Biotechnology, Section 1.
[10] Department of Agriculture and Agri-Food, Regulation of Agricultural Products of Biotechnology, Section 1, p. 2.
[11] Categories of livestock feeds include: complete feeds, supplementary and macro premixes, micro premixes, converter feeds, specialty feeds, and single ingredient feeds. Most biotechnology products can be found under specialty feeds.
[12] AAFC defines novel feeds as a feed comprising an organisms or organisms, or parts or products thereof, that has a novel trait, with novel trait referring to a characteristic that has been intentionally selected, created, or introduced into the feed through a specific genetic change and is based on valid scientific rationale, is not substantially equivalent, in terms of its specific use and safety both for the environment and for human and animal health, to any characteristic of a similar, currently available feed (taken from Canada Gazette, Part II, pp. 2813, 1993).
[13] Agriculture and Agri-Food Canada: CEPA Information Day. Held at the Citadel Hotel, September 23, 1996, p. 2.
[14] Plants with novel traits refers to "a plant variety/genotype possessing characteristics not normally present in the cultivated species." Human Resources Development Canada: *Road map for biotechnology regulations*, January 1996, Section 2.5.
[15] Agriculture and Agri-Food Canada: CEPA Information Day. Held at the Citadel Hotel, September 23, 1996, pp. 2–3.
[16] Agriculture and Agri-Food Canada: CEPA Information Day. Held at the Citadel Hotel, September 23, 1996, p. 5.
[17] Department of Agriculture and Agri-Food, BioInfo: Regulation of Agricultural Products. Revised, 1996.
[18] Agriculture and Agri-Food Canada: CEPA Information Day. Held at the Citadel Hotel, September 23, 1996, p. 6.
[19] Fisheries and Ocean Canada is not discussed, largely because it is only beginning to construct its aspects of the biotech regulatory system.
[20] Department of Agriculture and Agri-Food, BioInfo: Regulation of Agricultural Products. Revised, 1996, p. 2.

REFERENCES

Agriculture and Agri-Food Canada (1996). *Biotechnology, agriculture and regulation*. http://www.agr.ca/fpi/agbiotec/geninfo.html

Ayres, I., & Braithwaite, J. (1992). *Responsive regulation: Transcending the deregulation debate*. Oxford: Oxford University Press.

Baldwin, R. (1995). *Rules and government*. Oxford: Clarendon Press.

Biotechnology regulation in Canada: A matter of public confidence (1996). Third Report to the Standing Committee on Environment and Sustainable Development, November. Ottawa: House of Commons.

Breyer, S. (1993). *Breaking the vicious circle: Towards effective risk regulation*. Cambridge, MA: Harvard University Press.

Chemical and Bio-Industries Branch (1996). *Bioindustries*. Draft, July 15. Ottawa: Industry Canada.

Department of Fisheries and Oceans (1996). *Transgenic aquatic organisms: Policy and guidelines for research with, or for rearing in natural aquatic ecosystems in Canada*. Draft. Ottawa: Department of Fisheries and Oceans, Aquaculture and Oceans Science Branch.

Dewees, D. (Ed.) (1983). *The regulation of quality*. Toronto: Butterworths.

Doering, R. L. (1996). *Alternative service delivery: The case of the Canadian food inspection agency*. Paper presented to the Department of Justice Workshop on Alternative Service Delivery, Ottawa, November 25.

Doern, G. B. (Ed.) (1978). *The regulatory process in Canada*. Toronto: MacMillan.

Doern, G. B. (1981). *Scientific controversy in federal policy formation*. Ottawa: Science Council of Canada.

Doern, G. B. (1994). *The road to better public services: Progress and constraints in five federal agencies*. Montreal: Institute for Research on Public Policy.

Doern, G. B. (1995). *Fairer play: Canadian competition policy institutions in a global market*. Toronto: C. D. Howe Institute.

Doern, G. B. (1998). The interplay among regimes: Mapping regulatory institutions in the UK, US, and Canada. In: G. B. Doern & S. Wilks (Eds.), *Changing regulatory institutions in Britain and North America*, pp. 29–50. Toronto: University of Toronto Press.

Doern, G. B. (forthcoming). Institutional and public administrative aspects of regulation versus voluntary codes. In: D. Cohen & K. Webb (Eds.), *The role of voluntary codes*. Ottawa: Government of Canada.

Doern, G. B., & Conway, T. (1994). *The greening of Canada*. Toronto: University of Toronto Press.

Doern, G. B., Pal, L., & Tomlin, B. (Eds.). (1996). *Border crossings: The internationalization of Canadian public policy*. Toronto: Oxford University Press Canada.

Doern, G. B., & Wilks, S. (1998). *Changing regulatory institutions in Britain and North America*. Toronto: University of Toronto Press.

Francis, J. (1993). *The politics of regulation: A comparative perspective*. Oxford: Blackwell.

Grabosky, P. N. (1995). Using non-governmental resources to foster compliance. *Governance, 8*, 527–550.

Harris, R. A., & Milkis, S. M. (1989). *The politics of regulatory change*. New York: Oxford University Press.

Harrison, K. (1994). Prospects for intergovernmental harmonization in environmental policy. In: D. Brown & J. Hiebert (Eds.), *Canada: The state of the federation 1994*, pp. 68–87. Kingston: Institute of Intergovernmental Relations.

Harrison, K. (1996). *Passing the buck: Federalism and Canadian environmental policy*. Vancouver: UBC Press.

Harrison, K., & Hoberg, G. (1994). *Risk, science and politics*. Montreal: McGill-Queens University Press.

Health and Welfare Canada (1991). *Health protection and drug laws*. Ottawa: Minister of Supply and Services.

Hill, M. (1994). *The choice of mode for regulation: A case study of the Canadian pesticide registration review 1988–1992*. Ottawa: Carleton University, Department of Political Science. Unpublished doctoral thesis.

Hoberg, G. (1993). *Pluralism by design: Environmental policy and the American regulatory state*. New York: Praeger.

Hood, C. (1986). *Administrative analysis*. London: Harvester Wheatsheaf.

Human Resources Developement Canada (1996). *Road map for biotechnology regulations*. Ottawa: Human Resources Development Canada.

Hutter, B. M. (1989). Variations in regulatory enforcement styles. *Law and Policy, 11*, 153–174.

Jacobs, S. (1992). *Regulatory management and reform: Current concerns in OECD countries*. Paris: OECD.

KPMG (1995). *Improving Canadian biotechnology regulation: A study of the U.S. experience*. Prepared for the National Biotechnology Advisory Committee on the Chemicals and Bio-Industries Branch. Ottawa: Industry Canada.

Law Reform Commission (1986). *Policy implementation, compliance and administrative law*. Ottawa: Law Reform Commission of Canada.

Martin, J. K. (1995). *Regulating the regulators: The Canadian approach to implementing government-wide regulatory reform strategies*. Paper presented to the New South Wales Regulatory Review Conference, Sydney, Australia, June 20.

Mausberg, B., Winfield, M., et al. (1993). *Growing safety concerns about biotechnology and the regulatory process: A report to Agriculture Canada*. Ottawa: Canadian Environment Network Biotechnology Caucus.

McGarity, T. O. (1991). *Reinventing rationality: The role of regulatory analysis in the federal bureaucracy*. Cambridge, MA: Harvard University Press.

Mehta, M. D. (Ed.) (1996). *Regulatory efficiency and the role of risk assessment*. Kingston: Queen's University, School of Policy Studies.

Meier, K. J. (1985). *Regulation: Politics, bureaucracy, economics*. New York: St. Martins.

National Biotechnology Advisory Committee (1995). *Improving Canadian biotechnology regulation: A study of the U.S. experience*. Ottawa: National Biotechnology Advisory Committee.

Office of Consumer Affairs (1996). Biotechnology: What is it all about? *Consumers Quarterly, 1*(3), 1–4. Ottawa: Industry Canada.

Ogus, A. I. (1994). *Regulation: Legal form and economic theory*. Oxford: Clarendon.

Pildes, R., & Sunstein, C. (1995). Reinventing the regulatory state. *University of Chicago Law Journal, 62*, 1–129.

Schultz, R., & Alexandroff, A. (1985). *Economic regulation and the federal system*. Toronto: University of Toronto Press.

Schultz, R., & Doern, G. B. (1998). Canadian sectoral regulatory institutions in a North American context: No longer "governments in miniature." In: G. B. Doern & S. Wilks (Eds.), *Changing regulatory institutions in Britain and North America*, pp. 108–130. Toronto: University of Toronto Press.

Sheehy, H. (1996). *Consumers and biotechnology: A synopsis of survey and focus group research*. Ottawa: Industry Canada, Office of Consumer Affairs.

Skogstad, G. (1987). *The politics of agricultural policy making*. Toronto: University of Toronto Press.

Sparrow, M. K. (1994). *Imposing duties: Government's changing approach to compliance*. London: Praeger.

Stanbury, W. T. (1992). *Reforming the federal regulatory process in Canada, 1971–1992*. Published as Appendix to House of Commons, Standing Committee on Finance, Sub-committee on Regulations and Competitiveness. Issue No. 23, Ch. 5. Ottawa: Supply and Services Canada.

Standing Committee on Agriculture and Agri-Food (1994). *rbST in Canada*. Issue No. 13. Ottawa: House of Commons.

Standing Committee on Environment and Sustainable Development (1996). *Evidence from Meeting No. 34, 2nd Session, 35th Parliament, October 1, 1996*. Ottawa: House of Commons.

Swimmer, G. (Ed.) (1996). *How Ottawa spends 1995–1996: Life under the knife*. Ottawa: Carleton University Press.

Trebilcock, M., & Howse, R. (1995). *The regulation of international trade*. London: Routledge.

THE AUTHORS

G. Bruce Doern is a Professor, School of Public Administration and Carleton Research Unit on Innovation, Science and Environment, Carleton University, E-mail: bdoern@ccs.carleton.ca. Heather Sheehy is a policy analyst at Health Canada, P.L. 3007A, Holland Cross, Tower A, Suite 709, Ottawa, Ontario, K1A 0K9. E-mail: heather_sheehy@hc-sc.gc.ca. At the time of writing she was with the Office of Consumer Affairs, Industry Canada.

Christopher Green
The Industrial Economics of Biotechnology

ABSTRACT. At the core of the biotechnology sector are "new biotechnology firms" (NBFs). Many NBFs have important vertical links upstream to universities and research institutes and downstream to firms in the pharmaceutical, chemical, and agri-business industries. In addition to the industry's evolving vertical structure are two horizontal elements: the extent of economies of scale and scope and barriers to entry and survival. The survival of NBFs depend jointly on successful development of new ideas and on attaining necessary finance. The innovative nature of the industry may mean high short run profitability to the successful, but the threat of displacement by future innovators implies that long-lived monopoly power is likely to be minimal. These and other considerations suggest that there is currently little or no need for industrial policy or economic regulation of the sector.

PART I

Introduction

This paper tackles some of the industrial economics of the biotechnology sector. Modern biotechnology has been defined as "the application of scientific and engineering principles to the processing of materials by biological agents" (Sharp, 1991, p. 113). The biotechnology sector is made up of three important players or agents: (a) universities; (b) small to medium sized research intensive firms, or what are called New Biotechnology Firms (NBF); and (c) the large, "established" pharmaceutical, chemical, and seed (e.g., Pioneer Hybrid) firms that have created in-house research, production, and/or marketing activities related to biotechnology. This report focuses on the NBFs – the most dynamic, and least understood, of the three sets of agents. However, we cannot ignore the role of universities or the large, "established" firms, because the links between them and NBFs are both complex and important (Arora & Gambardella, 1990; Pisano, 1991).

Modern biotechnology, often treated as synonymous with "genetic engineering," dates from the early 1970's. Genetic engineering became possible as a result of the development of the recombinant DNA technique (rDNA) and hybridoma technology (Malerba & Orsenigo,

Journal of Consumer Policy **21**, 1998.
Bartha M. Knoppers & Alan D. Mathios (eds), Biotechnology and the Consumer.

1995). Each is traceable to the discovery by Watson and Crick in 1956 of the double-helix structure of DNA. The scientific revolution unleashed by Watson and Crick has its foundations in the field of molecular biology and biochemistry. Its commercial possibilities became evident in the 1970s.

The development of a biotechnology sector began with university-based research in the 1970s. It soon moved outside the university with the creation of numerous NBFs in the 1980s, often with the integral participation of university researchers and the financial support of large established pharmaceutical firms. By the late 1980s and 1990s, large established firms in the pharmaceutical industry were establishing an in-house presence in biotechnology. Their activities began to spread beyond complementary (to NBFs) research and development (R&D), production, and marketing activities to potentially competitive production and development activities.

The NBF Sector in Canada

As in the U.S., the NBF sector is made up of a large number of firms, almost all of which are very small enterprises. It is important to distinguish the NBF sector from what the Heller Report (1995) terms the Canadian Biotechnology Community (CBC). According to Heller, in 1993 there were 538 organizations in the CBC, a group which includes (in addition to NBFs):

- established corporations in the chemical, pharmaceutical, agricultural, etc., industries with biotechnology operations or links to them
- university departments of microbiology and related disciplines carrying out basic and applied research
- research institutes engaged in biotechnology research
- suppliers of equipment and biological reagents for rDNA and cell fusion research
- venture capital firms supplying substantial funding to NBFs.

Of the 538 organizations, 309 had 10 or fewer employees. Only 54 had more than 100 employees in 1993.

What proportion of biotechnology players are NBFs? Of the 538 Canadian organizations involved in biotechnology, perhaps around 100 fit the description of an NBF. The Heller Report (1995, p. 15) indicates that in 1993, there were 147 Canadian firms (94 of which were in the health field) engaged in rDNA, most of which would fit the definition of a NBF. On the whole, we can expect the NBFs to be among the smaller of CBC organizations. Very few, if any, are among the 54 largest. A distinguishing feature of NBFs is a high ratio of R&D expenditures to invested capital. This implies that the employees of NBFs are likely to be mainly scientists and technologists (S&Ts), unlike the other commercial components of the CBC.

Some support for these characteristics of NBFs is found in Table I. The table provides information on medical-related firms, many of them NBFs, in which the Canadian Medical Discoveries Fund (CMDF) has made investments. These are mainly companies with an "idea" in search of a "discovery" that will lead to a product. Of the 444 employees of the 19 NBFs in the CMDF portfolio in September 1996, 324 (73 percent) were S&Ts. The small size and nature of NBFs gives rise to complex links between them and other biotechnology institutions or agents (or, as Heller would put it, members of the CBC).

Specialization of and Links Between the Major Players

Despite increasing competition between NBFs and established pharmaceutical firms, the activities of these two important sets of biotechnology players reflect an important division of labour. This division is indicated in Table II where we cross-classify NBFs and "established firms" by two major activities – one "upstream" (R&D) and one "downstream" (marketing). It is clear from Table II that at least up through the early 1990s, NBFs specialized in R&D while the established firms specialized in marketing. The table shows that of 32 biotechnology products that had received or were awaiting U.S. Food and Drug Administration approval, established firms developed only three, and in each case marketed them. In contrast, NBFs developed 29 of the 32 products, but were responsible for marketing only eleven of them.

Left undisclosed by Table II is biotechnology manufacturing activity. Generally, manufacturing will be carried out by the agent that markets the biotechnology product. Available evidence suggests that while NBFs have traditionally done little manufacturing, there is an increasing overlap between NBFs and established firms in this respect. I will, however, leave to a later section the issue of vertical integration, in which NBFs are beginning to engage in manufacturing (forward integration) and established pharmaceutical firms in research

TABLE I
Canadian Medical Discovery Fund Portfolio, September 1996

Company	Location	Investment[a] CMDF/TOTAL	Staff[b] S&T[c]/TOTAL	1997 R&D Budget[a]
Aptogen	Ottawa	0.9/1.0	19/20	1.0
Terragon Diversity	Vancouver	0.2/1.0	16/20	1.0
University Medical Discoveries	Toronto	5.0/5.0	1/1	N/A
Anormed	Vancouver	3.4/19.7	15/22	3.3
Gylco Design	Toronto	1.3/4.5	20/23	3.5
LIFE Imaging Systems	London	1.4/4.5	16/20	1.0
Morphometrix Technologies	Toronto	2.0/2.5	19/21	3.5
Neuromotion	Edmonton	2.5/3.0	14/22	2.0
Neuroscience Pharma	Montreal	2.3/13.0	12/14	3.5
Osteopharm	Toronto	1.5/2.5	17/22	3.5
Bio-Cell Medical	Vancouver	3.0/8.0	6/14	1.4
Hemosol	Toronto	3.1/13.3	65/72	12.0
Micrologix Biotech	Vancouver	1.0/11.0	23/29	2.5
Nexia Biotech	Montreal	2.0/6.0	20/23	1.2
Seragen Biopharma	Montreal	1.0/13.5	1/3	2.5
Ultravision	Calgary	2.0/11.6	24/73	0.9
Vascular Therapeutics	Hamilton	6.2/17.0	12/12	1.7
Drug Royalty	Toronto	7.5/7.5	0/8	N/A
Stress Gen Biotech	Victoria	1.0/5.3	21/28	3.5
Total		47.3/149.9	324/444	47.9

[a] millions of dollars
[b] full-time equivalent
[c] scientists and technologists
Source: Canadian Medical Development Fund, Newsletter, Fall 1996.

TABLE II
Development and Marketing Activities of NBFs and Established Firms, 32 Biotechnology Products Launched 1982–1992

Development Activity	Marketing Activity NBFs	Marketing Activity Established Firms	Total
NBFs	11	18	29
Established Firms	–	3	3
	11	21	

Source: Pisano (1991, Tables 2, 3, 4, pp. 242–243).

and development activities (backward integration). At this point I will confine the discussion to the nature of links between "established" firms and the other two sets of economic agents in the biotechnology sector. The links between NBFs and universities will also be considered later.

There are numerous links between the large pharmaceutical and chemical firms and the other agents, NBFs and universities, in biotechnology. Arora and Gambardella (1990) have identified four types of links. First there are research and/or joint development links with other firms, usually NBFs. These tend to be either (a) product-specific, with the established firm focusing on "downstream" activities, as Table II suggests; or (b) aimed at developing and commercializing a particular discovery of the NBF. The Arora-Gambardella study is based on a sample of large U.S., European, and Japanese firms. The international reach of these firms implies that the linkages also apply to the Canadian biotechnology scene.

Second, the established firms have research agreements with universities. These agreements focus on basic research objectives and in many cases give the established firms first option on the licensing of new discoveries. Third, established firms have invested in the capital stock of NBFs. These minority positions have not only supplied the NBFs with an important source of finance, but have provided a means by which the established firm is able to monitor the internal research activities of NBFs. An example is the pharmaceutical firm, GLAXO, which has strategic alliances with and investments in a dozen NBFs (KPMG-DRI/McGraw Hill, 1996, pp. 15–16).

Finally, some established firms have acquired NBFs. Examples include the acquisition of Hybridtech by Eli Lilly in 1986; Hoffman-LaRoche's acquisition of Genentech in 1990; Bristol Myers' acquisition of Oncogen and Genetic Systems in the 1980s; Ciba-Geigy's 49.9 percent participation in Chiron acquired in 1994; and the acquisition by Angen, an industry leader in biotechnology, of financially troubled but product-rich Synergen in 1994. Acquisitions of NBFs by large established chemical and agri-industry firms which once were rare (Heller Report, 1995, p. 12) are now more frequent. Examples include acquisitions made by Monsanto and AgrEvo. There are undoubtedly numerous reasons why an "established" firm might acquire an NBF. Among the more important ones are (a) to add internal knowledge and (b) to provide a means of "catch up" for late entrants into the biotechnology business. That acquisitions are not more important is testimony to the fact that the other, less all-encompassing links, typically serve the established firms just as well, except where transaction costs are important (see below).

Now that we have established who are the players and the nature of the links between them, we are ready to turn to the structure, behaviour, and performance of the NBF component of the biotechnology sector. Part II of the report, by far the longest, follows traditional industrial organization lines in its subject matter. However, as we shall see, making headway with this non-traditional industry requires investigating some non-traditional approaches, especially with respect to the entry decision.

PART II

Structural Characteristics of the NBF Sector

Although we can broadly speak of a biotechnology sector, both officially and theoretically there is no biotechnology industry or sector as such. Statistics Canada uses a standard industrial classification (SIC) to organize firms into "industries" on the basis of similar production processes. Statistics Canada's Census of Manufacturers does not single out a biotechnology industry, much less a biotechnology sector. Moreover, the potential markets served by NBFs are too disparate in terms of consumer wants to constitute a theoretically meaningful industry. But in the past 10 to 15 years, the advance of science and its industrial applications have been such that quite a number of industries and sector have components which can be described as engaged in activities of a biotechnological sort. And taken as a whole, these activities are increasingly important economically, as well as scientifically, so that it is none too soon to begin to think about biotechnology from an industrial economic standpoint.

Because biotechnology does not yet constitute an industry, any attempt to describe much less analyze biotechnology from an industrial economics perspective encounters some difficulties. Among other things, it renders questionable the standard framework in which industrial organization economists relate industry structure, firm behaviour, and performance. True, there are many firms (or at a more basic level, establishments) engaged in biotechnology activities. But in many cases their activities, when viewed from an economic or commercial standpoint are too disparate to be viewed as competing – to be called an industry. Taken as a whole, biotechnology firms do not produce close substitutes – the usual demand side criterion for determining when a set of firms (or establishments) constitutes an industry. Moreover, on the production-supply side, what constitutes biotechnology involves a myriad of biological, chemical, and engineering processes. If there is any common denominator among biotechnology firms it is on the factor input-supply side. Biotechnology firms are heavily dependent on a certain type of knowledge capital – found at the intersection of the biological, chemistry, and chemical engineering disciplines, and additionally on a certain type of financial capital – venture – willing to bet on the commercial viability of the discoveries that may flow as output from this knowledge capital.

For the above and many other reasons which will become evident as the paper proceeds, it is virtually impossible to even roughly summarize the structure, much less the behaviour and performance, of a biotechnology industry. Any attempt at a snapshot description must also cope with the problem that no well-defined industry exists – that those captured in the picture are in such different stages of evolution as to make static analysis virtually useless. The problem is even greater when the setting is dynamic, with the biotechnology sector players on many different technological trajectories, including the production of biopharmaceuticals, transgenic drugs, enzyme discovery, genetic engineering of increased food production, the development of new diagnostic technologies, etc.

Nevertheless, it is possible to make some summary statements about "structure" that appear to fit much, if not all, of the biotechnology sector.

Size distribution of firms. The NBF component of the biotechnology sector consists almost solely of small firms – as we have seen. By inference, concentration by any measure is low (although no measure is very meaningful where an industry is so ill-defined). Even the well-established, large-scale firm, pharmaceutical industry has a relatively low concentration ratio. There are, to be sure, a few relatively largely NBFs such as Genentech (now acquired) and Amgen. In addition to NBFs, biotechnology activities are carried on within the divisions of some of the large, established pharmaceutical firms (e.g., Ciba-Geigy, Hoffman-Laroche, Lilly) and chemical firms (e.g., Monsanto, American Cyanamid).

Economies of scale. Here, the picture is complicated. Some economies of scale probably exist at the R&D level (if evidence from the pharmaceutical industry is relevant, Henderson & Cockburn, 1996), but there may be diseconomies of scale too, especially agency costs associated with the employment contract of scientists and engineers (Zenger, 1994). Whether economies of scale exist at the firm level (as opposed to the R&D level) is unclear. There is little evidence of production economies of scale, although to be sure, commercial success requires that the fixed costs of research and development be eventually spread over sufficiently large sales. Thus what production scale economies exist probably relate to length of production runs, i.e., they are "product specific" (Scherer, 1980) as opposed to plant or firm (multi-plant) specific.

In addition to economies of scale in R&D, there are economies of scope. An important empirical study of economies of scope in R&D of pharmaceutical firms shows them to be both large and important (Henderson & Cockburn, 1996). One source of economies of scope arises as pharmaceutical research has moved from "random" research to "rational" drug design. This change in research regime places an increasing premium on information exchange within the firm. Thus, Henderson and Cockburn (1996, p. 55) argue that the "primary advantage of size has become the ability to exploit internal returns to scope," that take the form of internal spill-overs of knowledge among products rather than economies of scale as such. To the extent that biotechnological products are likely to be specialized, with limited markets, the survival of NBFs may depend on their ability to capitalize on economies of scope in the development of several products.

Vertical integration. There is still limited vertical integration in the biotechnology sector. Table II above provides some evidence. But there is abundant evidence of extensive vertical links backward to sources of knowledge input suppliers (scientists). Thus many, if not most, biotechnology firms have professional contractual relationships with scientific members of research institutes and universities. Some NBFs engage in manufacturing, but there is little further integration downstream such as into marketing, much less into distribution. However, the importance of vertical integration is growing and may influence the eventual structure of the sector – a subject I take up later in the report.

Product differentiation. Differentiation is endemic, so much so that one might argue that each firm's product is essentially non-competing with that of other biotechnology firms. At the same time, there is competition among biotechnology firms in terms of research projects, and among those that succeed in terms of product success as well. However, product differentiation is probably not a very meaningful way of describing the output of the biotechnology sector. Further, neither of the two standard models of product differentiation, the Chamberlin and Dixit-Stiglitz (symmetric) model or the Hotelling ("location" or characteristic spatial) model is of much use in characterizing preferences for and competition among outputs of biotechnology firms. Either new products are non-competing or if they are substitutes for existing products they will tend to replace them in a "creatively destructive way." Thus most new products neither compete generally with existing products as in the symmetric model, nor do they fill in characteristic space as in the location model.

Barriers to entry. Barriers, in the sense of either Bain (1956) or Stigler (1968), do not appear important (Heller Report, 1995, p. 142). Bain found that both economies of scale and product differentiation constitute substantial barriers to entry into many industries. Stigler denied that either economies of scale or product differentiation constitute true barriers, arguing that if barriers exist they reside in absolute cost differentials of firms, irrespective of scale or the extent of product differentiation. Absolute cost advantages of incumbent firms could arise as a result of patents, access to scarce resources, or the strategic ability to raise the costs of newcomers. None of these appear important in the biotechnology sector. Nor do there appear to be reputational barriers (Demsetz, 1982). Reputations are made from successful innovation – and therefore are not barriers to someone with a good new idea. Yet the issue of entry, exit, and survival is an absolutely crucial and distinguishing part of the story of the industrial economics of the biotechnology sector, as we shall see below.

The Entry Decision

A key to understanding the industrial economics of the biotechnology sector rests on the entry decision (dealt with in this section) and the barriers to survival of NBFs (next section).

In order to analyze the entry issue, I focus on the new biotechnology firm (NBF) component of the biotechnology sector. NBFs are the most important industrial manifestations of the new (or second generation) structure of biotechnology. Explaining the entry decision by NBFs is an important factor in understanding the industrial economics of this sector. The approach taken here to the entry decision of NBFs, while not original, is somewhat novel. To appreciate the novelty and to provide a standard of comparison, it is useful to be somewhat more specific about the mainline model of the entry decision (Audretsch, 1995, pp. 45–46) which can be traced to Bain (1956) and was empirically estimated by Orr (1974).

In the mainline model, the entry decision depends on the difference between expected post-entry profits P^e of a potential entrant and the level of profits of an established firm that are protected in the long run by entry barriers B. According to Bain's limit price theory, some level of profits of established firms might be protected by economies of scale, a unique product or technology, and/or patents. If the difference between the expected profits of a potential entrant and barrier-protected profits of established firms is non-positive, then according to Bain's limit price theory, entry does not take place. Formally

$$(1) \qquad E_{jt} = 1(P^e_{jt} - B_j),$$

where E_{jt} is entry into industry j at time t; P^e_{jt} is expected post-entry profits in industry j at time t; B is barrier-protected profits; and 1 measures the speed with which entrants respond to excess profits. When such an equation has been estimated by adding a stochastic disturbance term u_{jt} to Equation (1), the statistical results are mixed in terms of the strength of the (positive) impact of (past) profits, with homogeneous product industries providing the best statistical fit.

A problem with the mainline approach to barriers is that it is a static formulation, when entry is inherently a dynamic process. This is doubly a problem when innovation is not only an important activity of incumbent firms, but is the raison d'être for entry, as is surely the case with NBFs. Moreover, it is not clear what are the barriers to entry for NBFs. And if there are none, there remains the question of why someone with new knowledge should choose to become an entrepreneur and enter an industry rather than sell his idea and/or become an employee of an established firm. To see the force of this point, it is useful to begin by comparing the NBF sector with the scientifically most similar industry, pharmaceuticals.

When one focuses on the structural characteristics of the new biotechnology firm (NBF) component of the biotechnology sector, there is something of a puzzle to be solved. The innovative activities of these firms – certainly their main raison d'être and defining characteristic – are in many respects similar to those of the pharmaceutical industry. Yet the NBF sector and the pharmaceutical industry are structurally dissimilar in terms of the size distribution of firms, vertical integration, product differentiation, the impact of economies of scale and scope on firm size, and the likelihood of new firm entry. The pharmaceutical industry is a loose oligopoly, dominated by large, typically world-scale firms which experience important economies of scale and scope in their R&D activities. These firms are vertically integrated from R&D through production to marketing, placing heavy emphasis on brand name (product differentiation) advertising and direct sales efforts with physicians.

In contrast, the NBF sector is dominated by small, relatively non-vertically-integrated, start-up firms, apparently capable of at least temporarily overcoming whatever economies of scale and scope exist, and experiencing limited product differentiation and brand-name exploitation.

A step toward solving the puzzle is to recognize that NBFs and pharmaceutical firms differ in at least one very important respect: the relative role of innovative activities. NBFs are very largely R&D firms – not just R&D-intensive firms. In contrast, the R&D-intensive pharmaceutical industry is heavily engaged in other activities as well, including production, promotion, and marketing, as well as innovative activities.

A further contrast is illustrated by the links of pharmaceutical firms and NBFs to their respective knowledge bases. Pharmaceutical companies have, at least until recently, limited their personnel links with university and other government funded research laboratories. In contrast, NBFs are strongly linked to university based research labs. The formation of NBFs has been described as typically "formed through the collaboration of a scientist and a professional manager backed by venture capital, in order to apply the new scientific discoveries [rDNA and hybridoma technology] to commercial product development" (Malerba & Orsenigo, 1996, p. 51). These discoveries have created great opportunities for innovation in a number of industries.

A likely explanation for the distinctive structure of the NBF sector is founded on the observation that in many, if not most cases, the great opportunities of commercially viable innovations made possible by the recombinant DNA and hybridoma technologies were not immediately foreseen or understood. There is, of course, nothing unusual in the idea that the economic possibilities from new scientific knowledge are highly uncertain or that substantial asymmetries exist across economic agents both between and within firms (Milgrom & Roberts, 1987). But it seems likely that in the case of the new generation of biotechnologies, the commercial possibilities were underestimated or undervalued by all but the scientists who were responsible for the new discoveries. If this is so, it would explain why the biotechnology sector is populated by very small, lab-type firms.

More than seventy-five years ago, Frank Knight (1921) observed that the productive choice facing an economic agent is whether to become an employee in an existing firm or an entrepreneur, i.e., start up a new firm. According to Knight, the choice depends, on the one hand, on the "wage" the agent can command as an employee and, on the other hand, the expected returns (profits) accruing from the establishment of a new firm (Audretsch, 1995, p. 49; Audretsch & Stephan, 1998). In more formal terms, the probability of (Pr) of entry (E) can be written as

$$(2) \qquad \Pr(E) = f(P^e - w)$$

where P^e is expected profits, w is the wage the economic agent can command in an established firm and the probability of entry, $\Pr(E)$, is positively related to the difference between the expected profit from being an entrepreneur and that wage or salary one can command as an employee.

Equation (2) differs from the mainline entry Equation (1) by making the entry decision depend on *two* variables that govern economic agent choice, P^e and w, and by ignoring exogenously determined barriers to entry, B. There are some compelling reasons why Equation (2) may be a more appropriate means of analyzing the entry decision of NBFs than is the mainline model summarized in Equation (1). These reasons are the following.

It is unclear what are the "natural" barriers to entry into the NBF sector, other than the need to have a new idea, some managerial help, and financial backing – none of which are the sort of barriers envisioned by the mainline model of entry. (There may, however, be "artificial" barriers in the form of regulation – or the fear of such.) Even sunk costs, which in traditional modeling constitute barriers to exit and thereby a barrier to entry, are mainly in the form of R&D expenditures, the value of which depends on the success of the innovation. Since that success is captured in the expected profitability variable, we need not consider it a barrier as such. But if there are no natural or artificial barriers, all that matters is post-entry profitability, which is likely to be highly uncertain in the case of biotechnology innovations.

On the face of it, if NBFs face barriers, they are more likely to be "barriers to survival," the subject of the next section, rather than barriers to entry.

The model of entry captured by Equation (1) works best in the context where entrants imitate the technology and products of incumbent firms. This model is not particularly appropriate where new entrants cannot hope to succeed unless they employ a new technology or offer a new product, which is surely the case with NBFs. In contrast, the model of the entry decision captured in Equation (2) more easily handles the introduction of new ideas or innovations.

The entry decision by NBFs is likely to be heavily influenced by the existence of asymmetric information. The expected value of a new idea may well be viewed differently by (a) the inventor of that idea or innovation, and (b) decision-makers in firms confronted with evaluating the innovation from a distance (Audretsch, 1995, p. 49). If the latter tend to initially undervalue the idea or innovation relative to the inventor, there is an increased probability that the inventor will consider becoming an entrepreneur, i.e., setting up a firm. This is precisely the choice captured in Equation (2) where the scientist-inventor's P^e is greater than that of decision-makers in established firms, and where w is what the latter would pay the scientist to work for or license his new idea to them.

If the preceding analysis is correct, the issue is not so much one of barriers to entry as it is the choice of entry in the first place. That is, in the biotechnology field, the key inputs are a new idea and the up-front finance necessary to pursue it. Leaving aside the issue of finance – to be discussed below – and the possibility of an "unfriendly" regulatory climate, the question is why a person with a new idea chooses to be an entrepreneur rather than an employee of an established firm. The tentative answer provided here is that NBFs are the result of some form of asymmetry which drives those with new ideas to prefer to enter (to become entrepreneurs) rather than throw their lot in with established firms. The newness of biotechnology and the uncertainty of consumer and social acceptability, as well as commercial opportunities, make them especially risky ventures. Established firms may be either too conservative or too fearful of cannibalizing their existing products, to place a high value on what seem like either far-fetched or profit-eroding ideas. In contrast, the scientist-inventor, after years of work in the field, may be reasonably certain of the scientific success of his idea. His conviction may be all that is needed to convince some venture capitalist(s) to provide the necessary financial support.

If this view of the entry decision of NBFs holds water, it suggests that we need to look beyond barriers to entry to the conditions of firm survival if we really wish to understand the roadblocks faced by NBFs. Moreover, we should anticipate that once the success of biotechnological innovations becomes routine and socially acceptable, we may see fewer NBFs and more initial effort carried out by relatively large established firms, whether they are former NBFs that have survived and expanded or are firms in related fields such as pharmaceuticals and specialized chemicals and fertilizers.

Barriers to Survival

The Heller Report (1995, p. 142) provides a table indicating the major barriers faced by a Canadian biotechnology firm "entering the market." Since the findings are based on respondents to a survey and the respondents were existing biotechnology companies, we can infer that (a) the firms initially overcame the barriers and entered and (b) the barriers are essentially viewed as barriers to the firms' continued survival. Overwhelmingly, the respondents cited lack of financing as the key barrier followed by lack of market acceptance. The only other frequently-cited barrier was Canadian regulation. The Heller Report (pp. 193–204) leaves the impression that environmental regulations applying to the Canadian biotechnological industry may be more stringent and costly than in other countries, particularly the U.S. The KPMG Management Consultants report (1995, p. 2) concluded that Canadian and U.S. biotechnology regulations are not "significantly divergent." However, KPMG found that Canada's system of biotechnology regulations, while more flexible and less complex, requires more data and is less supportive of commercialization than is the U.S. From the responses to the Heller Report survey, there was no evidence that Bain (1956) or Stigler (1968) type barriers are important or that sunk costs are barriers to either entry or survival. The survey is consistent with the analysis of the preceding section. But once entry has occurred, the issue is *survival* – and here there may well be barriers.

The question of what firms survive in the biotechnology sector is a crucial one with implications for public policy. Clearly, survival is partly based on the ability to make discoveries related to specific problems, and the allocation of venture capital is likely to reflect this ability. Nevertheless, if the lives of firms with good ideas are cut short by lack of finances before they have had a reasonable time to pursue these ideas, there is a potential social loss and there will be pressure for governments to provide financial help. But the exit of NBFs may be due to other failings than ability to attract finance, the latter often being a symptom rather than a cause of failure. To assess this possibility, we need some framework within which we can explain frequent entry followed almost as frequently by exit within a short (say 5 to 10 year) period of time.

A potentially useful framework is Jovanovic's (1982) theory of "noisy" selection. In Jovanovic's model, start-up firms are uncertain about their costs and managerial capacity. The decision to launch a firm is based only on the vaguest sense of post-entry performance. It is only after entry has occurred that the entrepreneur discovers his/her true ability to manage, to say nothing of the success of the new product. Thus in Jovanovic's theory, new, small firms are not deterred from entering even when economies of scale are important. The problem becomes one of survival.

Audretsch (1995) has argued that Jovanovic's theory helps to explain the abundant empirical evidence that, across almost *all* industries, most new firms are small, operating well below minimum efficient scale. This is certainly true of NBFs, although it is not yet clear that there are important economies of scale to exhaust. Following the logic of Jovanovic's theory, a dynamic view of the process of firm selection is that new firms typically start at a very small scale of output. Where small scale output is well below MES, the firm must eventually grow in order to survive. However, as applied to NBFs, survival would appear to depend more on finding sufficient finance to continue R&D and tide the firm over initially very short production runs of highly specialized biotechnological products.

In his study, Audretsch (1995, p. 156) found that the greater the degree of (a) technological change and (b) the extent of scale economies in an industry, the faster the rate at which new entrants will exit. Audretsch's empirical findings provide some support for Jovanovic's theory. For NBFs, rather than economies of scale, it is much more likely to be the rapidity of technological change, in conjunction with the need to finance highly risky R&D, which acts as barriers to survival.

To what extent does NBF survival depend on cooperative behaviour or the tactics of rivals? In an industry in which success depends on innovation, it is unlikely that the strategic behaviour we associate with static oligopoly theory plays much of a role. Neither collusion nor non-cooperative behaviour in the form of price wars make much sense where the struggle is to successfully develop a new product from the fruits of scientific knowledge. Nor are such tactics as predatory pricing and exclusion through raising rivals' costs relevant in an industry in which victory or defeat, success or failure, depend primarily on commercially successful innovation. Furthermore, since NBFs engage in limited production and virtually no marketing, these sort of strategies, if they are important at all, are played out by downstream firms who purchase from NBFs. However, legal strategies tied to regulatory provisions may be used, as the example of Avonex (described in a later section) suggests.

Performance of the NBF Sector

The standard way in which industrial organization economists empirically assess market performance is from data on price cost margins and profit rates. The theoretical logic here is that prices that are held consistently above (long run) marginal costs, or profits which are chronically in excess of the "normal" returns to (i.e., the opportunity cost of) capital signal the existence of market power and its effective exercise. However, neither of these measures is much good in the case of NBFs, and for two reasons. First, many NBFs have too few sales on which to make an assessment of price-cost margins. Second, for businesses which are so overwhelmingly driven by risky R&D, it is very difficult to calculate a normal rate of return. Certainly, comparison with most other types of business or sectors of the economy would be invalid given the unique characteristics of NBFs.

One possibly valid and insightful comparison might be made with experience in the pharmaceutical industry. Although, on average, the pharmaceutical industry has proved to be highly profitable over the past two or three decades, this performance is due to spectacular profits on just 10 percent of new chemical entities. In seventy percent of the cases, the new products yielded quasi rents lower than their R&D outlays (Scherer, 1996). Thus, high average profitability accompanied by a huge variance across new products may reflect nothing more than a normal return to highly risky R&D.

In order to usefully assess market performance in the NBF sector it is necessary to turn to dynamic measures of performance: innovation. Most economists now accept the Schumpeterian argument that innovation, in its many varieties, is the most significant indicator of industrial performance, if long run growth and welfare is one's ultimate interest. And clearly the NBF sector exists for and because of innovation. On the basis of casual empiricism, the biotechnology sector as a whole appears to have contributed its share of innovation if one measures by the number of genetically engineered products and the general world-wide growth of the sector. But can we say more about the relation between size/structure and performance?

One question we might ask is whether the small scale of all but a very few NBFs hinders the innovative effort? This question arises rather naturally out of the traditional Schumpeterian concern that small size and the absence of market power might not be conducive to innovative effort. (It should be mentioned in passing that this concern is a product of Schumpeter's later writings, especially in *Capitalism, Socialism and Democracy*, 1942, and is not found in his earlier, and arguably more incisive, *Theory of Economic Development*, 1911.) There is as yet very limited evidence of the impact of firm size on innovation in the biotechnology sector as such. There is much more evidence of the impact of size and market power on innovation across numerous industries, many of them high-technology ones. The most convincing study is that of Acs and Audretsch (1988), which uses a unique set of data on innovations compiled by the U.S. Small Business Administration.

Acs and Audretsch found that the number of innovations introduced by U.S. 4-digit SIC industries, in 1982, to be *negatively* related to concentration (a proxy for market power) and unionization (a proxy for internal rent seeking which reduces a firm's ability to appropriate rent from its innovations). Neither concentration nor unionization are important factors in the biotechnology sector. In contrast, innovation was found to be *positively* related to R&D and the relative importance of skilled labour. R&D and skilled (scientific) labour are, of course, salient characteristics of the biotechnology sector. But Acs and Audretsch also found that the greater the proportion of an industry consisting of "large" firms (defined as firms with more than 500 employees) the greater is innovative activity (1988, pp. 686–687). As we have seen, however, almost all NBFs are small by this standard. Moreover, Acs and Audretsch found that the possible negative impact of small size on innovative activity was most evident in industries with high R&D. These latter findings could be a hint – but only a hint – that the small size of NBFs is, or could eventually become, a factor that hinders the sector's overall innovative activity.

The links between large established firms and NBFs may be one means by which the small size of the latter is finessed, so far as innovative activity is concerned. Shan, Walker, and Kogut (1994) have empirically investigated the role of interfirm cooperation between large firms and innovation by smaller firms (NBFs) in the biotechnology field. The authors found that innovative output, in the form of patents, does not attract large firm relationships but rather depends on them. That is, the existence of interfirm agreements leads to innovation, but the reverse did not appear to be true, innovation does not lead NBFs and established firms to form agreements. At the same time, the size of NBFs appears to positively affect the level of NBF innovative output. But the size effect can be offset by interfirm cooperation initiated by the larger firm. Moreover, Shan et al. found that cooperation is more likely to occur if the NBF (a) has been able to attract equity capital through public offerings, and (b) has good network relationships with other biotechnology organizations.

The Shan et al. paper suggests that the relationship between size and innovation is more complicated than the Acs and Audretsch study might suggest, at least so far as the biotechnology field is concerned. Moreover, a general concern with the Acs and Audretsch study is that, as in most models used to estimate the impact of state variables on innovation, innovative activity is made *endogenous* to "structural" factors such as concentration, capital intensity, firm size, the advertising-sales ratio, and unionization. But in a dynamic setting, causation may be reversed, with market structure endogenous to innovation. This is especially likely to be the case where success in innovation breeds further success, with the result that a few unusually successful or "superior" firms come to dominate the industry. Only time will tell whether such a process will occur in biotechnology and influence the structure of the NBF sector.

What can we say about the relative performance of Canada's biotechnology sector? Here we can use as a yardstick the performance of other technologically advanced nations, in particular the U.S. The Heller Report (1995, Ch. 2) attempts a comparative cross-country assessment. There is some evidence that Canada has lagged behind the U.S. in the development of its biotechnology sector. There are various possible explanations for a lag, if it exists, including the fact that this tends to be true of most industrial developments here. But one factor that may be important is the availability of venture capital. The CMDF was set up to help remedy this problem. Another source of finance has been the federal government's National Biotechnology Strategy (NBS). If there is a barrier to survival which impinges negatively on NBF sector performance, it may well be on the financial side. But if financing is a barrier to survival of NBFs, it is not an easy matter to assess. Venture capital rationing may be the most efficient means of producing early exit of NBFs which have little likelihood of long run success.

Perhaps we can allow Jovanovich's theory of "noisy selection" to have the last word in our discussion of performance in the NBF sector. The ease with which an NBF can be set up once someone has a new idea and the uncertainty about the eventual commercial success of that idea place all the risk in the hands of investors. Consumers will benefit if the idea is successful, investors lose if it does not. Those with scientific ideas (the scientists) can escape with most of their human capital intact and in a position to return to the field another day. Thus the entry and exit of NBFs, as with many firms in the information technology field (especially software), reflects an industrial churning, almost chaotic in character. Some firms will fail, with the losses, if any, borne by investors. Of course, investor losses must on average be offset by successes, the high returns from which will often be of limited duration because of the capacity to genetically engineer new – and competing – products.

PART III

Dynamic Industry Structure and Evolution

To this point I have focused on the traditional elements of industrial economics in my assessment of the biotechnology sector: structure, conduct, and performance. To be sure, a relatively novel, and implicitly dynamic, approach was introduced in the discussion of entry conditions; one which among other things, pushed the consideration of "barriers" one step ahead to "barriers to survival."

Within this framework, I addressed what little is known about behaviour and performance within the NBF sector. But my discussion of traditional elements of behaviour is limited by the fact that NBF firms do not constitute a well-defined industry of competing firms. The NBFs are neither price takers (as in competitive models) nor strategic (oligopolistic) rivals, at least where *static* forms of behaviour (e.g., price, output) are concerned. Each NBF is working to follow-up and develop a new idea with potential commercial capabilities. Thus strategic rivalry, if it exists, is at the level of innovation. Only at the more downstream levels of manufacturing and marketing, mainly carried out by large established firms, is the traditional analysis of rivalristic behaviour and its impact on performance potentially relevant.

In this section, attention is turned to what are arguably more dynamic elements of industry structure, elements which seem particularly relevant to understanding the dynamic behaviour of the biotechnology sector and how the structure of the sector may evolve. The three elements are: (a) the knowledge structure of the sector,

(b) the location structure of NBFs, and (c) decisions by NBFs and established firms to vertically integrate. The knowledge structure, in which universities and independent (of firms) research labs play such an important role, commercially as well as scientifically, has influenced the location structure of the NBF component of the biotechnology sector. At the same time, attempts by established firms, particularly in the pharmaceutical industry, to build in-house R&D capabilities are likely to slowly but progressively alter the knowledge structure of the industry. Meanwhile, vertical integration downstream into production and marketing by larger NBFs, undertaken in order to protect their proprietary knowledge, is likely to introduce more traditional forms of competition into the biotechnology sector. The outcome of the forthcoming competitive struggle could have implications for the relative importance of the NBF sector one or two decades hence.

Knowledge structure of biotechnology. The knowledge structure of the biotechnology sector is both unique and influential. Biotechnology, at least in its present manifestation, is fundamentally dependent on new scientific ideas, resulting from basic scientific research. Much of the basic scientific research is carried on in universities and other publicly-funded research institutes. One result of the dependence of biotechnology on university research is a close relationship between university scientists and the businesses that their research has spawned. In some cases, universities have founded NBFs. In other cases the university scientists have, in partnership with managerial talent and financial support, founded NBFs. Many more scientists serve on the Scientific Advisor Boards (SABs) of NBFs. And as indicated above, some university-based biotechnology research is being funded by large, established pharmaceutical firms.

That there is a close relationship between science and the business of biotechnology has been documented by McKelvey (1996, 1997, 1998), who has described the "coevolution" of the institutions in biotechnology. Stephan and Everhart (1998) have provided empirical evidence on how biotechnology has changed the financial rewards to science. Many university-based scientists affiliated with NBFs have large share holdings, or options thereto, in these firms, as well as receiving consulting fees and salaries from the NBFs. Audretsch and Stephan (1998) and Audretsch and Feldman (1996) have shown that what university scientists contribute to NBFs is more than just "scientific" knowledge. They also contribute "tacit knowledge," i.e., knowledge whose transference depends on personal, face-to-face contact. The apparent importance of tacit as well as scientific knowledge in the business of biotechnology has influenced the locational structure of the NBF component of the biotechnology sector.

Locational structure of biotechnology. The Heller Report (1995, p. 10) provides some rough evidence on the provincial distribution of biotechnology "firms" (organizations) in 1993. These biotechnology "firms" are concentrated in Ontario, Quebec, and British Columbia. The distribution is approximately what one might have anticipated from the provincial distribution of GDP, with the exception of Alberta whose share of Canadian GDP far exceeds its tiny (2.6 percent) share of biotechnology "firms." However, the Heller Report does not appear to capture a "cluster" of approximately 30 agricultural biotechnology firms (as of early 1997) that has grown up in Saskatchewan, particularly around Saskatoon.

To get a better idea of how the role of scientists has influenced the location of NBFs, we need to look to evidence south of the border. Audretsch and Stephan (1996) have used a database that includes "virtually the entire population of biotechnology firms that prepared an initial public offering in the early 1990s" (p. 641) to investigate the way in which the location of scientists influences the location of NBFs. Table III, drawn from the Audretsch-Stephan paper, shows that NBFs are heavily concentrated in locales where university and other research lab scientists working on biotechnological subjects are concentrated. The concentration of NBFs in the San Francisco Bay, San Diego, Boston, and Philadelphia areas reflects the location of important centers of university scientific research in biotechnology.

Yet Table III also shows that only 31 percent of the contacts between university-based scientists and NBFs involve the same region. As Audretsch and Stephan (1996) have shown, the supply of biotechnology scientific talent is much less regionally concentrated than are NBFs. Since modern means of electronic communication virtually eliminate the problem of distance, Audretsch and Stephan search for an explanation of the particular locational structure of NBFs. Their answer resides at least partly in the role of "tacit knowledge." If tacit knowledge is an important source of the knowledge that NBFs gain from scientists, NBFs are likely to locate where there are pockets of biotechnology scientists. This is so even though many of

TABLE III
Location of Scientists With Links to Biotechnology Firms, By Region

Region (1)	Contacts (2)	Unique Firms (3)	Contacts in Region (4)	% in Region (4)./.(2)
San Francisco Bay Area	66	28	44	66.7
San Diego, CA	27	13	21	77.7
Boston, MA	69	26	41	59.4
Philadelphia, PA	17	14	8	47.1
New York, NY	36	22	9	25.0
Maryland	11	9	1	9.1
Research Triangle, NC	8	8	2	25.0
Los Angeles, CA	13	9	0	0.0
Dallas/Houston, TX	31	14	9	29.0
Other	167	108	3	1.8
Total	445		138	

Source: Audretsch & Stephan (1996, p. 643).

the scientist members of SABs come from outside the region in which the NBF is located.

A related factor influencing the location of NBFs is R&D spillovers. Audretsch and Feldman (1996) have used a large U.S. database consisting of more than 8,000 commercial innovations (some of which are biotechnological) to investigate the spatial distribution of innovative activity. Their main hypothesis is that innovative activity will tend to cluster in industries where new economic knowledge plays an especially important role. That is, they predict that in industries such as micro-electronics and biotechnology, where innovative activity is an important generator of new, commercially viable economic knowledge, firms will tend to cluster. Audretsch and Feldman found substantial supporting evidence for this effect. If the Audretsch and Stephan and Audretsch and Feldman results carry over to Canada, we can expect that most Canadian NBFs will, for the foreseeable future, be located in Toronto, Montreal, and Vancouver, which already have clusters of biotechnology firms, and in other areas which are able to successfully attract a cluster of firms as appears to be occurring in the agri-biotechnology firms in and around Saskatoon (KPMG-DRI/McGraw Hill, 1996). The Saskatoon experience raises an interesting question which cannot be addressed here: Is the biotechnology field particularly susceptible to planning by cities/regions willing to devote time and money to cluster industrial development?

Evolving vertical integration. The knowledge structure of the biotechnology sector is reflected in the sector's vertical structure and relationships. To summarize, as the industry developed, a number of university scientists who were engaged in biotechnological research, vertically "integrated" forward by helping to create NBFs. In many more cases, university scientists established vertical contacts with NBFs that were originally formed from initiatives taken outside the university. Realizing the potential of the new biotechnology, but unwilling to take a big initial leap, established firms in the chemical, agricultural, and particularly the pharmaceutical industries provided financial capital to NBFs and provided university scientists with research funding. In addition, the established firms undertook to provide manufacturing facilities and marketing services to NBFs that had succeeded in developing commercially viable products and the requisite regulatory (in the U.S., FDA) approval. Thus the three major components of the biotechnology sector developed a vertically efficient division of labour, aided and abetted by various links between them.

But as in any dynamic setting, it would be unrealistic to expect the vertical structure of the biotechnology sector to remain unchanged. Indeed it has begun, and will continue, to evolve – to slowly take on a different shape over time. As NBFs have succeeded in translating their R&D effort into a growing number of commercially viable products, some have begun to undertake downstream activities, particularly manufacturing. Why should the NBFs vertically integrate forward? Pisano (1991) suggests that NBFs may have transaction cost reasons for vertical integration. Among the more important transaction costs that might induce integration is the protection of proprietary know-how. This cost and inducement to integrate seems likely to increase.

As NBFs vertically integrate downstream, established firms, particularly in the pharmaceutical industry, have begun to build up in-house R&D capabilities and to directly finance university research under agreements giving priority to license. But why should the established firms integrate backward? One reason is to assure themselves a continuing supply of biotechnological products to manufacture and market. According to McKelvey (1997) most "pharmaceutical firms have now tried, more or less successfully, to integrate genetic engineering competence in house" as well as to tap external sources such as universities.

Although vertical integration of economic agents in the biotechnology sector will continue to increase, the process is likely to be a slow one. It takes time to build up in-house R&D capabilities, even for large, established firms. The acquisition route to rapid build-up of capabilities is a risky one, since most of the assets reside in scientists who may leave. For NBFs, capital constraints place a limit on the rate at which they can integrate downstream. Finally, the collaborative forms of agreements discussed at the beginning of the paper have generally served the players in the biotechnology sector well. Contractual integration, as an alternative to full vertical integration, is just as much an important (and efficient) option in biotechnology (perhaps more so) as in other industrial sectors of the economy.

In fact, for some observers, an evolution to vertical integration is not certain at all. James (1995) cites evidence of a proliferation in strategic alliances in the biotechnology field. A report in *The Economist* (Biotechnology mergers, 1995) indicated that in 1994 U.S. pharmaceutical firms entered into 117 ventures with biotechnology firms, 70 percent more than in the previous year. Magun (1996, pp. 24–25) provides some evidence of strategic alliance activity in the Canadian chemical, pharmaceutical, and biotechnology industry group. As long as the biotechnology field is in its development stage and NBFs continue to enter the field in large numbers, this "virtual integration" (James, 1995, p. 471) is likely to dominate vertical integration. Thus, for the foreseeable future, any strong trend toward the latter is likely to be masked by alliances and other forms of interfirm cooperation.

Finally, we should not overlook the fact that at least some NBFs will "horizontally" grow into large, established multi-product firms. For example, Amgen has succeeded in generating growth through a series of successful product launches, although in doing so, it has focused on only a few categories of medical ailments (Gertz, 1995). Another example is the British biotechnology firm, Chiroscience, which has evolved from a Cambridge University-based firm, making building blocks for other drug companies, into a major biotechnology firm with an interest in novel drug discovery (Gopinath, 1997). Chiroscience has on its own, or in collaboration with other firms, developed a number of new drugs, some of which are now undergoing clinical trials. By diversifying its product portfolio, the firm has enhanced its chances of success and the likelihood that it will become a leader in the biotechnology field.

PART IV: CASE STUDIES

The experience of the NBF sector is too recent, too varied, and too little documented to allow for useful summary. This is certainly the case where the focus is on the industrial economics of the biotechnology sector. In most respects, it is simply too early to generalize about the structure, behaviour, and performance of this unique sector, although we have been able to shed light on its current structure, especially on the vertical elements and entry conditions. To conclude this report, it seems useful, therefore, to abjure generalization and attempt to find insight in specific cases. The three cases described here are suggestive.

Avonex (Biogen)

In 1996, Biogen, one of the largest and oldest of the NBFs, received U.S. Food and Drug Administration (FDA) approval to market Interferon b-1a, a genetically engineered drug designed to treat multiple sclerosis (MS). Interferon b-1a will be marketed under the name Avonex. Biogen has the necessary manufacturing facilities in place to serve a market of 250,000 Americans afflicted with MS. But Avonex will have to compete with Betaseron which is sold under license by Berlex Labs, a U.S.-based division of the German drug producer Schering. Berlex Labs' license is from Chiron, an NBF, which introduced Interferon b-1b in 1993. In 1994, Chiron which was short of capital became substantially owned by the Swiss chemical and pharmaceutical giant, Ciba-Geigy (C&EN, 1996a). Such are the complexities of the organization of biotechnology firms.

Biogen's decision to manufacture and market Avonex represents a departure from past practice. Previous biopharmaceutical successes by Biogen, including a-interferon, hepatitis B vaccines and diagnostics, and interleukin-2, had been licensed to other firms generating sales of $1.8 billion in 1995 to the licensees, $152 million in revenue to Biogen, 89 percent of which came from royalties, and a profit to Biogen of $5.7 million (C&EN, 1996a).

Biogen's Avonex faces more than just a marketing battle with Berlex Labs' Betaseron for a market estimated at almost $1 billion. Berlex has gone further and has filed suit to halt the marketing of Avonex. Berlex charges that FDA approval violates the U.S. Orphan Drug Act which grants seven years of marketing exclusivity for products targeting rare diseases. Although Berlex lost the first round when a U.S. District Court turned down its request for a restraining order, it plans to continue its suit against the FDA. For its part, FDA considered Avonex different enough from Betaseron to warrant allowing Avonex to enter the market.

The Avonex example highlights several facets of the industrial economics of biotechnology. Even an industry leader such as Biogen has until now depended almost solely on royalty revenue from previous discoveries. Its decision to manufacture and market Avonex reflects its evolving vertical integration downstream. In so doing, it has come into face-to-face competition with another biopharmaceutical, Interferon b-1b, originally developed by another NBF, Chiron, with complex vertical and horizontal relationships to other pharmaceutical organizations. Chiron's licensee, Berlex Labs, represents more than just a market competitor for Biogen. The complexities of drug legislation and the vagaries of regulatory decision-making will be brought to bear on the outcome of the competitive struggle.

Extremozymes (RBI)

The creation of Recombinant Biocatalysis Inc. (RBI), which was originally incorporated as Industrial Genome Sciences in 1994, is an example of competitive entrance into a new biotechnology field. Extremozymes are the enzymes of exotic microorganisms that inhabit extreme environments such as hot springs, Arctic waters, and the like. These microorganisms, which are called extremeophiles because they thrive in extreme environments, are difficult to grow in laboratory cultures. What makes these microorganisms commercially interesting is that from them can be developed exotic enzymes that have the potential to replace chemical catalysts in industrial processing (C&EN, 1996b).

RBI is not alone in recognizing the market potential of extremozymes. Large suppliers of industrial enzymes such as Novo Nordisk Biotech and Genencor International have been stocking up on extremeophiles. Pfizer, the large pharmaceutical firm, has been adding extremeophiles to its inventory. Other NBFs such as Altus Biologies (Cambridge, Mass.) and Maxygen Inc. (Santa Clara, California) have been developing technologies that make enzymes more useful (C&EN, 1996b). Thus RBI is entering a new field into which both established firms and other NBFs are also moving. Taken as a whole, the activities of these firms are best described as a complex mixture of complementary and competitive developments in biotechnology.

As a new competitor in an advanced biotechnological field, RBI has many of the personnel characteristics so common among NBFs. One of its cofounders is a professor of biology at the California Institute of Technology. It has a Scientific Advisory Board which includes a professor of chemical engineering at North Carolina State. RBI which is located in Pennsylvania has 60 employees, 40 of whom are in technical positions.

Roundup (Monsanto)

In November 1996, Monsanto, the large St. Louis based chemical firm, announced that it was spinning off its US$2.7 billion chemical business. A few weeks later, it announced it would spend $1.02 billion U.S. to purchase two Iowa corn seed companies. Monsanto's moves are the latest in a series of steps the company has taken to transform itself from a traditional chemical firm into a leader in the agricultural biotechnology industry. Monsanto largely owes its dramatic moves to the success of Roundup, a potent and efficient herbicide that forced the firm to search for herbicide-resistant crops.

The Monsanto Roundup story is an example of how a chemical firm's innovative activities in one of its traditional markets, agricultural herbicides, forced it to look for answers in the biotechnology sector – specifically in genetically engineered varieties of corn and soybeans. Roundup generally kills any ordinary plant with leaves. Thus, without the development of herbicide resistant crops, Roundup's use would be equivalent to "throwing the baby out with the bath water." Genetic engineering was viewed by Monsanto as a means of developing herbicide-resistant crop varieties – a bet that appears to have turned out correct. The development of Roundup-Ready Soybeans (RRS), a Roundup resistant soybean variety, has been a success. Monsanto's decision to purchase Holden's Foundation Seeds Inc. (one of the Iowa corn seed companies), a producer of genetic material for corn hybrids, presumably reflects Monsanto's desire to extend its activities to herbicide resistant corn varieties.

At its inception, biotechnology was heralded as a means of making crops more disease and insect resistant, presumably making them less dependent on chemicals. The Monsanto Roundup example, however, seems to have turned an important field of agricultural biotechnology on its head, and in the process may eventually transform the structure of the biotechnology sector. The aim of Monsanto has been to develop genetically engineered crops resistant to its chemical herbicide. To do so, Monsanto has moved away from its chemical products base and into agricultural biotechnology. Monsanto has now become a large (agricultural) biotechnology firm, in a sector that has been almost wholly dominated by small startup firms.

Monsanto is not the only large chemical firm that is increasingly moving into the biotechnology business, although it is the only one to leave the chemical business in so doing. Ciba-Geigy, DuPont, Dow Chemical, Zeneca, and Hoechst have invested in genetically engineered crops. Ciba-Geigy has aggressively invested in a genetically engineered corn hybrid whose introduction in Europe has caused considerable controversy. DowElanco, a partnership between Dow Chemical and Eli Lilly, has a 52 percent stake in Mycogen Corp., a California-based agricultural biotechnology firm. Monsanto countered by raising its stake in Calgene Co., another California-based agricultural biotechnology firm, to a controlling 54.6 percent.

But the decisions of chemical firms to move into agri-biotechnology are not without risk. In addition to political resistance, particularly in Europe, some firms have suffered setbacks. A case in point is Monsanto's Bt (Bacillus Thuringiensis) cotton seed, Bollgard. The Bt toxin, harmless to humans, is supposed to kill three pests that attack cotton. However, following a large scale planting of this bioengineered, transgenic cotton seed, thousands of acres were attacked by cotton bollworms, one of the pests that Bollgard was supposed to control (Pets overwhelm Bt cotton crop, 1996). Nevertheless, it is reported that Monsanto's transgenic cotton seed may capture twenty percent of the market this year (Thayer, 1997). It is unlikely, therefore, that individual setbacks will be more than small bumps in a road that will lead many large firms in the chemical and pharmaceutical industries to seek to integrate biotechnology into their portfolio of production activities.

PART V: FINAL WORD

The Avonex, extremozyme, and Roundup examples are illustrative of the richness of the industrial structure and behaviour of the biotechnology sector. NBFs currently play the pivotal role in an industrial landscape that also includes university scientists and large pharmaceutical and chemical firms. The complex and changing relationships among the major players makes descriptive generalization difficult, but is reflective of the dynamics of the biotechnology sector. The sector is driven by a combination of scientific breakthroughs and commercial possibilities. The combination is too powerful to be inhibited by ordinary industrial barriers to new firms. But as biotechnology develops and its commercial possibilities multiply, the sector's structure is likely to slowly change – an industrial evolution which will probably leave its greatest mark on the NBF component. As surviving NBFs "mature," their size, integration, and behaviour will likely become more akin to that of their established counterparts in the pharmaceutical and chemical industries – perhaps even becoming integral parts of these industries. Moreover, as the commercial possibilities and attractions of genetically engineered products increase, their profitability is likely to lure established chemical and pharmaceutical firms into the biotechnology sector, even to the extent of forsaking their original base, as in the Monsanto example. If this occurs, the biotechnology sector will be transformed from its startup, NBF-dominated structure to one characteristic of the chemical and pharmaceutical industries.

The dynamic, evolving nature of the biotechnology sector makes drawing policy conclusions both tenuous and premature. Yet, my analysis of the industrial economics of the biotechnology sector suggests, albeit implicitly, that there is currently little need for industrial policy or economic regulation – and that there is little reason to believe this will change. The implicit message of my assessment of NBF sector performance is that its dynamism, the ease of entry and exit of firms, and the apparent limitless scientific possibilities, at least some of which can be carried to commercial success, virtually assures: (a) that the threat of long lived monopoly power is minimal, if not non-existent; (b) that failures are losses to investors, with the human capital able to move on to a new project; and (c) that finance is likely to be forthcoming as long as investors are able to reap the high returns from successes.

Although the structure of the biotechnology sector will evolve, its evolution is likely to be slow and its ultimate form remains problematic. However, one can conjecture that the important changes are likely to be mainly vertical in nature. Since vertical relationships are chiefly motivated by efficiency considerations, the structural changes that are likely to occur presumably will be, on the whole, more efficiency-enhancing then market power-creating. This is a further reason why there appears little need for government intervention, including subsidies or regulation, that would influence the economic structure, behaviour, and performance of the biotechnology sector.

REFERENCES

Acs, Z., & Audretsch, D. (1988). Innovation in large and small firms: An empirical analysis. *American Economic Review, 78*, 678–690.

Arora A., & Gambardella, A. (1990). Complementarity and external linkages: The strategies of the large biotechnology firms. *Journal of Industrial Economics, 38*, 361–379.

Audretsch, D. B. (1995). *Innovation and industry evolution*. Cambridge, MA: MIT Press.

Audretsch, D. B., & Feldman, M. P. (1996). R&D spillovers and the geography of innovation and production. *American Economic Review, 86*, 630–640.

Audretsch, D. B., & Stephan, P. E. (1996). Company-scientist locational links: The case of biotechnology. *American Economic Review, 86*, 641–652.

Audretsch, D. B., & Stephan, P. E. (1998). Knowledge sources in biotechnology through the Schumpeterian lens. In: G. Eliasson, C. Green, & C. R. McCann (Eds), *Microfoundations of economic growth: A Schumpeterian perspective*, pp. 169–187. Ann Arbor, MI: University of Michigan Press.

Bain, J. (1956). *Barriers to new competition*. Cambridge, MA: Harvard University Press.

Biotechnology mergers: Unseemly couplings (1995). *The Economist*, May 13, pp. 66–69.

C&EN (1996a). FDA okay's biogen's multiple sclerosis drug. *Chemical and Engineering News*, May 27, p. 6.

C&EN (1996b). Enzyme discovery heats up. *Chemical and Engineering News*, Oct. 14, pp. 31–33.

Demsetz, H. (1982). Barriers to entry. *American Economic Review, 72*, 47–57.

Gertz, D. (1995). Beating the odds. *Journal of Business Strategy, 16*, 20–24.

Gopinath, L. (1997). Chiroscience: From past to present. *Chemistry in Britain, 33*(2), 29–31.

Heller Report (1995). *Background economic study of the Canadian biotechnology industry*. Ottawa: Industry Canada.

Henderson, R., & Cockburn. I. (1996). Scale, scope, and spillovers: The determinants of research productivity in drug discovery. *Rand Journal of Economics, 27*, 32–59.

James, G. E. (1995). Strategic alliances as "virtual integration": A longitudinal exploration of biotech industry-level learning. *Academy of Management Journal, 38*, Best Paper Proceedings, 469–473.

Jovanovic, B. (1982). Selection and evolution of industry. *Econometrica, 50*, 649–670.

KPMG Management Consultants (1995). *Improving Canadian biotechnology regulation: A study of the U.S. experience*. Consultants Report Submitted to Industry Canada. Mimeo.

KPMG-DRI/McGraw Hill (1996). *Building technology bridges: Cluster based economic development for Western Canada.* Consultants Report Submitted to Industry Canada and four Western Provinces, Ottawa. Mimeo.

Knight, F. (1921). *Risk, uncertainty and profit.* New York: Houghton-Mifflin.

Magun, S. (1996). *The development of strategic alliances in Canadian industries: A micro-analysis.* Ottawa: Industry Canada. Working Paper No. 13.

Malerba, F., & Orsenigo, L. (1995). Technological regimes and firm behavior. In: G. Dosi & F. Malerba (Eds.), *Organization and strategy in the evolution of the enterprise*, pp. 42–67. London: Macmillan.

McKelvey, M. (1996). *Evolutionary innovations: The business of biotechnology.* Oxford: Oxford University Press.

McKelvey, M. (1997). Coevolution in commercial genetic engineering. *Industrial and Corporate Change, 6*, 503–532.

McKelvey, M. (1998). R&D as pre-market selection: Of uncertainty and its management. In: G. Elliason,C. Green, & C. R. McCann (Eds), *Microfoundations of economic growth: A Schumpeterian perspective*, pp. 188–211. Ann Arbor, MI: University of Michigan Press.

Milgrom, P., & Roberts, J. (1987). Information asymmetries, strategic behavior, and industrial organization. *American Economic Review, 77*, 184–193.

Orr, D. (1974). The determinants of entry: A study of Canadian manufacturing industries. *Review of Economics and Statistics, 56*, 58–66.

Pests overwhelm Bt Cotton Crop (1996). *Science, 273*, p. 423.

Pisano, G. (1991). The governance of innovation: Vertical integration and collaborative arrangements in the biotechnology industry. *Research Policy, 20*, 237–249.

Scherer, F. M. (1980). *Industrial market structure and economic performance* (2nd ed.). Chicago: Rand-McNally.

Scherer, F. M. (1996). *Industry structure, strategy and public policy.* New York: Harper Collins.

Shan, W., Walker, G., &. Kogut, B. (1994). Interfirm cooperation and startup innovation in the biotechnology industry. *Strategic Management Journal, 15*, 387–394.

Sharp, M. (1991). Pharmaceuticals and biotechnology. In: C. Freeman et al. (Eds.) *Technology and the future of Europe*, pp. 213–230. London: Pinter.

Stephan, P. E., & Everhart, S. (1998). The changing rewards to science: The case of biotechnology. *Small Business Economics, 10*, 141–151.

Stigler, G. (1968). *The organization of industry.* Chicago: Irwin.

Thayer, A. M. (1997). Betting the transgenic farm. *Chemical and Engineering News*, April 28, pp. 15–19.

Zenger, T. (1994). Explaining organizational diseconomies of scale in R&D: Agency problems and the allocation of engineering talent, ideas, and efforts by firms size. *Management Science, 40*, 709–726.

THE AUTHOR

Christopher Green is Professor of Economics at McGill University. Address: 855 Sherbrooke St. W. Montreal (Quebec) Canada H3A 2T7. E-mail: green@leacock.lan.mcgill.ca.

Claude Crampes and Abraham Hollander

Biotechnological Innovation and Industrial Performance

ABSTRACT. The first part of the paper gives an overview of the industrial structure of the biotechnology sector. It highlights the importance of partnership agreements between industrial firms and universities and examines issues of financing. The second part examines the impact of biotechnology on the two main user industries: agri-food and health. The final part investigates economic regulation, especially in the area of intellectual property protection.

Biotechnology is a vast array of techniques of a generic nature, driven by slow, empirical development over the centuries (OECD, 1989). In the broad sense it is defined as the industrial application of biological sciences. However, the term "biotechnology," as it is used today in the media and by the scientific world, covers only activities related to genetic engineering, molecular biology, and biochemistry.

The difference between biotechnology before and after 1970 is the same as that between analytical and synthetic chemistry. Formerly, chemistry focussed on the search for new uses for chemical substances and mixtures found in nature, but after the discovery of chemical synthesis it became possible to form compounds that did not exist in nature (Krimsky, 1991). Similarly, until the early 1970s biotechnology depended on the utilization of organisms that existed in nature but, as a result of the techniques of DNA recombination and cell fusion, it has undergone a transformation that makes possible a range of innovations likely to revolutionize certain industries such as agri-food and health. In February 1997 in Scotland, the first cloning of an adult mammal – a sheep – took place. This new advance in biotechnology suggests that know-how in the field will accumulate at an accelerating pace. It took 20 years from 1953, when Watson and Crick discovered the structure of the DNA molecule, to obtain the first successful manipulation of genetic sequences to modify the genetic code of an organism (by Boyer and Cohen in 1973). But as early as 1982 Genentech, the first pharmaceutical firm to specialize in genetic engineering, marketed human insulin obtained by recombination of

Journal of Consumer Policy **21**, 1998.
Bartha M. Knoppers & Alan D. Mathios (eds), Biotechnology and the Consumer.

deoxyribonucleic acid (rDNA), and in 1985 the U.S. Federal Drug Administration (FDA) authorized Genentech to market a growth hormone produced by rDNA.

The aim of this study is to review developments in biotechnology and determine their impacts on user industries. First, we give a preliminary quantitative evaluation of the biotechnology sector; in particular, we show the extent to which the players vary in status as well as in size. Second, we seek to highlight the importance of partnership agreements between industrial firms or with research units and universities. Third, we study the delicate issue of financing in an industry that has not yet reached its cruising speed. Notably, we consider the extent to which the dynamism of financial institutions is a prerequisite for the development of these types of activities, where commercial opportunities remain uncertain. Fourth, we examine the impact of biotechnology on the two main user industries, agri-food and health. Finally, we investigate the issue of economic regulation of bio-industries, that is, the necessary intervention of public authorities to deal with deficiencies in market mechanisms, especially in the area of intellectual property protection. We conclude the paper with a few remarks on the future of the sector.

RESEARCH, DEVELOPMENT, AND INDUSTRIAL STRUCTURES

The majority of U.S. biotechnology firms emerged from government-funded agencies conducting research in molecular biology. Only since the 1980s has private venture capital either replaced or complemented public financing. These small firms are responsible for the U.S. lead in biotechnology. The U.S. government financed the scientific base but did little to promote the industrial development and exploitation of biotechnology. Commercialization has instead been the result of a partnership between universities and venture capital (Senker, 1996). In 1994 the U.S. biotechnology industry contained 1,300 firms, of which approximately 265 were publicly-owned. Of these, 37% had fewer than 50 employees, 18% had between 51 and 135 employees, and 12% had between 135 and 299 employees.[1]

In contrast, the biotechnology industry in Europe is dominated by large firms, especially multinationals in the pharmaceutical and chemical sectors. Britain is an exception to this rule: There, many small biotechnology firms have emerged. Ernst and Young, the consulting firm, has estimated that in 1995 there were 400 biotechnology firms in Europe (Ernst & Young, 1995).

In Japan, small firms play an insignificant role for all practical purposes. Big Japanese firms, especially in fermentation, have adopted a strategy that involves forming partnerships with foreign firms of far smaller size, or simply positioning themselves downstream to produce and market innovations from abroad.

Table I clearly shows the U.S. domination of the biotechnology sector.

By itself, the European market for products that depended on biotechnology applications was worth around 38 billion ECUs in 1993–94: 15 billion in the sector of human and animal health, 10 billion in the food sector, 8 billion in diagnostics and equipment, and finally 44 billion in chemicals. In 1994, the industrial sectors that achieved significant growth with the help of biotechnology accounted for 9% of the gross value added of the European Union (EU), and 8% of EU jobs. Experts predict that the biotechnology market in Europe should top 80 billion ECUs by the year 2000 (Club de Bruxelles, 1996).

As for research and development (R&D), a Business Week survey (1997, March 10, pp. 78–84) found that 7 out of the 10 firms ranked as U.S. leaders in terms of R&D per employee are biotechnology firms. Ranked first was Biogen, with $208,724 per employee; last was Chiron, with $64,263 per employee.

In 1993, biotechnology firms spent an average of $59,000 on R&D per employee, as compared with an average of only $7,476 for U.S. firms in general. In 1994, R&D expenditures as a percentage of sales stood at 91% for biotechnology firms, but only 16% for the pharmaceutical industry.

Ernst and Young (1995) also reports that in 1994, R&D expenditures by U.S. biotechnology firms amounted to $7.7 billion, up by 28% from 1992 and by 10.5% from 1993. The same survey shows that, again in 1994, the U.S. federal government spent $4.3 billion on biotechnology research.

The U.S. government is present mainly in medical research. In 1991 it allocated to biomedical research over 80% of the total $3.5-billion biotechnology research grants budget. Most of the financing from the U.S. government has gone to basic research, and the main channel for distributing this assistance was the National Institutes of Health.

TABLE I
Percentage of Global Inventions per Country (With Applications in at Least Two Countries)

Sector	Ger.	Fr.	U.K.	E.U.	U.S.	Japan	Rest of world
	1986 and 1987						
Apparatus and devices for enzymology and microbiology	15.8	5.2	10.5	36.6	34.3	15.6	13.4
Micro-organisms, enzymes, cells	9.7	5.3	9.5	29.2	44.6	16.1	7.5
Fermentation	10.4	5.5	10.0	32.4	40.8	20.3	6.5
Measuring and testing methods based on enzymes or micro-organisms	13.1	4.9	8.2	29.3	49.4	14.9	6.4
Analysis of substances by testing their chemical or physical properties	12.7	6.0	7.1	29.6	45.6	15.2	9.3
Total (Double counts excluded)	11.6	5.8	8.7	31.0	41.8	13.4	8.8
	1988 and 1989						
Apparatus and devices for enzymology and microbiology	13.5	5.0	7.9	29.9	39.9	18.7	11.4
Micro-organisms, enzymes, cells	9.3	4.5	8.5	27.7	48.8	16.1	7.5
Fermentation	11.3	5.0	8.5	30.3	43.4	19.4	6.9
Measuring and testing methods based on enzymes or micro-organisms	10.0	5.1	7.7	25.6	6.2	9.3	7.8
Analysis of substances by testing their chemical or physical properties	13.0	6.5	7.7	31.3	45.7	14.1	8.5
Total (Double counts excluded)	10.5	5.5	8.5	29.7	45.9	16.1	8.3

Source: EPIDOS/INPADOC (1994), Ifo Patent Statistics, cited in Panorama of EU Industry.

TABLE II
Government Grants for R&D in the Health Field
(Millions of US$, 1985 Prices at Purchasing Power Parity)

	1975	1980	1985	1991
U.S.	4199	4895	5611	7753
U.K.	150	128	305	398
France	266	329	412	408
Canada	117	114	147	203

Source: OECD STIIU Database, 1993.

Table II shows funding allocated by various governments to medical R&D.

As for government expenditures on biology, for the year 1986 Coleman (1987) reports the following figures: $50 million for Canada, $100 million for France, $108 million for Britain, and $600 million for the United States.

BIOTECHNOLOGY PARTNERSHIPS

As the biotechnology industry developed, it became apparent that it would be increasingly difficult to group under one roof skills in basic and applied research, in manufacturing, in marketing, and in steering cases before regulatory authorities. To a large extent, R&D expertise lay within universities, research institutions, and biotechnology firms. To proceed from the research phase to the development phase, it is essential to have a thorough mastery of in-situ testing techniques, particularly clinical tests for medicines. Big firms, especially in the pharmaceutical sector, have the necessary knowledge and experience. They also have abundant financial resources and skill in steering cases before regulatory authorities. These resources are needed: In the health sector, 10 to 12 years may elapse from the first tests on a product until it receives final approval, and the process may cost from US$150 million to US$359 million (U.S. Congress, Office of Technology Assessment, 1993).

The firms faced problems as soon as they had to create an in-house environment conducive to basic innovation, however. Biotechnology requires new skills in the pharmaceutical field. Traditional pharmacology is based on the synthesis of new molecules. Chemistry laboratories within pharmaceutical companies are reservoirs of accumulated knowledge on synthesis techniques. Biotechnology, however, does not seek to produce a molecule by direct means. Instead it seeks to discover the means by which to induce a cell to produce a protein found in nature. The techniques used are very different from those in traditional pharmacology. Accumulated experience in traditional pharmacology is no longer an asset if the aim is to find a therapeutic treatment by means of biotechnology. This situation encouraged the development of partnership ties between small biotechnology firms and big firms.

Forms of Collaboration

Collaboration agreements have rapidly become a favoured form of industrial organization in biotechnology. These agreements are of various types but in general they fall into the following categories:

- research contracts and minority shareholdings that allow one of the partners to obtain information on the research undertaken by another firm;
- licensing and marketing contracts, which also give a firm the right to use a technology developed by the other partner;
- partnerships that involve share transfers;
- mergers and acquisitions.

It should be noted that the same company may have links with several other firms, involving very different types of collaboration (Roberts & Mizouchi, 1989). Table III shows some typical contracts

TABLE III
Collaboration Agreements in Biotechnology

Type of link	Type of partner
Joint R&D project	Another biotechnology firm, pharmaceutical company, research institution, university
Outside investment	Venture capital firm
Clinical test	Research hospital, specialized firm
Manufacturing	Pharmaceutical/chemical company
Marketing	Pharmaceutical company
Purchase of idea	University
Joint venture	Another biotechnology firm

Source: Powell, Koput, & Smith-Doerr (1996).

and indicates which partners are likely to be linked by these contracts.

The decision to proceed by means of collaboration or integration depends mainly on how the firm answers the following questions (Teece, 1996):

1. Is it possible to transfer know-how at lower cost when the production employees to whom it is transferred belong to the same firm, rather than when they do not?
2. How much protection do intellectual property rights provide for the technology to be transferred?
3. Is it possible to draw up a contract efficiently managing the transfer, particularly with regard to restrictions imposed on the recipient for the use of the technology?
4. Does the technology transfer provide access to complementary assets at lower cost than by purchasing those assets?

Collaboration Between Firms

Greis, Dibner, and Bean (1995) examined data from a survey of 244 U.S. biotechnology firms. The survey in question investigated business people's perceptions of the skills they consider essential to financial success, as well as the constraints that could slow the growth of their firms. Also studied was the relationship between the perceptions of company heads and their tendency to establish links with partners.

The first finding to emerge from the survey is that firms that have entered into at least one partnership agreement consider that they are exposed to more intense competition than do firms that have not formed any such partnership. Another finding is that the majority of business people thought that in 1990, the year when the survey was conducted, development and research were the factors determining success. However, these same people believed that expertise in production and marketing would become increasingly important over the coming years.

In addition, the tendency of firms to establish research partnerships correlates with the idea held by their directors of the protection offered by patents. Firms that regard the meagre protection provided by patent law as a barrier to innovation are much less willing to collaborate in the field of research. They are clearly worried about the use their partners could make of the information placed at their disposal.

The data are particularly revealing about the tendency to seek a partner for production. It is sometimes maintained that considerations of suitability are important in the decision to proceed with production via vertical integration or to outsource manufacturing. The survey indicates that the tendency to seek collaboration in production is instead linked with worries raised by financing and the difficulties of steering cases through the regulatory labyrinth. It also seems that many biotechnology firms seek to remedy their lack of management skills through association with a partner.

Marketing biotechnology products can also require considerable resources, above all in the health field. It has been estimated that, toward the end of the 1980s, marketing costs represented 24% of the receipts of pharmaceutical companies (The doctor's dilemma, 1990). Few biotechnology firms have a marketing team. What's more, because of the innovative nature of the products, a major task is providing information to physicians; this task requires large teams of sales representatives, such as multinational pharmaceutical firms alone possess. The survey data suggest that the wish to speed up marketing through a partnership arrangement is also associated with the degree of competition perceived by company heads.

It would be false to assume that a big firm, which generally is responsible for marketing, always holds exclusive production rights. The survey by Ernst and Young (1995) showed that in the 1980's roughly a third of biotechnology firms had production facilities but that in most cases these were merely pilot units. A study of a sample of firms by Pisano and Mang (1993) shows that, in nearly half the cases, collaboration agreements grant exclusive production rights to the partner that undertakes marketing. Nevertheless, in slightly under a third of the cases the biotechnology firm keeps exclusive production rights, while they are shared in 15% of cases (Pisano & Mang,1993). Figure 1 presents changes in the way manufacturing rights have been assigned over the years.

Other studies have examined the role of market structure in the decision to enter into a collaboration. Pisano (1990) thinks that when the number of collaborators is low, partners are exposed to the risk of opportunistic behaviour. He maintains that this risk would cause more research to be done in-house within big pharmaceutical companies. This observation agrees with theoretical teachings on the

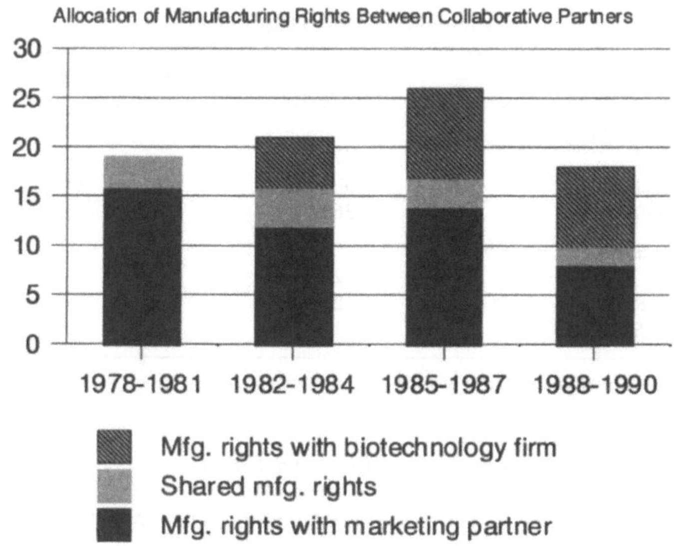

Figure 1. Allocation of manufacturing rights between collaborative partners.

economics of the firm: When specific investments are necessary for an economic activity, efficiency requires internalization that is partial (a long-term contract) or total (merger or in-house production). Otherwise, the fear of "expropriation" leads to less than optimal investments (Tirole, 1988, Introductory Chapter).

Last, certain authors maintain that in-house research and research collaborations are not substitutes but rather complements for each other. They stress that an internal research capacity is necessary to evaluate the research performed by others (Powell, Koput, & Smith-Doer, 1996). In this regard, a 1987 survey in the United States, examining big U.S. firms having major investments in biotechnology, indicated that nearly 96% of them performed in-house research while 83% outsourced research to other firms or universities (U.S. Congress, Office of Technology Assessment, 1988a).

Table IV is drawn from a survey of Canadian companies. It gives an idea of the type and scale of collaborations with other companies undertaken by Canadian biotechnology firms.

TABLE IV
Collaboration With Other Firms

	1993	Prediction 5 years later
Collaboration by field		
R&D	67%	–
Marketing	50%	84%
Manufacturing	36%	64%
Average number of partners		
R&D	2	
Marketing	2	
Manufacturing	1	

Source: Woiceshyn & Hartel (1996).

Collaboration Between Universities and Firms

Biotechnology has its roots in the university. The first experts in the field were professors. They were the first to explore the commercial potential of the new technology, and they founded the first biotechnology firms. In the United States, industry financed between 16% and 24% of university research in biotechnology – a proportion four to five times higher than the average for university research financed by industry (U.S. Congress, Office of Technology Assessment, 1988a). Nearly half of U.S. biotechnology firms subsidize university research.

University researchers contribute to biotechnology firms in three ways (Audretsch & Stephan, 1996):

1. they facilitate knowledge transfer;
2. they give a signal of research quality to the scientific community and the capital market;
3. they help set research direction.

Well before the emergence of biotechnology, the United States had a climate favouring commercial exploitation of university research, and links were rapidly established with the industrial sector. The American entrepreneurial spirit and the dynamism of the country's capital market are factors that strongly supported this development. In other countries, governments took measures to encourage the creation of links between universities and industry. A list of measures appears in OECD (1988). In Sweden, for instance, academic researchers are now legally entitled to remuneration for industrial research projects. Sweden has also given official standing to the position of adjunct professor, thereby enabling industry experts to teach part-time. In Germany, the Bundesministerium für Forschung und Technologie gives financial assistance to joint industry-university projects. In the Netherlands, the Ministry of Economic Affairs covers 45% of research costs once a university or research institution participates in a private-sector project.

In Japan and Switzerland, however, there is little joint industry-university research. In Japan the predominant form of collaboration involves industry and public research establishments. The lack of collaboration with universities is said to be due to a rule that bars professors from acting as consultants.

Audretsch and Stephan (1996) have studied the extent to which the links between university researchers and biotechnology firms are strengthened by geographic proximity. It seems reasonable to suppose that the tacit nature of some expertise requires personal contacts when knowledge is to be passed on, and that geographic proximity is then a factor promoting collaboration. However, the study by Audretsch and Stephan indicates that, while a large number of university researchers belong to local networks, nevertheless in the United States 70% of links between universities and firms are inter-regional. According to these writers, the importance of geographic proximity would depend on the type of service offered by the university partner. If the role of the academic collaborating with a firm is simply to give a signal of quality, there is less likelihood that this person will reside in the region where the firm is located. The likelihood is greater if the role of the academic is to contribute knowledge or to train students who will later be hired by industry.

Focusing more particularly on Canada, we note that nearly all firms have established research collaboration relationships with universities or government laboratories. They have also created many links with other small firms. Canadian firms have an average of 5.4 partnerships, with development being the aim of over two thirds of these. More than a quarter of the partnerships involve both basic and applied research, while only 6% of collaborations are devoted exclusively to basic research. Few Canadian biotechnology firms, however, have links with big firms. Those that are close to production and marketing stages tend to look for a partner to obtain expertise, such as for in-situ tests (Walsh, Niosi, & Mostar, 1995). Table V provides some additional information on the links between universities and Canadian biotechnology firms.

TABLE V
Collaboration With Universities

	Percentage of firms involved
Degree of collaboration on basic research	
Nil	11.6%
Very high	18.6%
Degree of collaboration on development	
Nil	16.3%
Very high	14.0%

Source: Woiceshyn & Hartel (1996).

FINANCING

The majority of biotechnology firms obtain initial financing through research contracts, private investments of venture capital, and government grants. Over the first years of its existence, the firm does not generate revenue. It can begin to do so after several years of activity, by granting licences authorizing other parties to use a technology that it has developed. Some firms can then develop final products, but only for markets of limited size. The last stage, reached by few firms, is that of launching a product sold on a large scale (Taunto-Rigby, 1988).

The fact that the great majority of firms are still in the start-up stage could explain the considerable losses recorded by the biotechnology industry. From 1990 to 1994, the industry suffered losses amounting to nearly $14 billion. On February 25, 1995, the British magazine *The Economist* wrote of the biotechnology industry as follows: "In America, where it first started, and which remains the world's leader, the sector is steadfastly losing money, $1 billion or so last year. Only 1% of the biotech firms make a profit. Of those that do, one company, Amgen of Thousand Oaks, California, makes more than all its leading competitors put together."

In the absence of profits, the major problem facing the industry is that of financing. The financing in question must come from equity capital since the main assets of the firm cannot be redeployed. Williamson (1988) explains that the decision in favour of financing by increasing capital or indebtedness is influenced by the ability of assets to be redeployed, that is, to be used for various purposes. Assets that can be allotted to several redeployed uses lend themselves better to debt financing. There are data that confirm this idea, showing that debt is a marginal source of financing. Among the 31 biggest biotechnology firms, nine had no long-term debt in the late 1980s; for five of them, long-term debt was less than 5% of total assets. The median percentage of debt financing was 13.2% (Avramovic, 1996).

Equity-capital financing sources are of three types: wealthy individuals, venture capital organizations, and the stock market. Venture capital organizations act as intermediaries between investors and companies. They do more than merely supply financing. They study the company, evaluate its chances of success, and provide management assistance. Often, the initiative for commercializing biotechnology has come from directors of venture capital firms. They are the ones who have identified the commercial opportunities and researchers, and they have subsequently arranged the link-up of capital, management, and scientific knowledge (Florida & Kenney, 1988).

In the United States, entrepreneurs in biotechnology have succeeded in selling shares to the public despite the losses accumulated by the great majority of firms and the fact that none of them pays dividends. The success must be attributed to investors' hopes of obtaining considerable profits in the future. However, it has often happened that company founders had to abandon a large part of their business to obtain funds enabling them to stay in operation.

In other countries, the financing experience has been different from that in the United States. A study undertaken in the late 1980s suggested that, in Britain, firm start-ups suffered as a result of the approach of venture capital organizations which, unlike their U.S. counterparts, did not become involved in the firms they financed and also offered financing only with very rigid conditions (Oakey, Faulkner, Cooper, & Walsh, 1990). As the London Stock Exchange did not allow companies less than five years old to be listed, the biotechnology sector faced a clear handicap.

In the United States new sources of financing became accessible in the 1990s when two companies obtained a listing on the National Association of Securities Dealers Automated Quotations. Out of fear of losing a lucrative market and under pressure from the industry, the London Stock Exchange altered its rules to make them less restrictive (Senker, 1996).

Canada's venture capital market is still less vigorous. Biotechnology financing comes mainly from big pharmaceutical, chemical, and agri-food firms that perform research either in-house or in collaboration with biotechnology firms. New share issues are infrequent occurrences. In fact, the Canadian market resembles Europe's more than that of the United States. Over the years 1989 to 1991, only 20 biotechnology firms are said to have obtained business capital from venture capital (Walsh et al., 1995). Canada's big commercial banks and trust companies did not invest in biotechnology.

Table VI shows the findings of a survey conducted in Canada on sources of financing.

It is, however, remarkable that financing does not appear to be a key factor in choosing a production site. A survey by Ernst and Young (1995) indicates that the most commonly cited factors guiding the choice of a site for a new biotechnology investment are:

1. regulations, especially those on product approval and patent protection;
2. access to qualified personnel, a concern that explains why firms seek partnerships with universities;
3. fears about public acceptance.

Some 80% to 85% of respondents considered the first two factors to be the most important in choosing a production site. In addition, 65% of the people surveyed cited consumer attitudes, while 55% mentioned public-sector support and interest.

TABLE VI
Financing Strategies of Canadian Biotechnology Firms

	First year	Currently (1993)
Status		
Privately owned, independent Canadian company	83%	63%
Independent division or subsidiary	7%	15%
Privately owned, independent foreign company	2%	5%
Publicly owned company	5%	12%
Main source of revenue		
Research contracts, grants, venture capital	58%	24%
Licensing of technology or products, sale of research products	19%	39%
Marketing of products	12%	37%

Source: Woiceshyn & Hartel (1996).

THE IMPACT OF BIOTECHNOLOGY ON USER INDUSTRIES

Should biotechnology be defined in terms of the technological processes involved, the firms that use these processes, the nature of the products manufactured, the needs that these products seek to fill, or the markets in which they are likely to find applications? The answer to this question is not merely academic because it has repercussions for the research strategies of firms that cannot easily identify future competitors. A vague answer also complicates the partnerships required for achieving industrial development and product marketing. Last, for its part, the still-strong uncertainty about the future of technologies does not ease the task of defining regulatory policies to be introduced by public authorities. One of the characteristics of biotechnology is that it involves major cross-sectoral interactions and effects: human and animal health; diagnostics for plants, animals and humans; and applications for agriculture, food and the environment. It is thus somewhat artificial to isolate the relations between biotechnology and a sector of activity defined by reference to traditional technologies.

Given the low price of oil and the lack of support for environmental protection campaigns, biotechnology has not yet had a decisive impact in the sectors of energy and environment (see Table VII). Despite certain hopeful prospects (for example, in pollution clean-up), we may suppose that the situation will remain the same over the next decade. In the longer term, however, it is reasonable to expect that genetically engineered organisms will appear allowing us to control air, soil, and water quality via the detection of pollutants or the industrial purifi-

TABLE VII
Distribution of Firms and Jobs in Biotechnology

	Firms (%)	Jobs (%)
Therapeutics	34.5	33.7
Diagnostics	18.6	4.6
Equipment	21.9	10.2
Agriculture	12.6	24.0
Chemicals	6.1	20.0
Food	1.2	2.0
Waste	3.5	2.3
Energy	1.3	3.5

Source: Swann & Prevezer (1996).

cation of biological substances and the monitoring of biological processes as they occur. In view of the uncertainty of these prospects, in this paper we shall discuss only the recognized impact of biotechnology on the agri-food and health industries. It would be still more difficult to attempt to explain the impact on the entire economy given the fact that, while the relative value of the agri-food sector is constantly dropping in most developed countries, related sectors are experiencing strong growth (e.g., the chemical sector upstream and the pharmaceutical sector downstream).

The Agri-Food Sector

So long as there are few experiments of real scope, estimating the agricultural consequences of the introduction of biotechnology is still a matter of predicting the future. Further, the farming sector remains heavily subsidized, with the result that failure or success in introducing biotechnological products or processes will depend greatly on the unfavourable or favourable attitude of public authorities. The final impact on the sector will thus be determined not only by scientific advances but also by the faith that private- and public-sector economic players have in the benefits of the new technologies.

The necessary conditions for introducing a new technology into an industry are its technical feasibility, profitability for the inventor and user, environmental and public health safety, and finally acceptance by producers and consumers. In all sectors, these conditions are obviously not independent of each other, but this is all the more true for agri-food. Further, the degree to which they are exogenous is highly variable, depending on the time frame selected. For example, environmental safety will probably be perceptible only in the long term; it thus does not represent a constraint so long as private- or public-sector decision makers are structurally short-sighted. On the other hand, it will probably take at least the equivalent of a generation for genetically engineered food products to win acceptance by the general public with full knowledge of the facts.

The result is that success in introducing a biotechnological innovation in agri-food depends mainly on the ability of developers to bring down the costs of implementation and exploitation, and on the possibility of convincing enough trend-setting consumers that the new products are improvements over the old ones. The alternative, which is to use aggressive marketing to raise the demand at a forced pace, could be counter-productive. The public today is still hesitant about food products of unknown origin. For example, the *Financial Times* (March 13, 1997) reports that a petition signed by 1.3 million people in Austria (or 20% of the country's voters) demanded a complete prohibition on the sale of genetically altered farm products, as well as an end to experiments outside the laboratory with genetically manipulated plants, and a ban on patents on genetically altered animals or plants. Organic foods account for a higher share of the market in Austria than in any other European country. Around 80% of its population say that they are opposed to any type of genetic engineering. Similarly, the German magazine *Der Spiegel* (April 7, 1997) reports that 75% of German consumers say that they will refuse to buy genetically altered foods. The refusal rate is higher for products of animal origin than for products of plant origin. However, according to experts quoted by the magazine, the growth rate for sales of genetically altered foods will rise to 30% over the coming years. The article also quotes the Senior Advisory Group of Biotechnology (SAGB), which estimates that the European market will expand from DM 10 billion to DM 30 billion by the year 2000. What's more, the SAGB estimates that it will become virtually impossible to buy foods that contain no genetically altered ingredients. As an example, it mentions that soybeans, which have been genetically altered, are used in the production of 30,000 food products.

To examine the current situation in agri-food, we shall review in turn the impacts on production and trade, on prices, on agricultural structures and employment, and last on international trade.[2]

Impact on production and trade balances. To analyze the effects of introducing biotechnology in the agri-food sector, we should distinguish between the following:

a: product innovations:
- a1: new products: corn with high concentrations of tryptophan, industrial soybeans, firm tomatoes;
- a2: substitutes for existing products (proteins of unicellular origin);
- a3: alternative uses for existing products (bio-ethanol from grains);

b: process innovations:
- b1: protection enhancers (diagnostic kits, vaccines);
- b2: performance enhancers (bovine growth hormone, Rhizobia);
- b3: improved agricultural products and livestock (disease resistance).

With regard to products, categories a1 and a2 will introduce competition in sectors that are already relatively fragile, even though products in category a1 are differentiated from existing products. Differentiation can be an excellent marketing tool so long as the proposed food products suffer no loss of taste or colour, and are guaranteed to pose no health danger. But it is unlikely that the emergence of these new products will create new needs. Accordingly, differentiated or not, they should intensify competition. In contrast, category a3 allows new uses of grains, thereby offering an additional opportunity to producers. But this category will generate competition with other markets, such as that for ethanol. We may suppose that substitutability for older technologies is risky, especially for process innovations. Some, such as categories b1 and b3, make it possible to reduce losses caused by disease and epidemics. Others, such as b2 and b3, have the effect of improving plant and animal productivity. In sum, we may expect new processes that reduce costs while at the same time increasing yields.

The U.S. Congress, Office of Technology Assessment (1991b) has tried to estimate the impact of new technologies on the annual growth rate of certain crops in the United States. Some of its findings are presented in Table VIII.

With the exception of cotton, the predictions are thus for slower annual growth than in the past despite the introduction of new technologies. However, these should have a considerable effect, as can be seen by comparing figures in the two right-hand columns of Table VIII.

In contrast, in animal production the prospects for development through biotechnology are more clear-cut, notably because of the use

TABLE VIII
Impact on Annual Growth Rates in the United States

	1962–1982	1982–2000 without new technologies	1982–2000 with new technologies
Corn	2.6	0.5	1.2
Cotton	0.1	0.3	0.7
Rice	1.2	0.2	0.9
Soybeans	1.2	0.8	1.2
Wheat	1.6	0.7	1.2

Source: U.S. Congress, Office of Technology Assessment (1991b).

TABLE IX
Annual Growth Rates in Yield of Milk Cows in the United States

	1962–1982	1982–2000 without new technologies	1982–2000 with new technologies
Milk (per cow per year)	2.6	1.4	3.9

Source: U.S. Congress, Office of Technology Assessment (1991b).

of growth hormone. As an example, Table IX shows how the yield of milk cows has changed in the United States (average annual growth rates).

For animal products, it is thus clear that the new technologies could worsen the problem of surpluses. This is one of the main stumbling blocks that biotechnology development is likely to face in the medium term. The other is the hesitation of consumers, mentioned earlier.

Impact on prices. An advance in technology ought to bring down production costs. Biotechnology can have this effect in the agricultural sector, first by replacing old production methods; this would make it possible to reduce the time needed for regeneration of plants and animals, and thereby to lower costs. Next, it will offer savings on operating expenses. For example, by using plants resistant to insects and pesticides, farmers will save on plant health products. On this point, it is clear that savings by private operators are far outweighed by the gains to the community since the products for which we may hope to see lower consumption by farmers are pollutants that are particularly harmful to the environment. It would therefore be efficient to assist the development of "environmentally friendly" plants by distorting their relative prices, either through subsidies promoting their use or through taxes on pollutants. Last will be savings via lower prices for agricultural inputs once biotechnology has swelled the supply of such inputs.

But the beneficial effects can be achieved only in a competitive environment. Biotechnology processes are protected by property rights giving considerable market power to their holders. At the same time, it is not unusual to see partnerships or financial unions between seed and herbicide producers, for example: The farmer may thus be offered a joint product consisting of a herbicide-resistant plant as well as the herbicide in question. In the long term, the regulatory mechanisms of the market should lead to lower prices for food products as a result of the contribution of bio-industries; nevertheless, it is likely that certain producers will temporarily drive up prices for their products and will receive monopolistic rents.

Impact on agricultural structures and employment. The historical trend in the agricultural sector has been a decline in the number of farms, an increase in the size of surviving farms, greater specialization, and the development of links with other sectors of the economy. In the United States, it is estimated that the total number of farms should drop from 2.2 million in 1982 to 1.2 million around the year 2,000 (U.S. Congress, Office of Technology Assessment, 1988b). The number of large operations with turnovers above $200,000 will increase. The move toward consolidation will gather headway; it is estimated that 4% of the largest operations (with a turnover exceeding $500,000) will control 75% of all farm sales. Small holdings requiring full-time work seem doomed to disappear, but small farms worked part-time could very well persist. Is the emergence of biotechnology likely to affect this shift?

The new biotechnology should intensify the vertical integration of agriculture with input and output sectors. But it is not clear whether biotechnology will accelerate or slow down the phenomenon of horizontal consolidation; if the new technologies are inherently likely to introduce savings on work, we may suppose that they will also allow savings on capital. Accordingly, with a capital-labour ratio little affected by the technological shift, changes in agriculture structures should continue to be determined by changes in demand and in the lifestyle of farmers.

The same may be said of employment: It is unlikely that biotechnology will greatly change the downward trend in the proportion agricultural employment represents of total employment, a trend that has been observed since the beginning of the century. On the other hand, we may suppose that biotechnology will trigger a redistribution and redefinition of jobs. A share of formerly agricultural jobs will be redefined upstream as industrial jobs. Work will be reorganized within real "agri-refineries," where all the components and all the potentials of plants will be put to use. Another share of jobs will be redefined as service jobs, for example, because of growing management and marketing requirements.

Impact on international trade. Biotechnology is likely to affect global agricultural trade for three reasons:

1. because it is becoming possible to grow plants outside their original geographic environment – for example, we can extend corn production northward by introducing a more cold-resistant variety;
2. because we can now make plants diverge from their original characteristics – for example, we can produce hemoglobin with transgenic tobacco; and
3. because fractionation allows us to employ industrial methods in order to extract all the components of plants and thus to expand the spectrum of possible uses, as is already the case with the cracking of petroleum, for example.

As long as farmers in underdeveloped countries cannot acquire knowledge (i.e., access to patented products/processes at reasonable prices or indeed without charge) as well as know-how (i.e., access to bio-industrial production methods), it will be difficult to avoid their being placed at a disadvantage by the new biotechnology. An often-cited example is the production of isoglucose, a sweetener with high fructose content, from corn: From 1975 to 1985, its production increased by a factor of 12, displacing cane and beet sugar. In contrast, efforts to replace traditional animal feed proteins with proteins from isolated cells (large-scale yeast or bacterial cultures) have failed, notably because of the low price of soybeans. Accordingly, we cannot yet say that biotechnology has disrupted international trade in agricultural products. However, the status quo may not survive into the future, and we may well see disruption triggered by surplus production of certain commodities.

Developments. Plant engineering is taking off and is expected to speed up plant growth as well as develop varieties resistant to disease, microorganisms, insects, herbicides, salinity, and climate extremes. Grains (corn, wheat and rice) are the main focal points of research, but other plants raise hopes for the development of "pharming" – for example, the production of human hemoglobin through the cultivation of genetically manipulated tobacco plants. The production of transgenic animals has become a reality, with results superior to those obtained through conventional selection. Production and breeding may now be encouraged, while preventing and controlling certain types of disease. We may thus expect impacts on everything related to the area of human diet.

In quantitative terms, the new tools correspond to traditional approaches in agricultural research, which seeks to increase production and reduce costs. For livestock raising, researchers aim to reduce losses caused by disease, or else to increase production. For cultivated plants and silviculture, researchers seek to protect harvests and achieve mass production of clones in order to standardize plant species. These quantitative effects could well worsen the problem of farm surpluses if the countries concerned do not change their policies and systems. In qualitative terms, aside from possible improvements in food taste, biotechnology is expected to produce healthier foods containing less residue of pesticides and other agri-chemicals, and it is also expected to allow production to be better differentiated to meet specific needs.

Health Industries

Paradoxically enough, the future of biotechnology seems less clear in health industries than in the agri-food sector, not because doubts persist about its scientific value but because of the difficulties in setting a timetable for carrying through innovations. The needs to be met, competitors, and regulations are constantly changing and upset sector professionals' decisions about investment and current management.

Changing markets. The radical changes brought about by biotechnology and the difficulties of defining its limits in the health sector can be illustrated by the example of growth hormone.[3]

Human growth hormone (hGH) extracted from the pituitary gland has long been the only treatment for pituitary dwarfism. Production of hGH is limited by two totally independent factors: the number of pituitary glands available and the quantity of growth hormone extractable from each. In the 1980s, hGH availability changed drastically from scarcity to abundance with the production of growth hormone through genetic engineering techniques (recombination). In these circumstances, the supply is no longer limited only to uses authorized by government agencies and by economic regulation of prices. The potential interest in expanding the market is thus understandable, for example, by redefining dwarfism to cover children who previously were simply regarded as short. The problem of market definition is complicated by the fact that the scientific community is not very sure about the limits of therapeutic applications of growth hormone: It has been claimed that hGH conserves muscle mass in the elderly, is a treatment for osteoporosis, provides dietary assistance to obese people, is a treatment for burns and ulcers, and (on the fringes of medicine and ethics) increases muscle mass in athletes.

This example is quite typical of the vagueness that still colours hopes for biotechnology products, whatever the sector concerned. In the health field, an extremely limited number of products have made it to market and achieved indisputable industrial and commercial success. Insulin and growth hormone are notable exceptions. As we await the introduction of medicines that are targeted, effective, and without secondary effects, we may nevertheless find encouragement in the steady development of immuno-diagnostic tests based on monoclonal antibodies, and of gene probes. Before long, vaccines produced by DNA recombination will surely lead the next wave of products brought to market.

For the moment, the impact of biotechnology is thus still weak in terms of products in the health sector. This may be explained by the fact that pharmaceutical industries operate in a market with particularly high risks.

A 1990 survey (Ernst & Young, 1995) showed that 80% of the medications that reach the stage of clinical trials are produced by only 3% of firms in the sector. The survey also showed that 10% of the firms received 70% of financing in 1989, a finding that confirms the lack of uniformity in the sector. Only 1 medication in 10 that undergoes clinical trials is subsequently commercialized; of this proportion, only 30% are capable of generating returns allowing producers to recover R&D costs (T. Coperman, as quoted in U.S. Congress, Office of Technology Assessment, 1991a). This risk, along with increasing pressure by public authorities on prices, limits initiatives to develop new products and spurs company mergers.

Less recent research, however, has created new opportunities, and since 1986 we may observe a slight acceleration in the introduction of new products: insulin, growth hormone, interleukins, and growth factors. But priority is given to diagnostics, prevention, and immunology with the arrival on the market of a growing number of immuno-diagnostic tests using monoclonal antibodies, biosensors, and gene probes.

Restructured competition. The impact of the new technologies is making itself felt through a redefinition of the roles of firms in the sector. For example, according to OECD (1989, p. 81), for some time the pharmaceutical industry has been undergoing change: from suppliers of products (mainly medicines), it is on the way to becoming a "function industry" or a health care industry (i.e., a supplier of a wide range of therapeutic products, diagnostic products, ancillary materials, equipment and machines, as well as biomedical techniques and systems). The scientific base is increasingly interdisciplinary: This branch of activity finds support not only from chemistry, biology, and medicine but also from physics, electronics, computer science, lasers, etc. These changes have already given rise to problems having to do with internal reorganization.

Further, Pisano, and Mang (1993) stress that more and more pharmaceutical companies are integrating biotechnology departments within their organizations. A few years ago, the majority of firms in the sector still considered biotechnology as a method of developing medicines based on proteins, that is, based on large molecules that were too complex to be produced artificially, for which we had to put to use the "skills" of living organisms. Increasingly, however, they make use of biotechnology to complement their skills in conventional pharmacology. Many of them use biotechnology to improve their knowledge in designing small molecules created by organic chemistry.

If, as is likely, diagnostic tests continue to be developed on a large scale, the current transformation of the industry will make necessary the adoption of multidisciplinary organization and methodology. Now that the general public has testing tools, biological tests must be used and computerized equipment must be developed, along with software for processing data. The trend is thus toward greater integration of biotechnology, micro-electronics, and telecommunications.

The result is that, in the sector of medical technology, traditional pharmaceutical companies could find themselves competing with electronics firms. A new test is thus emerging for the pharmaceutical industry. The decisive factor in meeting this new challenge probably lies more in the capacity to run complex systems than in biotechnological knowledge itself. It is thus possible that in this battle the electronics industry will have an advantage over the pharmaceutical industry.

This shift has profound organizational and institutional repercussions. It offers privately owned firms the opportunity to extend their activities to services, to directly manage diagnostic centres, and thus to take the place of the public health system. This new trend could become more widespread, especially in countries where the public sector faces severe difficulties in adapting to technological change: Public authorities might have to seek organizational support from private industry (OECD, 1989, p. 81).

ECONOMIC REGULATION OF BIO-INDUSTRIES

As in most other activities arising out of a technological revolution (for example, the internal combustion engine or electronics), development of the earliest biotechnology products was based on their scientific and technical feasibility, without taking into account their economic cost and without the slightest marketing analysis. This period seems now to have come to an end and concerns about profitability are taking priority in the process of defining research programs. The result is that established firms have adopted strategic behaviours; for example, they are seeking to erect entry barriers, particularly by applying for patents.

Necessity of Regulation

Can an activity as innovative as modern biotechnology be completely abandoned to market mechanisms? Its basic principles as well as the variety of its applications suggest the need for a certain measure of public control. This can be technical, to meet existing quality and safety criteria; it can be economic, to limit extraction of monopolistic rents; but it can also be ethical, because of technologies that affect the integrity of the cell and its nucleus. In this paper we shall examine problems of economic regulation alone. For the moment, economic regulation of the sector basically involves controlling the use of intellectual property rights in order to offer strong research incentives to firms without, however, completely allowing them free hand to demand the rents made possible by their monopolistic position.

As proof of the importance of regulation of intellectual property, we need simply recall the role of research activities in this industry. In the United States, since the inception of the industrial phase of biotechnology in the late 1970s, firms in the sector have invested over $10 billion in long-term R&D. During the year 1991 alone, they spent $3.2 billion. In other words, the industry re-invested more than 47% of its 1991 sales in R&D. To grasp the importance of biotechnology research, we need only compare this figure with the 14% re-investment rate in the traditional pharmaceutical industry, itself considered one of the sectors undertaking the most research (Raines, 1991/1992). Similarly, for the years 1989 to 1993 in Canada, it is estimated that R&D efforts increased by an average of 41% per year (Heller Report, 1995).

Unlike other industries, where heavy industrial equipment is essential (e.g., the "white rooms" in the computer chip industry), the most important inputs for biotechnology research are intangible goods (OECD, 1989, p. 76): scientific knowledge, an information base, and know-how. We therefore understand why most large firms in the sector have felt the need to establish partnership agreements with universities (e.g., Kenney, 1986), and also why they attach so much importance to conditions on database access and to the legal conditions that will prevail in order to protect their innovations.

Protection of Biotechnological Products and Processes

In this type of activity with a high content of endogenous innovation and based mainly on intangible assets, we see the conflicting demands of static efficiency (which requires the market power of firms to be limited, particularly by prohibiting any monopoly) and dynamic efficiency (which requires protection, and indeed reinforcement, for activities characterized by increasing cross-time yields, if necessary sheltering them from any competition). The latter naturally takes precedence but the former is not completely forgotten, as the following example shows.

In 1983, the U.S. Congress passed a law to encourage development of medicines intended to treat rare diseases affecting fewer than 200,000 people. This Orphan Drug Act found a perfect field of application with recombined growth hormone. The Act gives a seven-year monopoly to pharmaceutical companies producing medicines intended for small groups of patients. By passing the law, the U.S. Congress sought to give these firms the opportunity to amortize their research expenses in a market that was insufficiently broad; the solution was to extend the market in time through the entry barrier established by the Act. In October 1985, Genentech was granted the status of Orphan Drug (OD) for its product, Protropin, on the U.S.

recombined growth hormone market. In 1987, however, well before the end of the seven-year monopoly granted to Genentech, the U.S. Food and Drug Administration approved marketing of a product from the Eli Lilly laboratories, also awarding it OD status on the grounds that the product differed from Protropin. Contrary to its expectations, Genentech thus found that it had a competitor in the U.S. market. We may regard this situation as a good example of the principle of "dynamic inconstancy," according to which decisions of the regulator do not have the value of a commitment. That being so, now that years have passed it is in the regulator's interest (and the public interest) to reconsider such decisions. The immediate risk was that Genentech would sue the FDA for breaching the 1983 law. The dynamic risk is that future research candidates would no longer attach any value to FDA commitments, with the result that research incentives would be weakened. But the company did nothing, perhaps finding consolation partly in the rapid growth of the U.S. market and partly in the effectiveness of its marketing: Between 1986 and 1995, its sales rose by over 400% from $41 million to $216 million.

Patents obviously represent a more traditional type of legal protection than that we have described. After some token opposition in the 1980s, reliance on patents has become a common practice in the biotechnology sector. It should, however, be noted that many biotechnology products cannot be patented because they do not meet the requirements of being new. Under U.S. law, the three prerequisites for a product to be eligible for a patent are that it be new, not obvious, and useful. Since the product already existed in nature, even though it is now obtained via a bio-industrial process, it is against the law to place it in the sphere of private property. Accordingly, the first medicines produced by biotechnological means, such as human growth hormone, human insulin, and erythropoietin or EPO (a hormone produced by the kidneys) could not be patented since they had already been used for years, the source being blood or cadavers.

For innovators, the alternative is then to patent the process instead of the product (as Genentech did for insulin in 1978) or to obtain the special status of Orphan Drug (as Genentech again did in 1985 for recombined growth hormone).

In these conditions, applying for patents is a particularly strategic move. When, after three years of effort, the firm Amgen succeeded in producing erythropoietin (EPO) by recombination techniques, it submitted patent applications for:

- the host cell and its vector (the DNA segment that surrounds the transplanted EPO gene and transports it into the host cell);
- the biotechnological process of producing EPO;
- EPO itself.

The application thus covered the full range possible: the process, the tools, and the product itself. Only the patent on the tools (host cell and vector) was granted in October 1987. The application for a patent on the process was refused on grounds of lack of novelty, and Amgen was also refused a patent on the product; that had been granted three months earlier to another company, Genetics Institute (GI), despite its clear inability to produce EPO industrially. To counter GI's attempt to have its EPO produced by a Japanese firm (Chugai) and have it imported and distributed in the United States by Upjohn, in 1983 Amgen applied for and obtained the status of Orphan Drug for its own product. The target population is chronic anemia patients; erythropoietin stimulates production of red corpuscles. GI's efforts to overturn this step finally failed in 1991. It was thus under the protection of this special status and not of a patent that Amgen was able to expand its production of EPO.

Biotechnology and Diversity

Diversity of products is desirable because consumer preferences are heterogeneous. Accordingly, if consumers are well informed and, particularly, if public authorities or professional associations guarantee simple labels identifying products of biotechnological origin, we may suppose that market mechanisms will ensure a proper balance between natural and genetically engineered products.

Concerning the diversity of technologies used, the situation is less clear but the problem is not specific to biotechnological activities. The variety of technologies offers insurance against technical and economic risks. By not putting all their eggs in one basket, producers are able to respond to unexpected price hikes for certain inputs, to breakdowns or disease, to regulatory changes. and to the predatory behaviour of agents controlling certain essential inputs (for example, through patent licences). The risk of loss of diversity was brought up particularly in early 1997, when researchers in Scotland transferred a nucleus to produce a cloned lamb. According to experts, however, this risk may be no greater in the case of cloning than that presented by widespread adoption of the practice of artificial insemination (Colleau, 1993). At the level of individual producers, the risk of loss of diversity can be covered by an insurance contract or mutualized within professional associations requiring their members to use a set variety of technologies. At the level of the community, public authorities must finance the conservation of plant and animal collections; they represent public goods that must be made available to all in case of need.

Prices and Trade

Price control is not an issue specifically for biotechnological products. The authorities must simply see to it that firms enjoying a monopoly (whether natural or legal) do not profit thereby in a way that is inefficient for society. It is by monitoring patents on basic processes and access to databases that monopolistic positions can be identified most easily. In particular, national and international authorities must quickly clarify the exact status of databases lacking proper legal protection in terms of either copyright or royalties. And if any price should be set by regulation, it is that for free access to biotechnological data.

Trade in biotechnological products can be limited for health-related reasons. But product regulation that leads to trade restrictions can be a violation of international rules on the free movement of goods. In Austria, for example, we noted earlier the widespread rejection of food products of biotechnological origin. A few months ago the country's government banned genetically modified corn produced by Novartis (the company created by the merger of Sandoz and Ciba), but this step has put it in conflict with European Union regulations. Brussels is demanding that the prohibition be lifted unless Austria can demonstrate that the banned corn poses health risks.

CONCLUSIONS

In industrial economics, the structural parameters characterizing a sector are, as a general rule, considered to be fixed in the short term. Horizontal concentration, vertical integration, the regulatory framework defining or limiting entry barriers, but also technical production conditions and the state of demand are all exogenous from the individual calculation. It is within these constraints that already established firms make their short-term decisions. In the biotechnology industry these structural characteristics are not generically different from those that may be observed in other sectors, but they also bear the stamp of the immature nature of the industry. For example, much more importance is given to R&D activities than in most other industries; markets are not clearly defined; and agri-food and health firms string together founding, buyback, and merger operations driven by the fear of missing the revolution in controlling the production of living organisms.

Partnerships are important in biotechnology because, at least at present, there is only very partial application of the concept of firms representing clearly defined entities that compete with each other according to rules found in the classic texts on micro-economics. What we find are groups of networks that partly overlap, with shifting scope and with components that simultaneously co-operate and compete with each other (Greis et al., 1995). Modern industrial economics has developed tools for analyzing these fluctuating entities of research, production, and marketing (Tirole, 1988, Ch. 10).

The arrival of new, genetically engineered products will not necessarily lead to the disappearance of the products they were intended to replace, at least not in the short or even the medium term. Some consumers do not want to use products created through biotechnological processes, and it will thus be necessary to maintain product differentiation at least for a generation. In some cases, however, the new product may be a clear improvement and we may expect that the product it has superseded will be retired. For example, the relative superiority of recombined human growth hormone over hormone extracted from pituitary glands became absolute, indisputable superiority on the April 1985 day when the U.S. Food and Drug Administration announced that a patient treated with pituitary hGH had died of Creutzfeldt-Jacob disease.

In a sector so avid for intangible assets, clearly public action is most urgently needed on the issue of access to databases, and it is obvious that the solution will be international in scope. For both ethical and economic reasons, what is required is a hierarchy of databases, from the most restricted to the most public. The problem is how to provide financing, and access fees will have to be charged even for databases open to all.

NOTES

[1] "Patents on Biotechnological Processes" and "To Authorize Use by Regulation the Presentation of 'Woodsy Owl'," Hearing before the Subcommittee on Courts and Intellectual Property of the Committee on the Judiciary. U.S. House of Representatives, HE 1269 and HR 587, Serial No. 16, March 29, 1995.

[2] Our main bibliopgraphical source for studying the agri-food sector is OECD (1992, Ch. VIII).

[3] Most of the following discussion of growth hormone is taken from Kelvey (1996).

REFERENCES

Audretsch, D. B., & Stephan, P. E. (1996). Company-scientist locational links. *American Economic Review*, *86*, 641–653.

Avramovic, M. (1996). *An affordable development? Biotechnology, economics and the implications for the Third World*. London: Zed Books.

Burgelman, R. A., & Rosenbloom, R. S. (1993). *Research on technological innovation. Management and Policy*, Vol. 5. London: JAI Press.

Club de Bruxelles (1996). L'avenir des biotechnologies en Europe – De la recherche-développement à la compétitivité de l'industrie. Brussels: Club de Bruxelles.

Coleman, R. (1987). National policies and programmes in biotechnology. Paper presented at the Canada-OECD Joint Workshop on National Policies and Priorities in Biotechnology. Reprinted in: OECD (Ed.), *Biotechnology and the changing role of government*, pp. 88–101. Paris: OECD.

Colleau, J. J. (1993). *Les Cahiers Agricultures*. Paris: Institut National de Recherche Agronomique.

The doctor's dilemma (1990). *The Economist*, January 27, p. 314.

Ernst & Young (1995). *Biotechnology's economic impact in Europe: A survey of its future role in competitiveness*. Toronto: Ernst & Young.

Florida, R., & Kenney, M. (1988). Venture capital, financial innovation and technological change in the USA. *Research Policy*, *17*, 119–137.

Greis, N. P., Dibner, M. D., & Bean, A. S. (1995). External partnering as a response to innovation barriers and global competition in biotechnology. *Research Policy*, *24*, 609–630.

Heller Report (1995). *Background economic study of the Canadian biotechnology industry*. Ottawa: Industry Canada.

Kelvey, M. D. (1996). *Evolutionary innovations: The business of biotechnology*. Oxford/New York: Oxford University Press.

Kenney, M. (1986). *Biotechnology: The university-industrial complex*. New Haven: Yale University Press.

Krimsky, S. (1991). *Biotechnics and Society: The rise of industrial genetics*. New York: Praeger.

Oakey, R., Faukner, W., Cooper, S., & Walsh, V. (1990). *New firms in the biotechnology industry: Their contribution to innovation and growth*. London: Pinter.

OECD (1988). *Biotechnology and the changing role of government*. Paris: Organisation for Economic Co-operation and Development.

OECD (1989). *Biotechnology: Economic and wider impacts*. Paris: Organisation for Economic Co-operation and Development.

OECD (1992). *Biotechnology, agriculture and food*. Paris: Organisation for Economic Co-operation and Development.

Pisano, G. (1990). The R&D boundaries of the firm: An empirical analysis. *Administrative Science Quarterly*, *35*, 153–176.

Pisano, G., & Mang, P.Y. (1993). Collaborative product development and the market for know-how. Strategies and structures in the biotechnology industry. In: R.A. Burgelman & R. S. Rosenbloom (Eds.), *Research on technological innovation. Management and Policy*, Vol. 5. London: JAI Press.

Powell, W. W., Koput, K. W., & Smith Doerr, L. (1996). Interorganizational collaboration and the locus of innovation: Networks of learning in biotechnolgy. *Administrative Science Quarterly*, *41*, 116–145.

Raines, L. J. (1991/1992). Protecting biotechnolgy's pioneers. *Issues in Science and Technology*, *8*(2), 33–39.

Roberts, E., & Mizouchi, R. (1989). Inter-firm technological collaboration: The case of Japanese biotechnology. *International Journal of Technology Management*, *4*, 43–61.

Senker, J. (1996). National systems of innovation, organizational learning and industrial biotechnology. *Technovation*, *16*(15), 219–229.

Swann, P., & Prevezer, M. (1966). A comparison of the dynamics industrial clustering in computing and biotechnology. *Research Policy*, *25*, 1139–1157.

Taunto-Rigby, A. (1988). Biotechnology companies as pharmaceutical competitors: Their marketing strategies may hold the key. *Spectrum*, *1*, 1–21.

Teece, D. J. (1996). Firm organization, industrial structure, and technological innovation. *Journal of Economic Behavior and Organization*, *31*, 193–224.

Tirole, J. (1988). *The theory of industrial organization*. Cambridge, MA: MIT Press.

U.S. Congress, Office of Technology Assessment (1988a). *New developments in biotechnology: U.S. investment in biotechnology*. Report OTA-BA 350. Washington, D.C.: U.S. Government Printing Office.

U.S. Congress, Office of Technology Assessment (1988b). *Technology, public policy and the changing structure of American agriculture*. Report OTA-F 285. Washington, D.C.: U.S. Government Printing Office.

U.S. Congress, Office of Technology Assessment (1991a). *Biotechnology in a global economy*. Report OTA-BA-494. Washington, D.C.: U.S. Government Printing Office.

U.S. Congress, Office of Technology Assessment (1991b). *The U.S. dairy industry at a crossroad: Biotechnology and policy choices*. Report OTA-F-470. Washington, D.C.: U.S. Government Printing Office.

U.S. Congress, Office of Technology Assessment (1993). *Pharmaceutical R&D: Costs, risks, and the rewards*. Report OTA-H-522. Washington, D.C.: U.S. Government Printing Office.

Walsh, V., Niosi, J., & Mustar, P. (1995). Small-firm formation in biotechnology: A comparison of France, Britain and Canada. *Technovation*, *15*(5), 303–327.

Williamson, O. (1988). Corporate finance and corporate governance. *Journal of Finance*, *43*, 567–591.

Woiceshyn, J., & Hartel, D. (1996). Strategies and performance of Canadian biotechnology firms: An empirical investigation. *Technovation*, *16*(5), 231–243.

THE AUTHORS

Claude Crampes is Professor of Economics at Université des Sciences Sociales de Toulouse, Institut d'Economie Industrielle, Place Anatole France, F-31042 Toulouse Cedex, France. E-mail: crampes@cict.fr. Abraham Hollander is Associate Professor of Economics at Université de Montréal, Centre de Recherche sur Transports, CP 6128, Succursale Centre-Ville, Montreal, Quebec, H3C 3J7, Canada. E-mail: hollande@poste.umontreal.ca.

Jeffrey G. MacIntosh and
Douglas J. Cumming

Consumer Controversy and the Funding of Biotechnology Research

ABSTRACT. Manufacturers of the products of biotechnology confront a variety of risks in making decisions regarding how to allocate research and development budgets. This paper explores the importance of one such factor: the prospect of consumer resistance based on real or imagined harms to consumers or the environment, and/or philosophical opposition to particular biotechnologies. A survey of the Canadian biotechnology industry provides evidence that a majority of manufacturers take potential controversies into account in funding decisions. However, even in those industries in which controversies are most likely to arise, potential controversy plays a minor role compared to other determinants of research and development budgets.

The pace of scientific change has increased at a staggering clip in the twentieth century. Biotechnology is a crowning example. It was barely more than 20 years ago that biotechnology got its start, with the founding in April 1976 of Genentech Inc., a company now worth over $6 billion in market capitalization (Hall, 1996). Optima Consultants (1994) reports that by 1994, field trials for over 300 genetically engineered crops had been approved in Canada. In 1995, biotechnology sales of Canadian firms amounted to nearly $3.5 billion, and as of April 30, 1996, approximately 12 varieties of genetically altered foodstuffs had been approved for sale (Canadian Biotechnology, 1997). In the United States, 29 varieties of genetically altered foodstuffs have been approved for sale, and nearly 7% of all agricultural acreage is devoted to producing genetically altered food (I'm modified, 1998). In addition, despite long product development cycles, the U.S. genetic engineering industry had 39 drugs on the market in 1996 (K. Day, 1996). Other significant developments have been realized or are in progress not only in food products, but in health and environmental applications as well.

The promise of biotechnology is truly staggering. As a broad generalization, the bulk of the benefits appear to fall within three main subject areas: food and agriculture, the medical sciences, and the environment. In food and agriculture, biotechnology has already made

Journal of Consumer Policy **21**, 1998.
Bartha M. Knoppers & Alan D. Mathios (eds), Biotechnology and the Consumer.

advances in controlling both weeds and pests. For example, through gene transplantation, scientists have succeeded in causing a number of food crops to produce a protein that is indigestible to a variety of plant pests (Estruch et al., 1997). Such advances hold out the promise of significantly enhanced crop yields, benefitting developed and undeveloped countries alike.

Genetical alterations also hold out the promise of improving food products in significant ways. For example, various crops can be engineered to ripen more quickly or more slowly (depending on what produces a superior product), to have a longer shelf life, look better, taste better, exhibit superior nutritional profiles, reduce frost damage, mitigate freezer burn, enhance drought-resistance, and so on (Bohrer, 1994). Indeed, genetic engineering has already reached the stage at which virtually any gene that can be identified can be transferred to another organism without respect for the natural barriers that ordinarily inhibit gene transfer (Pan et al., 1994; Ronk, Maryanski, & Thompson, 1990; Yom & Bremel, 1996).

Biotechnology also holds immense promise in the medical sciences. This promise includes the production of novel medicines and improved ways of producing old ones. For example, since the early 1980s, recombinant DNA technology has been used to produce human sequence insulin on a scale, and at a cost previously unheard of (Patrick & Williams, 1993). This has been a life-saver for many of the tens of millions worldwide who suffer from diabetes (Nilsson, Jonasson, & Samuelsson, 1996). Many other novel medicines are on the drawing-board: In 1995, there were 771 biotechnology drug projects known to be in development worldwide, including gene therapy and recombinant DNA projects (Schlumberger & Stradler, 1997).

Biotechnology may also revolutionize the theory and practice of environmental science. Bioremediation of oil spills is an example. Oil spills can cause enormous damage to marine environments, often killing virtually all flora and fauna within affected areas, particularly at shorelines. Minimizing environmental damage on the occurrence of an oil spill requires a fast-acting and highly effective means of converting spilled oil into substances less toxic to the marine environment. Bioremediation utilizes naturally occurring hydrocarbon-degrading microorganisms to clean oil spills. Affected waters may either be seeded with microbial hydrocarbon-degraders, or fertilized to enhance the oil-consuming abilities of naturally occurring microbes

(Atlas, 1995). Bioremediation was successfully used, inter alia, to clean up the Exxon Valdez spill in Prince William Sound in Alaska in 1989 (Atlas, 1995).

Despite its enormous promise, biotechnology is not uncontroversial. The first and most widely publicized North American controversy over the use of biotechnology concerns the use of rBST (Bohrer, 1994). This recombinant version of a naturally occurring cow hormone is manufactured and injected into cows to boost milk production. Early reports claimed that rBST increased milk production by up to 20 or 25%, although more recent reports suggest that rBST's success in boosting milk production is in fact more modest (rbST Task Force, 1995). Despite anticipated benefits, some have suggested that the use of rBST poses a danger either to treated herds or to humans who consume cows' milk (Bueckert, 1996; P. Day, 1994). While rBST has now been approved for use in the United States, it remains under study in Canada.

That some consumers find various biotechnolgies unpalatable is well illustrated by the reaction of the citizens of various European nations to the use of genetically engineered food products. Environmental activists, consumer groups, and others have vigorously protested the importation from North America of genetically engineered crops such as corn, soybeans, and tomatoes (Ibrahim, 1996; O'Neill, 1997; Weiss, 1996). These protests have had a major impact in keeping many of these products off the European market (I'm modified, 1998).

As discussed below, there are non-trivial risks that inhere in the use of biotechnologies, whether in the food, health, or environmental fields. There are also imaginary risks with little or no basis in scientific fact, which arise out of a fear or mistrust of science that extends well beyond the confines of biotechnology. But whether the risks are non-trivial or imaginary, they represent a potential cost for companies doing research in biotechnology. For example, a company might spend tens or hundreds of millions of dollars (and many years) developing a genetically engineered soybean, only to find the market evaporate overnight because of vigorous opposition from environmental activists – or as a result of regulatory disapproval spearheaded by interest group opposition.

An ideal capital budgeting program for companies engaged in biotechnology research will take these prospective costs into account. In theory, projected market demand should be adjusted to take account of any future controversies. Admittedly, however, such risks are hard to quantify. First, the strength of any adverse consumer (or interest group) reaction and its affect on demand must be estimated. Second, the likelihood of controversy must be evaluated, and the projected cost of the controversy discounted by a factor reflecting this likelihood.

Obviously, making these evaluations will be extremely difficult, particularly when corporate decision makers are presented with a variety of competing projects in very early stages of development. At an early stage, the nature of the products that will ultimately result from various alternative projects may not be fully known. Moreover, a myriad of factors that are only vaguely understood enter into the determination of consumer and interest group reaction.

Despite the difficulties of evaluating the risks of a consumer controversy, the potential for adverse consumer reaction to biotechnology products like rBST and genetically altered food products suggests that manufacturers should make at least a qualitative evaluation of potential controversies on future demand. The purpose of this paper is to determine whether, and to what extent, that is in fact done. To that end, questionnaires were sent out to every firm in Canada engaged in biotechnology research. A copy of the questionnaire is available on request.

THE ORIGINS OF CONSUMER CONTROVERSIES

Because biotechnology is a broad-ranging field that embraces a wide variety of different applications, the possibilities for controversy are as limitless as the myriad products that may emerge from biotechnology research. Consumer controversies may obviously arise as a consequence of various adverse impacts of the new technologies. These adverse impacts may be spread widely, or concentrated among a few. It is thus useful to think of the risks created by biotechnology as either risks facing society at large, or risks faced by particular individuals (or discreet sub-populations of individuals). For example, should biotechnology accidentally produce a class of "superweeds" that resist conventional herbicide treatments, it may be necessary to use *greater*, and not lesser, quantities of herbicides to protect crops from weeds, resulting in enhanced herbicide-related pollution of the environment. Since the impact of an outcome of this nature is spread widely, this adverse impact is more societal than individual in nature.

By contrast, should genetically altered soybeans contain allergens that have a potentially devastating impact on a very small part of the population, this impact is less societal than individual in nature. The benefits associated with products developed through biotechnology can be similarly classified.

Controversies often arise not simply as a result of concerns about adverse impacts of particular technologies, but about related moral and ethical issues. Genetic testing, for example, poses little direct physical health risk, but raises important ethical concerns associated with how test results are used. For example, will test results be used by prospective parents as a basis for aborting fetuses with "undesirable" characteristics? Will they be used to facilitate attempts to create a master race?

Indeed, in many cases, concerns about adverse impacts are difficult to disentangle from ethical concerns associated with particular technologies. In the above noted examples, the potential impacts of the use of particular technologies are regarded by some as adverse precisely because these impacts are considered to be morally repugnant (rather than directly harmful to the health of some or many individuals). There will be cases, however, in which moral or ethical concerns are partly or wholly divorced from questions about adverse impacts. Genetic engineering, whether of foodstuffs or otherwise, is a prime example. Many oppose genetic engineering not because it may lead to the spread of allergens in food products, create a danger of ecological disasters, or lead to other particularized societal or individual harms. They oppose genetic engineering because, in their view, altering the genetic code of living things for any purpose is morally wrong.

In addition to adverse societal impacts, adverse consumer impacts, and moral and ethical concerns, controversies may arise out of a misperception of actual risks associated with particular technologies. Food irradiation, while not strictly a biotechnology, supplies a good example. First approved in the United States in 1990 for some (but not all) foodstuffs (McCann, 1995), food irradiation consists of exposing foodstuffs to gamma radiation (X-rays), usually from cobalt-60 isotopes (Nooitgedagt, 1997). Such exposure reduces the number of pathogenic and spoilage micro-organisms in food, generally without causing any harmful side effects (such as adversely affecting nutritional quality or taste). The result is a safer food product with lower levels of preservatives, greater transportability, and a longer shelf life (Nooitgedagt, 1997). The safety and the efficacy of food irradiation have been widely recognized by scientists and by reputable bodies such as the World Health Organization (Loaharanu, 1996; Nooitgedagt, 1997).

The process of irradiation involves the use of X-ray technology that leaves no residual traces of radiation in food. Many consumers, however, mistakenly believe that food irradiation makes food radioactive. Consumer resistance to purchasing irradiated foods has thus been considerable (Sparks & Shepherd, 1994). Unfortunately, despite dissemination of information concerning the safety of irradiation by both industry and governmental organizations, manufacturers have found it difficult to overcome these persistent consumer prejudices (McCann, 1995; Perry, 1995).[1] Many consumers continue to associate food irradiation with negative images gleaned from such events as Three Mile Island, atomic bombs, and painful cancer treatments (McCann, 1995). These false and groundless images of the dangers of food irradiation serve as a potent reminder that misperceptions of the nature and consequences of particular technologies may keep valuable innovations out of the marketplace.

EXAMPLES OF ACTUAL AND POTENTIAL CONTROVERSIES

As discussed, the most controversial aspect of biotechnology in North America is the controversy associated with the use of rBST, a synthetic hormone used to boost milk production in cows. It is not difficult, however, to find other examples of adverse consequences potentially associated with biotechnology that either have generated controversy or may generate controversy in the future. For example, many have suggested that genetic engineering has the potential to create "superweeds" that will wreck havoc on the environment. A superweed might be created, for example, by conferring upon a field crop genetic resistance to a particular pest or variety of pests. This genetically modified crop might escape into the environment, outcompeting local plant populations and altering the delicate balance of the local eco-system (MacIntosh & Cummings, 1996).

The introduction of new plant species may alter surrounding ecosystems in other ways. For example Nature Mark's New Leaf potato has been genetically altered so that it produces a bacterial protein that is indigestible to the Colorado Potato Beetle. This allows for much lower use of pesticides to control these pests. Animal or plant species that feed on the beetle, however, may find themselves without a food source. Those that feed on these feeders may in turn be adversely affected, with a cascade effect reverberating throughout the entire ecosystem.

Biotechnologies may also create adverse consequences for individual consumers. Genetic engineering may result in the transfer of an allergen from one food to another. For example, Pioneer (a company engaged in agricultural biotechnology research) attempted to enhance the protein quality of soybeans by implanting a gene from the brazil nut. Unfortunately, the transfer of genetic material resulted in a transfer of the brazil nut's allergenicity. Pioneer abandoned the project because of the threat that the altered soybean would pose a danger to those with severe, and often life-threatening allergies to brazil nuts (Nordlee, Taylor, Townsend, Thomas, & Bush, 1996). Just as genetic manipulations of foodstuffs may transmit allergenicity from the donor to the donee organism, so it may create toxicity in foods where none existed before. By affecting the combination of proteins, carbohydrates, and lipids in food substances, genetic engineering can also alter the digestibility or nutritional characteristics of foods.[2]

One of the more intractable problems associated with biotechnology is the risk of adverse long-term consequences that do not show up during product testing. While the testing process for new biotechnologies will typically last many years, it is possible that adverse consequences will show up only after a decade or more, and hence will go undetected in the approval process. An example of unanticipated long term consequences (although admittedly not from the sphere of biotechnology) is the deleterious consequences associated with the use of asbestos. Asbestos, in various forms, was used for decades before associated health hazards became known (Lee, Van Orden, Corn, & Crump, 1992). Because biotechnology is so young, it is difficult to evaluate the danger of long-term unanticipated affects. Potential long-term affects, however, properly belong in the category of "non-trivial" risks, rather than "imaginary" risks.

The introduction of new biotechnologies designed to reduce risks (particularly those in the health fields) may bring about changes in behaviour that tend to offset the risk mitigation. For example, increased resources devoted to mountain rescues have induced a greater number of inexperienced climbers to climb mountains, resulting in more rather than fewer deaths (Clark & Lee, 1997). It can be anticipated that similar responses may occur following the introduction of various products of biotechnology. For example, the introduction of genetic therapies to address heart disease may lead to changes in life style and diet that partially (perhaps even completely) offset the benefits.

Such changes in behaviour may indeed heighten various risks for those who either do not consume biotechnology products, or consume them only indirectly. For example, increases in auto safety have led to more aggressive driving, resulting in increased risks to pedestrians (Peltzman, 1975). In like fashion, some biotechnology products are likely to directly change activity levels in ways that result in increased harm to consumers or to the environment. Bioleaching, for example, promises to lower the cost of mineral extraction. Because lower costs mean higher profits, this will almost certainly lead to a higher aggregate level of activity in the mining industry. Many resource properties that cannot currently be mined at a profit will be brought into production, and the scope of operations that are currently profitable will be expanded. This has obvious implications for the environment.

IMAGINARY RISKS

Uninformed critics sometimes imagine risks that can only be described as "imaginary," given that their associated probabilities are either close to zero or zero. For example, a gene that yields resistance to ampicillin (an antibiotic) has been used as a marker gene to separate corn that has been genetically altered from corn that has not. Some have raised the spectre of the gene jumping from the corn to cattle fed with corn, and from cattle to humans who consume beef from the cattle. Cross-species gene jumps, however, appear to be sufficiently unlikely to assign this fear to the "imaginary risk" category (Weiss, 1996).

As discussed above, the experience with food irradiation amply illustrates the potential consequences of imaginary risks. Food irradiation appears to be extremely safe. Nonetheless, there is an imagined and fanciful perception that food irradiation results in making food radioactive. This has kept the public from enjoying the many benefits that might result from more widespread use of food irradiation.

WHAT FACTORS MEDIATE CONSUMER REACTIONS TO BIOTECHNOLOGY?

There are a host of factors that influence consumer perceptions of biotechnology products. A study of these factors suggests that consumer perceptions vary substantially from product to product based on a wide variety of considerations. This is vitally important. In assessing the risks of potential consumer controversies, manufacturers must make a case-by-case determination based on the unique characteristics of each product in question.

These factors – discussed in more detail in MacIntosh and Cummings (1996) – can be summarized as follows.

Outrage Factors

There is evidence that certain product characteristics are more likely than others to produce consumer "outrage," and increase the likelihood of a controversy. For example, Douthitt (1995) notes that:

> There is . . . growing theoretical literature positing that *outrage factors* mediate risk perceptions. Outrage factors and perceived risk are positively correlated; that is, consumers underestimate hazards of risks that are low-outrage and overestimate the hazard of risks that are high-outrage. Risk psychologists believe that consumers are more likely to express outrage with (1) involuntary (versus voluntary) exposure to risk, (2) exotic (versus familiar) products, and (3) products or services that have salient characteristics of deep moral significance (versus insignificance) (p. 329).

Grobe and Douthitt (1995) suggest that protests over the use of rBST are a product of the high outrage factor produced by this product. In particular, the consumer has no choice about whether the milk she purchases comes from treated cows. Nor will she even know whether the milk is derived from treated cows. As highlighted by at least one survey of Canadian consumers (Ekos Research, 1996) the absence of knowledge and choice appears to be a particularly potent "hot button" for some consumers.

Product "Purity" and Interference With the "Natural Order"

One of the findings to emerge from a study of consumer reactions to rBST is that consumers are less tolerant of alterations to products that are perceived to be "natural" or "pure" (Busch, 1991; Douthitt, 1995). Many consumers perceive milk to be a "pure" food product, and regard the use of rBST as unnatural and interfering with purity of milk. This is true despite the fact that cows naturally produce BST. Biotechnology, by its very nature, produces "unnatural" results; i.e., results that are not found in nature. Douthitt (1995, p. 329) suggests that "biotechnology is considered by most people to be an exotic, unfamiliar technology . . . [S]ome consumers are likely to hold deep, moral beliefs that human interference with the 'natural order' is wrong."

Interference with the natural order is thus an overarching factor that colours consumer perceptions of biotechnology. However, some products are seen as interfering more with the natural order, or product purity, than others. This finding emerges not only from a study of the rBST case, but from surveys of Canadian consumers as well (Ekos Research, 1996).

Benefits to Consumers

Consumers are more likely to accept biotechnological innovations if they perceive that significant benefits will result to themselves. The Calgene Flavr Savr tomato, for example, results in a firmer, more nutritious, more colourful, and better tasting tomato. The use of rBST, by contrast, results in a product that is indistinguishable from the milk produced by untreated herds. As a consequence, the Flavr Savr tomato is more readily accepted by consumers (Douthitt, 1995).

Benefits to Society

Canadian surveys of consumer reactions to biotechnology suggest that consumers are far more likely to be accepting of a particular technology if they perceive a resulting benefit to society, and not simply to consumers. A survey by Ekos Research (1996) specifically investigates the relative weighting of personal and societal benefits in consumer evaluations of biotechnologies. Ekos used focus group methodology to determine consumer reactions to two products: the Calgene Flavr Savr tomato and Nature Mark's New Leaf potato. The tomato, which is better tasting and more nutritious, was chosen by the survey researchers as paradigmatic of a *personal*, rather than a *societal*, benefit. The potato, on the other hand, is engineered so as to be naturally resistant to the Colorado Potato Beetle, allowing for significantly reduced use of pesticides. The potato was thus included in the survey as paradigmatic of a *societal*, rather than a personal, benefit.

The Ekos study offers confirmation of the view that consumers value societal benefits over personal benefits; consumers exhibited a fairly strong preference for the potato. This preference was based partly on the reduced pesticide use associated with the potato. That was indeed thought by consumers to result in a benefit to society. In contrast, the tomato was not considered an important or particularly desirable innovation because of the absence of a societal benefit.

Optics

It is clear that consumer reactions sometimes depend not merely on an understanding of the dangers and benefits associated with particular products, but on visceral reactions. As Creative Research (1996) notes:

> Semantics are extremely important. Jargon breeds suspicion. "Bio" sounds friendly. "Genetic" conjures up images from mutant-filled horror movies. "Bioremediation" and "biolixiviation" are impenetrable. The public is more comfortable with "biofiltering" and "biorestauration" because they recognize parts of the terms (p. 7).

Consumers sometimes respond in a visceral way not only to labels attached to products or processes, but to the origins of the product. For example, consumers appear to react negatively to a tomato that has been rendered frost resistant by the insertion of a fish gene (Optima Consultants, 1994). While many people are quite happy to eat fish, or indeed fish with a tomato salad on the side, the direct combination of tomato and fish genes is uncongenial for many people. Thus, it is clear that manufacturers must realize that optics will play a significant role in consumer acceptance of various biotechnologies.

Attitudes to Science Generally

One of the more consistent findings in the literature assessing consumer attitudes towards biotechnology is that these attitudes are not divorced from attitudes towards science in general. In particular, past and highly visible scientific failures tend to create mistrust that spills over into concerns about biotechnology. As the study by Creative Research (1996) notes:

Where knowledge is lacking, reactions are generally characterized by suspicion. Thalidomide, breast implants, asbestos and Urea Formaldehyde insulation are all fodder for this distrust of new technologies (p. 6).

Mistrust of science based on past mishaps was observed in all three Canadian surveys of consumer attitudes reviewed for this paper. Anecdotal evidence suggests that the same is true in Europe, where public nerves about genetically altered crops have been rattled by recent controversies over mad cow disease, nuclear power, and overhead power cables.

Mistrust of Private Actors

Consumer surveys also demonstrate that the public has a deep-seated mistrust of the private actors who produce biotechnology products. The survey by Creative Research (1996) found that "those with the least credibility on biotechnology issues include religious leaders, corporations and industry associations (p. 10)." The survey by Ekos Research (1996) found that participants expressed scepticism and concern about the ability of governments and corporations to "do the right thing." With respect to government, there was also a concern about whether it could exert enough control when challenged by multinationals, who were perceived to be aggressive and self-serving by some participants.

Mistrust of Government

Consumer surveys and focus group discussions have also disclosed that many people mistrust government actors. Government approvals of particular products are not sufficient to allay concerns over those products. According to Optima Consultants (1994), this stems in part from hi-tech mishaps, in which government has been implicated, and where public health was endangered (e.g., thalidomide and silicone breast implants). Ekos Research (1996) also found that "in the view of participants . . . government regulators were not infallible when it came to product safety decisions." The Ekos survey also suggests that some people feel that government can be pushed around by aggressively self-serving multinational corporations.

Ethical Concerns

Opposition to genetic engineering is not based simply on health concerns over the use of genetically altered products. It is also based on a conviction, deeply held by many, that genetic engineering will lead humanity down a path that it should not travel. In general, the public is acutely mistrustful of any research involving genetics. For example, Creative Research (1996) found that "virtually all participants were uncomfortable with technologies that involve genetically altering organisms." Fears of what might come of genetic research have been fed by genetic experiments with animals (the most dramatic being the cloning of a sheep by Scottish researchers). If animals can be genetically engineered, then clearly so can human beings.[3] One concern that many have is that genetic engineering will be used in attempts to develop a "master race" of people. Another is that fetuses with genetic flaws that can be detected before birth will be routinely aborted. Such "flaws" may, for example, include fetuses that show a genetic predisposition to homosexuality. And even if genetic manipulations such as cloning are confined to animals, some have raised important concerns about animal welfare (Gordon, 1997).

The Luddite Factor

Some people will oppose new technologies simply because they do not like change, or are deeply suspicious of science and technology. Ekos Research, for example, found that:

> Biotechnology was also viewed by some people as generating problems rather than solving problems. Biotechnology had the capability, for example, to eliminate jobs and erode a traditional way of life (e.g., farming).

The Ekos Research survey also disclosed that some people "objected in principle to the addition of a foreign gene" to a food product, even though genetic interchange and mutation lies at the heart of the natural evolutionary process that has produced all of the animal, plant, and microbial life on earth. Some of the more activist anti-biotechnology groups, especially in Europe, seem to fall into the Luddite camp. Karacs (1996) notes the following about German activists:

> In a country obsessed with the environment – witness the Brent Spar controversy – eco-zealots have seized on DNA as the symbol of Man's destructive tinkering with

Nature. Diagrams of the double helix adorn Green stalls on the main squares of Germany's towns. The tomatoes, sheep-goat chimeras, and now the killer maize lurking on the country's borders are the new totems of the Green movement; the ecologists' final frontier.

Such activists are unlikely to be persuaded by any amount of reason or science that biotechnology products can be used safely. As the German experience indicates, however, these activists can have a large impact on popular debates about biotechnology.

OTHER FACTORS THAT MAY RESULT IN OPPOSITION TO BIOTECHNOLOGIES

A variety of other factors may play a role in creating or deepening controversies. For example, local producers, fearing the incursion of foreign competitors with a leg up on particular biotechnologies, may disguise anticompetitive motives by raising ostensibly substantive concerns about particular biotechnologies. Other political concerns may intrude. An example is the Human Genome Diversity Project, in which "researchers would compare genes from hundreds of ethnic groups in order to map relationships, track prehistoric human migrations and perhaps make medical discoveries" (Hanley, 1996). A concern which some have expressed is that the project has the potential to upset native myths and oral histories, especially concerning the origins of aboriginal peoples.

CONTROVERSY AND CONSUMER DEMAND FOR THE PRODUCTS OF BIOTECHNOLOGY

Food biotechnologies have been the focal point for controversies about biotechnology. These controversies, particularly those related to genetically engineered food products, have clearly had an impact on the demand for these products. The biggest impact has been felt in Europe, where such foods are extremely controversial. In some cases, grocery stores and other businesses have responded to pressure from consumers and others by indicating that they will not carry products resulting from biotechnology. As reported by Ibrahim (1996), grocery chains in England such as Sainsbury, Tesco, and Safeway P.L.C. have demanded that American producers both separate and label genetically altered soybeans. In Germany, Unilever and Nestle Deutschland A.G. have pledged not to use genetically altered soybeans.

Other commercial interests have demanded labelling of genetically altered foods. In England, for example, London restaurateurs have supported a ban on unlabelled genetically altered foods, as has the U.K. Consumers' Association (Neale, 1996). EuroCommerce, a trade association representing one third of the European Union's food wholesalers and retailers, has also demanded labelling of genetically modified crops. Because of difficulties associated with keeping genetically altered crops separate from conventional crops in the distribution process, demands for separate labelling are often thinly disguised campaigns against the use of genetically altered crops. As a result of these pressures, some American food distributors have agreed to give undertakings to their European customers that their crops have not been, and will not be, genetically modified (Neale, 1996).

While most critics of biotechnology make their views known peacefully, this is not true of all environmental groups. While anti-technology terrorists like the Unabomber are clearly outliers, many consumer and "green" groups in Europe have resorted to more than merely speechifying to make their point. Neale (1996) reports that European activists, for example, commandeered dockside cranes in an attempt to prevent the docking of a ship from the United States carrying genetically altered soybeans. In the U.K., activists have distributed leaflets calling on "Earth and Animal liberation groups to target biotechnology and genetic engineering, this including both animal and food research stations." Research farms in the Netherlands have been attacked. Karacs (1996) notes that in Germany, "research institutes have been forced to erect top-security fences, not so much to confine their creations to the lab than to guard against terrorist attacks." Neale (1996) reports that even in the U.S., Greenpeace activists have sprayed dye on genetically modified crops. Weiss (1996) reports that an international coalition of health and environmental activists has also spearheaded the charge for a consumer boycott of 10 U.S. corporations that either use or intend to use genetically altered crops in their products.

THE INTERPLAY BETWEEN CONSUMER CONTROVERSY AND THE FUNDING OF BIOTECHNOLOGY R&D: SOME EMPIRICAL FINDINGS

The purpose of this section is to utilize survey data to establish empirical regularities (or "stylized facts") regarding the factors that play a role in determining the funding of biotechnology research and development. In support of establishing such stylized facts, references are made to respondents' descriptive commentaries relating to the due diligence process employed in allocating research and development funds. In addition to a variety of sub-issues, the following principal questions will be addressed herein:

1. Do firms take into account current or potential consumer controversies in their due diligence review in deciding whether to fund particular R&D projects?
2. What are the financial and other characteristics of firms that address consumer controversies in their R&D expenditure decisions?
3. In which sectors of the biotechnology industry is the funding of R&D most influenced by consumer controversies?
4. Would greater consumer acceptance of biotechnology lead to more funds being directed towards research and development?
5. How do current and potential consumer controversies compare to other factors that influence the funding of research and development in Canadian biotechnology firms?

Description of the Survey

A questionnaire on research and development in biotechnology firms (referred to below as "the questionnaire") was sent to 716 Canadian biotechnology organizations on May 1, 1997. The questionnaire was the product of the following process. First, a preliminary draft was formulated. Comments on this draft by various constituencies connected with the Industry Canada initiative were collected by Industry Canada. Further comments were collected from about a dozen people involved in the commercialization of biotechnology. Appropriate revisions were made. In some cases, people in the latter group furnished us with a second set of comments after the preliminary draft had been revised. These survey data were married with data from Contact Canada which collects data on the biotechnology industry in Canada.

Data from Contact Canada (*Canadian Biotechnology, 1997*) indicate that, at the time the questionnaire was sent out, Canada's biotechnology industry consisted of 14 non-profit organizations, 169 government and university organizations, and 533 public and private firms. Although the questionnaire was primarily designed for the public and private biotechnology firms, it was sent to all biotechnology organizations in Canada in order to obtain the most feedback possible.

The questionnaire was divided into three parts: (A) "Identification," (B) "Firm Characteristics," and (C) "Correlations Between Research and Development Expenditures and Consumer Perceptions of Your Firm's Activities." In part (A), the name and address of the respondents were requested in order to keep track of replies and to be able to contact the respondents in the event of a query with their responses. The fact that the questionnaire was sponsored by Industry Canada and that the names of all respondents would be kept confidential was noted at the beginning of the questionnaire. Part (B) inquired about the respondents' characteristics including questions about the financing stage of the firm and the firms' balance sheet. Part (C), the most important part of the questionnaire, had a total of 15 questions of both a quantitative and qualitative nature. Questions 1 through 5 explored research and development expenditures and the due diligence process whereby various prospective projects are evaluated and eventually funded. The first question sought to elicit the quantum of research and development expenditures as a percentage of total expenditures. Five choices were offered: (a) > 80%, (b) 60–79%, (c) 40–59%, (d) 20–39%, and (e) < 20%. Question 2 asked the respondents to describe the internal process whereby R&D budgets are determined in the respondent's firm, and how R&D dollars are allocated between various competing projects. Questions 3 and 4 asked whether future product demand was accounted for in this "due diligence" process, and if so, whether estimates of future demand took into account potential consumer controversies that might emerge from the research. Question 5 asked the respondent to describe the details of this due diligence process.

Question 6 asked the respondents to rank on a scale of 1 to 10 the importance of the following factors in determining their firm's level of R&D expenditures: taxation factors, economic climate, availability of capital generally, availability of capital for the project, market demand for particular types of products, product push (an idea for an innovative product without any clear indication that market demand exists), demand pull (market demand for a product to perform a particular function), ability to protect the product by means of a patent, copyright, or trademark, whether the technology is a platform technology,[4] whether the technology is innovative, how many competitors presently compete in the market, the relevance of the product to the established expertise of the company, potential consumer controversies that might inhibit demand, whether manufacturing the product will reduce dependence on suppliers, the availability of a strategic alliance with another company, government, or university, or some other factor left open for the respondent to specify.

Question 7 asked if the respondent's firm would spend more on research and development if there were greater consumer acceptance or awareness of biotechnology products in general.

Questions 8 to 10 focused on research and development expenditures and consumer controversies. Question 8 asked firms to indicate whether they had ever had a controversy regarding research and development activities (or whether a close competitor had experienced a controversy regarding a similar product). If the answer was yes, respondents were asked to describe the nature of the controversy. Of those that have had a controversy, Question 9 sought to determine whether the controversy influenced their firm's decision to pursue research and development in the area of the controversy, and Question 10 asked whether the controversy influenced the firm's decision to pursue research and development in general. Descriptive explanations were also requested in Questions 9 and 10.

Question 11 asked the respondents to specify their firm's typical planning horizon for research and development projects. Questions 12 to 15 focused on firms' exposure to the general public. Respondents were asked whether their company had recently been in the news. Those who answered in the affirmative were asked to indicate the nature of the coverage, and whether the respondent viewed this exposure as positive, negative, or neutral. The respondents were also asked to rank the overall general awareness of the firm's products on the part of the general public. Finally, respondents were asked whether the respondent's firm participated in efforts to promote biotechnology.

Profile of the Respondents

As of June 19, 1997, there were a total of 70 respondents from 716 questionnaires distributed on May 1, 1997. Four of the respondents were eliminated from the data pool because each incorrectly assumed that consumer controversies included intellectual property disputes. This left 66 respondents, consisting of 15 public corporations, 39 private firms, 2 non-profit organizations, 6 government organizations, and 4 university-affiliated research institutes that responded to the survey. Though we have data on only 66 firms, these firms tend to match, on key elements, other industry wide data sources. Given that there has been a large amount of work on consumer surveys but almost no systematic work on surveys of producers this data collection is a first step in understanding how firms react to consumer controversies in the marketplace.

The geographic distribution of respondents parallels the distribution of biotechnology organizations across the provinces. The majority of respondents were from Quebec and Ontario, although there were respondents from every Canadian province except New Brunswick. The respondents represented every sector of biotechnology, and their sectoral distribution parallels the distribution reported in other sources. Nearly one third of the respondents were in the health field. Horticulture, forestry/pulp, aquaculture/veterinary, and food/beverage/fermentation accounted for another third, while environment, energy/mining, biomaterials, cosmetics, research, and software accounted for the remaining third. Twenty-four of the respondent firms were established in the 1990s, 22 in the 1980s, 9 in the 1970s, 3 in the 1960s, and 8 before 1960. This reflects the fact that biotechnology is a new industry; the vast preponderance of firms were created in the 1980s and 1990s.

Despite their relatively recent vintage, respondent firms were fairly sizeable on average Excluding non-profit organizations, and government and university facilities, forty-nine had their own biotechnology labs, and 38 had manufacturing facilities. The average number of employees was 153. Average estimated revenue in 1995 was $102.5 million. Average biotechnology revenues were $24.4 million. The average estimated R&D expenditures were $3.7 million. Average biotechnology R&D expenditures were $3.0 million (Canadian Biotechnology, 1997). Though there were some obvious inaccuracies in the balance sheet information (left and right side did not balance in some cases), average current assets were $39.8 million, average total assets were $377.2 million, average current liabilities were $124.1 million, average total liabilities $207.4 million, and average shareholders' equity was $251.39 million.

With respect to the financing stage the greatest number of respondents were in the development and expansion stage (40.9%); 13.6% were in the start-up stage; 15.2% were publicly traded with more than $100 million of capital; 13.6% were publicly traded with less than $100 million of capital; and 4.5% were in the seed stage. The capital structure of the respondents' firms (current and totat debt-equity ratios) are analogous to typical Canadian firms at each stage of financing. The current and long-term debt-equity ratios were 0.01 and 0.01, respectively, for seed firms, 0.65 and 0.69 for start-ups, 0.19 and 0.32 for development and expansion firms, 0.46 and 0.56 for mezzanine firms, 0.12 and 0.17 for public firms with less than $100 million market capital, and 0.24 and 0.35 for public firms with more than $100 million market capital.

R&D Expenditures and Consumer Perceptions of Firms' Activities

The purpose of this section is to ascertain whether certain factors influence firms' R&D expenditures. It is primarily descriptive in nature; multivariate statistical approaches are available in MacIntosh and Cummings (1996).

Part C of the questionnaire was entitled "Correlations Between Research and Development Expenditures and Consumer Perceptions of Your Firm's Activities." It asked each respondent to indicate the approximate amount of its research and development expenditures as a percentage of the firm's total expenditures. The most striking result is that 62.1% of all respondents allocated either less than 20%, or more than 80%, of their budgets to R&D, with the rest falling in the mid-range. The population of respondents may thus be thought of as consisting of two subpopulations of firms; those whose primary mission is research, and those in which research is a subsidiary part of the business. Discussions with industry players suggest that firms from each of these populations will be contractually linked. Instead of (or in addition to) in-house R&D, many firms will outsource their R&D, hiring a specialist R&D firm typically referred to as a "Contract Research Organization," or "CRO," to do some or all of its research.

Determination of R&D budgets. Respondents were asked to "describe the internal process whereby R&D budgets are determined in their firm, and how R&D dollars are allocated between various competing projects (i.e., the due diligence process whereby various prospective products are evaluated and eventually funded)." Virtually all 66 respondents had comments to offer. Some of the more interesting responses are as follows:

"R&D budget is established by the Sr. VP Operations and approved by the Parent Co. CEO / Board of Directors. Criteria are (1) requirements of current business, (2) opportunities for short-term growth and (3) long-term (> 2 year) opportunities in that order. Projects are allocated on the basis of ROI and relevance to strategic business."

"Time to market, novelty, market size, fit with our capabilities, etc."

"[Firm Name] is a custom manufacturer of pharmaceutical actives. The R&D carried out is proportional to the size of the customer's project. Much of our R&D is charged to the customer on a contract basis."

" a) Identification of potential markets
  b) Identification of resources needed
  c) Identification of internal expertise
  d) Determine priority in the actual portfolio
  e) Simulation analysis to predict the outcomes and impact on the other R&D programs"

"Projects evaluated and compared based on ROI factoring in market data on 'delta revenue' or manufacturing data on 'delta cost of sales' as a result of new development. Rank crude projects of approximately equal risk according to payable time. Choose top ranking projects. Projects that are speculative (high risk/long term) are funded out of separate 'slush fund')."

" a) detailed market research and evaluation
  b) detailed assessment of competition (if any)
  c) evaluation of potential marketing and distribution partners
  d) evaluation of technical feasibility
  e) costs to develop and launch product vs. revenues
  f) liability issues associated with product's use"

"New projects are selected based on: the willingness of industry to co-fund; priorities developed by advisory committee, other clients, etc.; projected return on investment of research project; and public good generated by research."

"A research project needs specific funding, e.g., with the help of a government grant or done as a paid contract. We don't have an R&D budget as such."

" 1. Corporate objectives set.
  2. R&D input translates objectives into action plan and decides if programs will be implemented intra or extramurally.
  3. Research budgets set.
  4. Research budgets reviewed and approved.

5. Using internal due diligence which includes benefit, cost, risk, scientific analysis. Scientific analysis is done intra- and extramurally."

"R&D expenditures are allocated in priority to products under development which are the closest to commercial licensing and approval."

" 1. Fit with corporate strategy / focus.
2. Market need.
3. Technical feasibility (cost/speed of development).
4. Whether technology / product can be protected by patent."

"General Manager and Manager of Production Plant decide which are deemed to have the most chance of commercialization and these are given priority."

"R&D budgets allocated to research networks; within network's allocations according to priorities/request as per client's advice or needs."

"The research and development budget is funding driven. If the funding levels are low, then the subsequent R&D spending is also low. [Firm Name] applies to funding sources such as IRAP and ACOA for consideration. Due to the development nature of these government funding sources, the funding requests that are submitted are very market driven. [Firm Name] focuses on products in our area of expertise (shellfish toxin detection) where there are few competitive products. [Firm Name] manufacturers very specific products and the R&D expenditures are used to expand this product 'family'."

"[Firm Name] uses gating process which is a series of technical and marketing reviews. R&D funds are then allocated based on estimated ROI. The gates are prospecting, applied research, development, prototyping, trial manufacturing, pre-commercialization. Each gate requires a greater certainty of technical and business feasibility."

"Ongoing R&D budget, supplemented by obtaining government subsidies for individual projects. Generally we look for a payback time of 1–3 years for small projects, 3–5 years for major projects."

The range of responses indicates that there is no standard method for determining which projects to fund; a heterogeneous variety of methods are used.

Potential consumer controversies. A major focus of this study is whether biotechnology firms take potential consumer controversies into account in deciding which projects they will fund. Almost all (65 of 66 respondents) indicated that potential future demand for a product is taken into account in the due diligence process that is used to decide whether to fund a particular project. Of the 65 respondents whose firms did take future demand into account, 45 (or 69%) indicated that their firms considered current or potential consumer controversies concerning any products that might emerge from the research. This constitutes prima facie evidence that Canadian biotechnology firms, in the main, are not ignoring potential controversies in allocating research funds.

Those respondents who took potential controversies into account were asked how this was done. Some of the more interesting responses included the following:

"[Firm Name]'s product line includes supplies to the biopharmaceutical industry – primarily related to cell fermentation media, growth factors, supplies, etc. Project decisions must take into account key customer positions on controversial products and materials. Thus, R&D is significantly "customer" (Biopharmaceutical Co.) driven."

"We try and anticipate expert opinion and public acceptance of product based on current attitudes. In fact, we design some of our new services to avoid or circumvent current controversies. For example, we have developed non-radioactive tests where only radioactive tests currently exist."

"Consultation with end-users (groups), government, industry allies, and internal to address issues. Data presented to the Board of Directors for discussion and approval."

"Products made by recombinant DNA technology face regulatory and customer acceptance hurdles, which vary widely by industry and geographical area. These points are reviewed or determined for all rDNA-based project proposed, which are typically maybe 1/3 of the projects. The information is gathered primarily by sales staff through discussions with customers. Depending on the situation, the project can be cancelled (if the response is highly negative), delayed (to allow time to present a case to address the customers concerns), or go ahead (if the concerns will be addressed with certificates of analysis, etc.)."

"Our products are medical devices. We consider the potential of product failure, reliability, and ease of use to the customer as potential 'consumer controversies.' We also evaluate the current regulatory environment as well as the need for qualified medical advice in conjunction with use or interpretation of results. If a potential product is highly controversial or embroiled in political/regulatory confusion – we avoid it until things are clarified. For example, we were asked to construct a diagnostic device for rBST. We turned down the request because the regulatory issues were highly political and quite unclear. We declined despite the promise of a large world market. Without a clear mandate for testing – there is no market."

"In case of serious opposition, may delay R&D progress."

"Market and clinical [people] meet and evaluate research, market, and regulatory trends. Risk/reward is assessed."

"Our research is twofold. Firstly, we look at the technical aspects of our business. Secondly, even if the technology is sound, the public fear of permitting may be too expensive or risky to proceed."

"Particularly with regards to genetically modified micro-organisms (GMO) to be used as foods or food ingredients. No major investment will be made if we have the evidence that the GMO will generate significant public controversy."

"Major research expenditures will only go into areas which management feels will result in "significant" profit margins. Trends in other industries and in related organizations are evaluated, with emphasis on what the consumer response typically has been. Generally, unless there is a strong indication that 'potential controversy' will not arise and affect market ability, our organization will choose to not proceed."

"We have access to information from various sources including suppliers, etc., of a global nature. In addition we actively participate in a number of industry based associations. In terms of such controversies we often do not act alone and make a decision on our best assessment of the situation. There is no major formula."

"Knowledge of market environment; customer sensitivities; use of client advisors."

"Consumer controversies are marketing development concerns. Should they feel that such controversies are bound to affect product or technology implementation, they will modify their needs for such development. However, if the controversies are not based on facts but only on a false perception of the reality, the development may still be required technically but implementation may never take place."

"Difficult to answer without getting into specifics of individual products. 1. Blood screening product: consumer controversy only likely to yield more demand since nothing else available for fatal disease diagnosis (CJD). 2. Transgenic (mouse) animal: (not yet sold to consumers) controversy unlikely to affect pharmaceutical licensing deals."

"[Firm Name] estimates of future demand do take into account consumer perception regarding our product. This is done from a marketing and sales perspective. The future demand projections are based on the assumption that the output of the R&D effort will be known. [Firm Name] does not plan for controversies concerning products that might inadvertently be discovered during the R&D exercise."

"Our products are medical. Accordingly, we try to gauge trends in the usage of these products, competitive break-through potential, impact of demographics, etc. As much information as possible is used to make determinations of likely demand. Ultimately, where novel products with new, innovative modes of action are involved, one is guessing!"

"Our business is bioherbicides, biofungicides, bioinsecticides, and biofertilizers. Issues of environmental concerns and sustainability are currently contributing to a good public image of the products. We must ensure that all products are safe and effective to maintain this image."

"Our approach to value added cereals is mainly through genetic engineering. Public understanding is usually low and acceptance varies depending on countries (high in N.A., low in Europe, etc. . . .) Our primary market will be Canada and U.S. where max customers (i.e., farmers) are favourable."

"Controversy may be genuine or perceived; in either case would be a risk factor for demand for a product. My understanding is that generally large companies look for established (assured) markets for major future products, and prefer to avoid consumer controversies. If consider controversy to be perceived only (unreasonable) try to assess likelihood that perception may be changed. Often there is a race to be second in market for controversial product; i.e., avoid being first."

Clearly, just as there is no standard means for allocating scarce research resources, there is no standard means by which potential controversies are accounted for. It is interesting to note, however, that a number of respondents classified potential controversies – as we did even prior to sending out the questionnaire – as either real or "perceived" (i.e., based on non-substantive factors).

Types of firms and sectors which express concern for consumer controversies. Eliminating all sectors with fewer than 10 responses, the following sectors had the highest proportion of respondents indicating that they took consumer controversies into account: food/beverage/fermentation (70%), research group/CRO (Contract Research Organization)[5] (68%), agriculture (67%), health (diagnostics) (63%), and health (therapeutics) (62%). If all sectors related to food production are combined (agriculture, food/beverage/fermentation, horticulture, aquaculture/fisheries, and veterinary biotechnology), then 34 of 47 (or 72%) of the respondents reported taking potential consumer controversies into account.

By contrast, if environment and forestry/pulp/paper are combined, then 16 of 18, or 89% of all respondents, indicated that they took consumer controversies into account. The comparative number in the three health care sectors was 28 of 48, or 58%. Based on the review of controversies presented in the earlier sections we would expect that agriculture and food companies would be the ones most concerned about controversies. In fact, in this sample, biotech companies engaging in environmental research and forestry/pulp/paper were the most concerned, followed by companies engaged in food and agriculture, and then health care. The high number of respondents in the environmental and forestry/pulp/paper sectors evincing a concern for consumer controversies may be an artifact of the relatively small number of responses from firms engaged in these sectors. It may also constitute evidence, however, that widespread popular sensitivity to environmental issues (rather than controversies about biotechnology in particular) has made companies in these sectors more cognizant of potential controversies arising out of their work.

Those firms that were concerned about potential controversies were more leveraged than other firms in the sample. The average ratio of current liabilities to current liabilities and equity was 0.236 for firms that were concerned about consumer controversies versus 0.168 for firms that were not concerned. From a theoretical standpoint, this is initially a puzzling result. Higher leverage results in greater financial risk, and hence greater systematic risk. In theory, changes in the firm's systematic risk should not influence the managers' choice of investment opportunities. In all cases, managers should choose any prospective project with a zero or positive net present value. Whether a prospective project has a high or low likelihood of a consumer controversy should not matter, save as it alters the firm's projections of the future net cash flow to be derived from a particular project. It is true that increasing the probability of a consumer controversy will increase the unsystematic risk of any particular project. However, unsystematic risk is diversifiable risk, and should not therefore enter into a calculation of the net present value of the project.

In practice, however, managers are often under-diversified, given that a disproportionate part of their investment portfolios will be tied to the fortunes of the firm that employs them. In this respect, the manager's salary can be thought of as a financial asset. When added to stock in the firm, stock options, and other incentive schemes whose value is tied to the fortunes of the firm, under-diversification results (Coffee, 1986). As incompletely diversified agents (who are exposed to at least part of the unsystematic, as well as systematic risk of the firm's portfolio of business projects), their investment choices will tend to be skewed towards projects with low unsystematic risk. This will be the case even if these projects do not have the highest net present value of all opportunities available to the firm (Coffee, 1986). It is very likely that the higher the firms' systematic risk, the less likely the managers will feel comfortable with projects exhibiting a high level of unsystematic risk.

The importance of consumer controversies in determining R&D as compared with other factors. Respondents were asked to rank 19 alternative factors on a scale of 1–10 in their importance in determining their firms' level of research and development. Market demand was the most important factor in determining R&D across the respondent firms. Two respondents provided the following comments which in part explain this result:

> "Market demand and innovative technologies are most important. Patents, trademarks, and copyrights are not relevant because people steal things in any case."

> "Economic climate '10' in that it dictates level of government funding. Competitors are not relevant because we have a monopoly."

Although 69% of the respondents indicated that current or potential consumer controversies are taken into account in estimating market demand, this factor ranked only 15th of the 19 alternatives specified. This suggests that potential controversy was not a factor that was weighted particularly heavily. Despite this, there is evidence (discussed below) that when firms had direct experience with a controversy, this experience was likely to have an impact on the funding of R&D projects in the area of the controversy.

Table I presents the average rankings of the factors that influence R&D listed in Question 6 on the basis of the biotechnology sector in which the respondents' firms operated. "Market demand for product" was ranked highest in every sector except health and software/training. In the health sector, "ability to protect by copyright" was ranked first, but "market demand for product" was a close second. In software/training "relevance to company expertise" was ranked the highest.

"Concern about consumer controversies" was ranked highest in the cosmetics sector, followed by research group/CRO (such units are found in all sectors of biotechnology), environmental/forestry/pulp and paper, software/training, health care, food/beverage/fermentation, agriculture/horticulture, aquaculture/veterinary, and energy/engineering/mining/biomaterials. The high concern amongst those working in the cosmetics area probably relates to the considerable controversy that has arisen in relation to testing new products on animals, which is not a concern that is unique to products produced using biotechnology. The result relating to software/training may also be biased by controversies not uniquely related to biotechnology – in particular controversies related to intellectual property disputes. While we attempted to cull from our data sample any responses that listed intellectual property disputes as "consumer controversies," it may be that some of the responses from software firms were coloured by previous experience with such disputes.

Grouping the remaining sectors into the three main categories of biotechnology (food related, health, and environmental), the environmental sector came out on top, followed by health care and food related biotechnology.

Table II categorizes the average rank of the factors listed in Question 6 on the basis of firm ownership and stage of financing. Public firms with assets under $100 million, and firms in the mezzanine stage of financing, gave consumer controversies a much higher ranking than public firms with assets in excess of $100 million. Firms in the mezzanine financing stage are typically in the last stage of

TABLE I
Average Rank of Factors in Question 6 in Each Biotechnology Sector

Question 6. Factor	Biotechnology Sector								
	Agriculture/ Horticulture	Food/ Beverage/ Ferm.	Cosmetics	Aquaculture/ Veterinary	Health – Diagnostics Therapeutics Vaccines	Environ./ Forestry/ Pulp/ Paper	Research Group/ CRO	Energy/ Engineering/ Mining/ Biomaterials	Software/ Training
Taxation Factors	4.50	4.30	1.00	5.43	4.30	4.88	4.13	5.00	6.00
Economic Climate	5.04	4.70	5.00	4.86	3.97	4.06	4.80	4.14	4.50
Availability of Capital Generally	4.88	5.10	7.00	4.14	6.21	5.41	5.63	7.43	3.50
Availability of Capital for Project	5.13	5.30	9.00	5.14	6.39	5.58	5.93	8.00	5.50
Market Demand for Product	7.21	8.30	10.00	5.57	7.33	7.12	7.70	9.00	4.75
Product Push	4.96	4.80	5.00	5.00	4.15	5.00	5.07	5.00	5.25
Demand Pull	6.33	7.30	9.00	4.29	6.36	6.24	6.60	6.00	4.00
Ability to Protect by Patent	5.92	6.10	2.00	3.57	6.06	5.65	6.10	5.29	3.25
Ability to Protect by Copyright	3.21	3.50	2.00	3.57	7.76	3.82	3.57	2.29	3.00
Ability to Protect by Trademark	3.92	4.10	3.00	3.57	3.67	3.88	4.17	1.86	4.50
Platform Technology	3.00	4.50	5.00	5.29	6.42	3.65	5.33	4.57	5.25
Technology is Innovative	6.08	6.50	8.00	5.29	6.61	6.35	7.00	5.42	5.50
Competitors in the Market	5.96	5.50	8.00	3.43	6.15	5.18	5.57	5.00	3.75
Relevance to Company Expertise	6.17	7.00	10.00	5.29	6.36	6.00	6.70	5.29	6.75
Consumer Controversies	4.00	4.10	8.00	3.57	4.42	5.12	5.30	3.43	4.75
Reduce Dependence on Suppliers	4.04	4.80	2.00	4.43	4.09	3.53	4.20	2.86	4.50
Strategic Alliance with Another Co.	5.83	6.20	6.00	4.14	6.67	5.35	6.27	5.14	3.00
Strategic Alliance with Gov't or Univ.	5.50	6.00	4.00	4.14	4.24	4.88	5.40	4.57	2.75
Other	0.00	0.00	0.00	1.86	0.18	0.00	0.43	0.00	0.56

TABLE II
Average Rank of Factors in Question 6 on the Basis of Ownership and Financing Stage

Question 6. Factor	Ownership					Financing Stage					
	Public	Private	Non-Profit	Gov't	Univ.	Seed	Start-up	Development & Expansion	Mezzanine	Public, Capital < $100m	Public, Capital > $100m
Taxation Factors	4,60	5.31	2.50	2.00	0.25	3.67	5.89	4.70	3.00	4.78	4.78
Economic Climate	4.73	4.26	6.50	3.00	4.75	6.67	4.11	3.63	4.62	5.11	4.22
Availability of Capital Generally	6.93	5.46	6.00	4.83	5.50	6.33	7.33	4.41	6.75	8.00	5.11
Availability of Capital for Project	6.40	6.08	7.50	4.33	5.75	8.33	7.89	4.30	6.87	7.00	6.56
Market Demand for Product	7.93	7.36	9.00	7.17	6.50	9.33	9.33	6.00	8.50	8.44	5.11
Product Push	4.73	4.54	5.50	4.00	3.00	5.33	4.11	4.37	3.63	4.00	5.11
Demand Pull	7.00	6.26	8.00	6.00	6.50	8.00	7.89	5.15	8.50	7.22	5.89
Ability to Protect by Patent	6.07	5.38	9.00	4.67	4.75	9.33	4.78	5.22	4.50	7.33	5.33
Ability to Protect by Copyright	4.73	3.10	2.50	3.67	3.75	2.33	2.67	3.44	3.13	6.56	2.56
Ability to Protect by Trademark	5.20	3.31	4.50	3.67	3.75	5.67	3.00	3.41	3.25	7.00	2.78
Platform Technology	6.80	4.85	2.50	3.33	5.00	1.67	5.22	5.04	5.75	6.89	6.56
Technology is Innovative	7.13	6.10	8.50	5.67	6.25	8.33	5.77	6.11	6.75	7.44	6.11
Competitors in the Market	6.47	5.77	8.50	3.50	4.25	8.33	6.11	4.81	6.75	7.66	5.67
Relevance to Company Expertise	6.73	6.15	8.50	6.17	7.00	9.00	7.89	4.89	7.50	8.22	4.67
Consumer Controversies	5.07	3.74	5.00	4.67	6.25	5.33	4.22	3.11	5.75	6.33	2.78
Reduce Dependence on Suppliers	5.07	3.95	2.50	3.00	5.25	2.33	4.33	3.67	3.75	4.89	5.89
Strategic Alliance with another Co.	6.53	5.56	7.00	5.83	6.00	7.33	7.11	4.70	6.38	7.33	5.44
Strategic Alliance with Gov't or Univ.	5.27	4.67	7.00	4.33	5.50	7.33	5.22	4.07	4.38	4.56	6.00
Other	0.00	0.49	0.00	0.00	0.00	0.00	1.56	0.19	0.00	0.00	0.00

private financing before going public. Thus, firms that anticipate raising public funds in the near future, as well as smaller public firms (which may be high growth firms that also anticipate going back to the public market in the near future), are more concerned about consumer controversies than are larger public firms.

This may reflect the relatively great dependence of small, high growth firms on the continued availability of capital in the public market. Such firms may believe that the adverse publicity associated with consumer controversies will jeopardize the firm's ability to tap into the public market. Private firms (those in the seed, start-up, and development and expansion stages) ranked consumer controversies as somewhat less important than firms in the mezzanine stage of financing and small public firms, but as more important than large public firms did. The ranking declined from firms in the seed stage of financing, through those in the start-up stage, to those in the development and expansion stage. This may be because in the seed stage, the product has not yet been marketed; indeed, at this stage, the firm has typically not yet constructed a prototype. Uncertainty about consumer acceptance of a given product will be high. As the firm moves through the start-up and development and expansion stages, however, uncertainty about the likelihood of a consumer controversy will tend to be resolved, leading to a lower level of concern. While this would predict that concern will be lower still in the mezzanine and small public stages of financing, it may be that concern about consumer controversies reasserts itself when the need to tap the public markets is imminent.

University affiliated biotechnology organizations are nearly as concerned about consumer controversies as small public corporations in their funding of biotechnology research and development. This may be because universities, as publicly funded institutions, are more subject to criticism from interest groups concerned with biotechnology than are organizations in the private sphere. The relatively high probability of receiving adverse criticism may serve to sensitize university researchers to issues associated with consumer controversies.

Firms in the seed stage of financing tend to be cash-strapped firms with minimal complements of researchers. This is reflected in the survey responses. Seed firms were very concerned about the availability of capital generally, and the availability of capital for the particular project. Seed firms also ranked "relevance to company expertise" very highly, reflecting the absence of a large research complement. Seed firms also indicated a preference for developing innovative products for which there was a demand pull, few competitors, and the ability to protect the technology by means of a patent. The ability to consummate a strategic alliance with another company, the government, or a university was also ranked highly. Given the minimal resources of seed firms, the latter result is not surprising.

Firms in the start-up stage of financing were also motivated by the availability of capital and market demand. However, some differences may be noted. Start-up firms were generally less concerned than seed firms about developing an innovative technology, the ability to protect oneself by a patent (perhaps because patent protection had already been secured), and the existence of competitors in the market. They were more concerned, however, about developing platform technologies and reducing dependence on suppliers. Both seed and start-up firms ranked consumer controversies as a factor to be considered, but one of secondary importance.

Firms in the development and expansion stage were far less concerned than early stage (i.e., seed and start-up) firms about the availability of capital. Presumably this is because in order to grow to this stage, the firm will already had secured significant capital commitments. It will also have a proven product, making it easier to tap into additional sources of funding. Such firms were also less concerned than early stage firms that there be existing demand for the product and less concerned about the existence of competitors in the market. This may be a natural product of the maturation of the firm; in the development and expansion stage, the firm will presumably already have found a market niche in which it had already found that it can compete. Such firms were also less concerned than early stage firms about degree of relevance to the company's expertise. This is difficult to explain. Firms in the development and expansion stage will tend to be larger and better financed, and hence may have an enhanced ability to hire people when expertise is lacking. However, if this were the case, then firms in the mezzanine and public financing stages would also show relatively little concern about relevance to company expertise. In fact, they rank this factor quite highly.

Another way of overcoming a lack of expertise in-house is for the firm to strike a strategic alliance with another firm; for example, a Contract Research Organization or university. However, development and expansion firms ranked strategic alliances as being of much lower importance than did early stage firms. However, firms in the development and expansion stage did not rank strategic alliances very highly. And again, mezzanine firms and small public firms ranked strategic alliances more highly.

Interestingly, once firms reached the mezzanine stage and public stages of financing, the availability of capital, market demand, and number of competitors in the market assumed a level of importance similar to that of early stage firms. It may be that this reflects the need of firms nearing, or at the stage of making a public offering to grow rapidly. Indeed, there is good evidence (MacIntosh, 1994) that initial public offerings (and especially those of technology companies) are driven by investor expectations of high growth. In order to satisfy the demands of mezzanine and public investors for high growth, the firm will need to aggressively expand into new markets. In addition, IPO markets are notoriously cyclical. Firms about to go public should therefore be concerned about the availability of capital.

Small public firms (i.e., those with less than $100 million in capital) are the ones most concerned about intellectual property protection (patent, trademark, and copyright). Why might this be the case? Going public will tend to attract the attention of competitors to the firm's existence and the nature of its products. Thus, it may be that the likelihood of challenges to the firm's intellectual property assets will be highest soon after the firm goes public.

There appears to be a tendency for "reducing dependency on suppliers" to be a factor of increasing importance as one proceeds to later stages of financing. This makes sense in that the more products the firm is producing, the greater will be its dependency on suppliers, raising the likelihood of a bilateral monopoly arising as between supplier and buyer. This enhances the likelihood of opportunistic conduct by a supplier, resulting in the reduction of dependency on suppliers assuming relatively great importance.

Table III categorizes the average rank of the factors listed in Question 6 on the basis of the firm's R&D expenditures as a percentage of total expenditures. Firms with less than 20% of their budget spent on R&D ranked availability of capital generally, market demand for product, and strategic alliance with a government or university organization as more important than other firms. Firms with 20–39% of their budget spent on R&D ranked taxation factors as high as firms in the 60–79% category, and higher than firms in the other categories. Firms in the 40–59% category ranked product push, demand pull, and ability to protect by copyright higher than firms

TABLE III
Average Rank of Factors in Question 6 on the Basis of R&D Expenditures/Total Expenditures

Question 6. Factor	Question C.1. R&D Expenditures as a Percentage of Total Expenditures				
	< 20%	20–39%	40–59%	60–79%	> 80%
Taxation Factors	3.61	5.00	4.71	5.00	4.61
Economic Climate	3.72	4.44	3.71	4.11	5.09
Availability of Capital Generally	6.61	4.44	4.57	6.44	5.70
Availability of Capital for Project	6.44	5.11	5.14	5.33	6.57
Market Demand for Product	8.28	7.33	6.71	7.11	7.26
Product Push	4.06	2.78	7.00	4.89	4.52
Demand Pull	7.00	4.56	8.29	6.44	6.26
Ability to Protect by Patent	6.44	4.78	6.00	7.56	4.22
Ability to Protect by Copyright	3.94	2.33	4.14	3.89	3.39
Ability to Protect by Trademark	4.28	2.33	3.71	4.33	3.91
Platform Technology	4.11	5.22	4.86	6.78	5.22
Technology is Innovative	6.33	6.11	6.86	7.67	5.87
Competitors in Market	6.28	4.00	3.86	7.00	6.00
Relevance to Company Expertise	7.39	5.00	4.14	7.78	6.35
Consumer Controversies	4.78	3.89	2.43	5.33	4.30
Reduce Dependence on Suppliers	3.61	4.56	3.43	3.89	4.47
Strategic Alliance with another Co.	6.33	3.89	5.43	7.89	5.65
Strategic Alliance with Gov't or Univ.	5.33	4.00	4.14	4.44	5.30
Other	0.44	0.00	0.00	0.88	0.48

in the other categories. Ability to protect by patent, ability to protect by trademark, platform technology, innovative technology, competitors in the market, relevance to company expertise, consumer controversies, and strategic alliances with another companies were ranked highest by firms with 60–79% of their budget allocated to R&D. Firms with more than 80% of their budget allocated to R&D ranked availability of capital for project and economic climate higher than other firms. Market demand for product was the most important factor for firms with 39% or less and firms with more than 80% of their budget allocated to R&D. Firms with 40–59% of their budget allocated to R&D ranked demand pull as the most significant factor. Firms with 60–79% of their budget spent on R&D ranked strategic alliance with another company as the most important factor.

Impact of a growth in consumer acceptance/awareness of biotechnology. Would greater consumer acceptance or awareness of biotechnology products in general lead firms to spend more on research and development (Question 7)? Only 12 of the 66 respondents indicated that this would be likely. Among these 12 were the following comments:

"Probably small increase [in R&D] in some areas."

"Yes [more would be spent on R&D], in that more projects would be funded through cooperation with industry."

"Yes [more would be spent on R&D], very general linkage, related to investment climate."

Among the comments from those who indicated that no more would be spent on R&D were the following:

"Not relevant. Using narrow spectrum, popularist definitions in vogue."

"No, our research budget is based on our cash flow."

A further breakdown of responses indicates that only firms in the research group/CRO sector would be more likely than not to increase their funding of biotechnology R&D if there was greater consumer acceptance of biotechnology in general.[6] Non-profit organizations, government, and universities appear to be as likely as not, and private and public firms appear unlikely to increase R&D funding in the event of greater consumer acceptance of biotechnology products in general.

Examples of actual controversies. Question 8 sought to determine whether the respondent's firm had ever had a controversy regarding its research and development activities (or whether such a controversy had arisen in relation to a close competitor, regarding a product similar to that produced by the respondent). Only 9 of the 66 respondents responded affirmatively. These comments are illuminative of how respondents interpreted the word "controversy."

"New medical diagnostic product – controversy of whether its the best diagnostic test."

"We began operating a central treatment facility for the bioremediation of petroleum soils. In 1995 we switched to a thermal process due to the limitations of bioremediation. To date other bioremediation facilities continue to operate, yet we are the only one meeting the terms and conditions of operating permit. In short, bio-remediation does not work for this application, yet regulations, consumers turn a blind eye and allow "controlled dumping" with no field demonstrated results."

"Falsely accused of causing morbidity due to improper formulation. Firm was *completely exonerated.*"

"Yes [we have had a controversy], in the area of food irradiation in the late 80s."

"RR soybean varieties – export"

"Regarding transgenic viruses to control forest pest: The organization carried out a "risk communication" process to better understand the arising controversy and communicate to the scientific community and public at large."

"Concern over dark colour of dust suppressant for roads. Product is water soluble, after rain, run-off was dark, lay in pools on road surface, ran into storm sewers then lakes."

The controversies arose in the following sectors: 2 in health care, 2 in environmental, 1 in forestry, 2 in food products, and 1 in cosmetics. One firm that preferred to keep the nature of its controversy confidential was active in both environmental and agricultural applications of biotechnology.

In response to Question 9, eight of thirteen respondents indicated that the controversy that they had experienced influenced their firm's decision to pursue research and development in the area of the controversy. Some of the comments were as follows:

"We did some additional research on bio-remediation but primarily our focus changed to R&D on low temperature thermal desorption.

"Product liability risk was high. This risk contributed to decision to divest production of product line. We now distribute product of a strategic partner rather than make our own."

"The area was almost abandoned, i.e., limited to the irradiation of packaging materials."

"Export food products to Japan."

"Intensified research related to environmental impact analysis; gathering information for adequate communication activities."

"Firm looked at toxicity of this product to fish and other organisms to confirm suitability of product as dust suppressant."

In explaining why the controversy did not influence their decision to pursue R&D in that area of the controversy, some of the respondents commented:

"We knew this would become the accepted technology."

"We were determined to continue R&D since we were convinced that we would win any litigation."

"The company was successful in defending its position with regard to the controversy."

A further breakdown of responses indicates that firms in the health-therapeutics and research group/CRO areas are the only sectors in which controversies are unlikely to affect research and development in the area of the controversy. Controversies surrounding R&D in the agriculture, cosmetics, environment, and food/beverage/fermentation sectors are likely to affect firms' R&D expenditures in the area of controversy. Government-sponsored organizations and firms with less than 20% of the total expenditures on R&D are also most likely to alter their funding of a controversial product than other firms.

Only 1 of the 9 respondents (a government organization in the research group/CRO sector) indicated, as a response to Question 10, that the firm's experience with a controversy had affected its R&D budget in general. It commented:

"Yes [it affected R&D in general because it] helped to identify the gaps in knowledge, better evaluate the risks, and improve development strategies."

Some of the respondents who indicated that it did not affect its R&D budget in general explained that:

"Our company's future is based on R&D, so we continue to support and to increase our efforts to develop new efficient drugs to treat important diseases that affect an important percentage of the population."

"Our business (generic drugs) is highly litigious. We cannot afford to waiver in the face of threats from the foreign-owned multinational."

"No, the company continues to conduct R&D at an increasing level."

"[The controversy] stimulated R&D in other areas."

Media coverage. Question 12 asked respondents to indicate whether their organization had recently been in the news. Twenty-nine of the respondents indicated "yes." The responses indicated coverage of a broad nature unrelated to consumer controversies, such as:

"A variety of articles ranging from new products, to stock value (possible takeover rumours), environmental issues, charitable initiatives."

"[Firm Name] has received exposure in trade journals (customers), local newspapers and regional magazines. The "nature" of the coverage was informative. The articles were describing the company, the product, and the marketing focus of [Firm Name]."

"Patent issues; expansion projects; science promotion."

"European patent received: Decontamination system for products. U.S. Patent received: Antisense therapy (AIDS)."

"Announcement of a new transgenic animal model of Alzheimer's disease."

"Collaboration with another company for development of diagnostic products. Our holding company – [Firm Name] – Aids therapeutics."

"Focus on adding value to agricultural commodities."

"Press conference . . . Introduction of new enzymes for pulp bleaching."

"We were interviewed by local TV and radio stations because we are developing test for antibiotic resistance, and there was a local outbreak."

"Interest in recent development in cloning or cattle embryos following release of report from research in England."

"Colorectal cancer screen test; scar prevention cream."

"Technology transfer to agricultural producers."

"Significant controversy over the Federal Government's refusal to issue a Notice of Compliance for our lead product, despite being instructed by the Federal Court of Appeal to do so."

In Question 13, the respondents were asked to indicate if the press coverage was positive, negative, or neutral. Thirty indicated that it was positive, 34 indicated it was neutral, 2 indicated that it was both positive and negative, and none of the respondents indicated that it was strictly negative. Some of the "positive" comments included:

"Positive – media gave positive spin to creating high-tech jobs and products."

"Positive because it increased public awareness of our company and the market we are in."

"Positive. Company activities have been accurately reported to the general public. Generally the news media seems to be more professional than it was several years ago."

"Positive. Gives visibility and attracts potential collaborations."

"Positive, a model was needed to develop better drugs faster. Positions company as innovative, capable, beating out competition."

"As a young biotechnology firm our marketing focus is one of building primary demand. As such, any exposure that creates awareness in the marketplace is desired."

"Very positive. Company strategy of buying up old plants shut down by multinationals, refurbishing, providing local jobs, requiring creativity and innovative firm employees (labour as well as management) at same time possibly slightly less remuneration. This comes across well generally to public."

Firms that indicated that media coverage was neutral were less concerned about building sales through direct contact with the public:

"Neutral – sales to the public are not the focus of the business."

Firms that indicated they had both a positive and negative experience commented:

"Negative in that it was bad for employee's morale. Positive in that it highlighted the unfairness of the Regulations and may set a precedent for repeal."

"(a) odour generation (we did not generate) – negative; (b) odour control project with UBC / Science Council – positive."

Public awareness of the company's products. In Question 14, respondents were asked to rank the public's general awareness of their products. Only 3 respondents felt that the general public was "very aware" of their products; 17 "moderately aware"; 19 "vaguely aware"; and 23 "hardly aware at all." The following were representative comments:

"[The general public is] not aware at all of our products in their R&D phase."

"[The general public is not aware because] we are not a direct consumer company."

Promotional activities. Thirty-six of the 66 respondents indicated that their firm participated in efforts to promote biotechnology. Comments on the nature of such promotion included:

"Member of the different biotech organizations, participant in various events related to the biotech sector (conferences, symposiums, etc.), science promotion in high schools and universities related to biotech, etc."

"Participation in Government committees, provincial and local groups, dialogue with end users, publications (lay press and scientific), annual report."

"Participation in conferences, workshops, and forums. Discussions with government agencies concerned with economic development, registration, policy."

"Promoting safety of genetically modified foods through BIO, Washington lobbying, European initiatives."

"Active in BIO Organization (CEO is a long-time BIO Officer – currently secretary). Exhibits at international and domestic biotechnology meetings, etc."

"Ottawa Life Sciences Council – Very active participation. Canadian Society of Microbiology – participate in sessions on industrial biotech."

"Toronto Biotechnology Initiative; Connaught Student Fair."

"Member of Biotechnology Working Groups (Nova Scotia). Former Treasurer of BWG."

"Previously Quebec Biotechnology Association. Now Association Quebecoise des Bio-Industries. Table de Concertation en Biotech. Seminars in schools, universities, etc."

"Build awareness through meetings with our members. B of D of As Work Biotech [sic] and CSTA which has biotech committee. Supported ABIC/NBAC."

Some of the respondents that do not promote biotechnology commented:

"No, not enough cash flow."

"No, but would like to do more."

"We attempt to maintain a low profile so as to discourage misinformation."

The need for promoting biotechnology. The responses to the questions on media coverage, public awareness, and promotion suggest that most respondents felt that the general public is not very aware of their firm's activities. However, many firms invest both time and money promoting biotechnology. This suggests that even though most firms see the public as possessing only a dim awareness of what they do, they are of the view that public reaction to biotechnology is important to their economic future.

Most respondents regard news on their company's activities as favourable or at least neutral. This may help to explain why the potential for consumer controversies was not ranked as high as a variety of other factors in most respondents' decisions regarding the allocation of funds between various R&D projects.

Econometric Modelling of Firms' R&D Expenditures

In an extended draft of this paper, we used OLS and multinominal logit and probit models to relate firms' R&D expenditure decisions to their ranking of factors influencing R&D decisions, as well as to their total assets, planning horizons, and debt-equity ratios. The regression results are summarized briefly below.

In the surveys, the respondent firms ranked market demand as the most important factor in influencing R&D decisions. However, our multivariate analysis indicated that a firm's ranking of market demand was not important in explaining that firm's ratio of R&D expenditures to total expenditures, relative to other firms. The availability of capital was also statistically insignificant; this may reflect the fact that in recent times an abundance of capital has been made available to Canadian biotechnology firms.

The product push coefficient was statistically insignificant in the models. We hypothesized that the product push coefficient would be negative because projects are riskier where market demand for a project has not been demonstrated, all other things being equal. This effect, however, may be offset by the possibility that product push may sometimes be associcated with particularly innovative products whose market demand has not been established simply because the product anticipates, rather than follows, the market. In such cases, the developing firm may anticipate an extremely high upside profit potential and invest heavily in R&D. If the presence of product push sometimes deters some firms from investing heavily in R&D, but encourages others, it would not be surprising that the coefficient for product push is statistically insignificant.

The ability to achieve patent protection was one of the most significant factors in explaining expenditures on research and development. The increasing prominence of intellectual property concerns has been noted by industry analysts (Goudey & Nath, 1997). Firms that pursue products with a view to obtaining patent protection tend to spend a higher percentage of their total expenditures on research and development. The pursuit of a platform technology which can be licensed to other developers/users involves basic research. Such research is less expensive and implicates fewer regulatory hurdles than shepherding products through to the marketing stage (Goudey & Nath, 1997).

While the multivariate estimates suggested that firms in pursuit of platform technologies tend to allocate a smaller portion of their total expenditures to R&D, this result was not statistically significant at conventional levels (with p-values of approximately 0.15). This result may be a product of the fact that firms seeking to develop platform technologies may simply choose to pursue a larger number of lower cost R&D projects, leaving their total R&D budgets relatively unaffected.

In our regressions, the respondents' rankings of the importance of developing innovative technologies was not a significant factor in explaining differences in R&D expenditures, nor was the number of competitors in the market. This is a somewhat surprising result. Perhaps the threat of competition stimulates R&D in firms that are already active in a particular biotechnology sector (in order to stay competitive), while inhibiting R&D on the part of new entrants (who are likely to search for market niches in which there are few or no competitors).

Although the consumer controversies coefficient had the expected sign (negative), it was not statistically significant. This is an important result. As indicated above, consumer controversies ranked lower than many other factors in determining R&D expenditures. The regression results reinforce the view that differences in the relative degree of concern about consumer controversies do not explain differences in relative levels of R&D expenditures. Indeed, even if consumer controversies had been a statistically significant variable, the coefficient was small, adding further support to the view that the degree of concern over consumer controversies does not have a significant impact on R&D expenditures in Canadian biotechnology firms.

The coefficient for reducing dependence on suppliers was not significant, nor did it have its expected sign. Both strategic alliances with another firm and alliances with government/university had their expected sign, but only strategic alliances with another firm was a significant factor (with a p-value of 0.05) in explaining expenditures on R&D. Total assets, R&D planning horizons and debt-equity ratios were significant in explaining R&D expenditures at the 0.10, 0.05, and 0.01 levels of significance, respectively. This suggests the importance of scale economies and financial risk in explaining firms' R&D budgets.

In sum, our multivariate estimates suggest that the availability of patent protection and low debt-equity ratios are the most influential factors in determining R&D expenditures (as a percentage of total expenditures) across Canadian biotechnology firms. These results were robust to alternative econometric specifications. Strategic alliances with another firm, R&D planning horizons, and total assets also appear to be significant factors. Concerns over current or potential consumer controversies were not a significant factor in explaining R&D differences across Canadian biotechnology firms in any of the econometric models.

CONCLUSION

Canada's biotechnology industry has tremendous promise to make vast improvements in agriculture and food and beverage products, health care, and the environment, among other things. Nevertheless, developments in the biotechnology industry have been and will continue to be the subject of consumer controversies. These controversies will relate to many matters that are substantive in character, and a few that we have classified as "imaginary." The objective of this study was to examine the extent to which current and potential consumer controversies are a factor in determining R&D expenditures by Canadian biotechnology firms.

The interplay between consumer controversy and the funding of research and development in Canadian biotechnology firms was perhaps best summarized by one respondent that stated:

> "Products . . . face regulatory and customer acceptance hurdles, which vary widely by industry and geographical area. The information is gathered primarily by sales staff through discussions with customers. Depending on the situation, the project can be cancelled (if the response is highly negative), delayed (to allow time to present a case to address the customers concerns), or go ahead (if the concerns will be addressed with certificates of analysis, etc.)."

Our survey of the interplay between consumer controversy and the funding of biotechnology research and development in Canadian biotechnology firms suggest a number of empirical regularities:

1. Current or potential consumer controversies are taken into account by over two-thirds of firms in deciding whether to fund particular R&D projects.
2. The prospect of controversy tends to be more important when firms are in the mezzanine stage of financing or have recently experienced an initial public offering.
3. Controversies appear to be of greatest concern in the environmental and forestry/pulp/paper area, followed by agriculture and food, and then health (contrary to our working hypothesis that such controversies would be most important in the agriculture/food products area).
4. Firms that are more leveraged are more likely to be concerned about current and potential consumer controversies.
5. If a firm (or a competitor) has had a controversy surrounding its R&D, it is more likely than not that the controversy has

influenced the firm's funding of R&D in the area of the controversy, but not of its funding of R&D in general.

6. Firms in the health/therapeutics and the research group/CRO are the only sectors in which controversies are unlikely to affect research and development in the area of the controversy.
7. Controversies surrounding R&D in the agriculture, cosmetics, environment, and food/beverage/fermentation sectors are more likely than not to affect a firm's R&D expenditures in the area of controversy.
8. Controversies are more likely to affect R&D in public firms than private firms. Government-sponsored organizations and firms spending less than 20% of their total expenditures on R&D are also more likely to avoid funding a controversial product than other firms.
9. Only firms in the research group/CRO sector would be more likely than not to increase their funding of biotechnology R&D if there was greater consumer acceptance of biotechnology in general.
10. Non-profit organizations, government, and universities appear to be as likely as not, and private and public firms unlikely to increase R&D funding in the event of greater consumer acceptance of biotechnology products in general.

NOTES

[1] Irradiation has in fact been banned in three states of the United States (New York, New Jersey, and Maine) (McCann, 1995; Perry, 1996).

[2] U.S. Food and Drug Administration, "Statement of Policy for New Varieties of Food Plants", 57 Fed. Reg. 22,984 (1992), at 22,993.

[3] The cloning of "Dolly" has resulted in stepped-up efforts to prevent the cloning of human beings, particularly with the announcement by a researcher in Chicago that he intends to clone human beings (A slapdash proposal on cloning, 1998; Marwick, 1997; Wade, 1998).

[4] A platform technology is one that may be used to launch a range of different products. This is generally distinguished in the trade from single-use technologies that do not possess the potential to spawn further products in the future.

[5] For a description of the CRO, see Wendy Porter, "The CRO: Partnering for Successful Clinical Development", and Bruce L. Maloff, "Insourcing: A New Perspective on Contract Research", both found at http://ContactCanada.com/articles2.html, for a discussion of CROs.

[6] Since CROs are often externally funded on contract, this may evidence an expectation that the amount of contract work would increase if there were greater consumer acceptance of the products of biotechnoogy.

REFERENCES

A slapdash proposal on cloning (1998). Editorial. *New York Times*, February 10, 1998, p. A20.

Atlas, R. M. (1995). Bioremediation of petroleum pollutants. *Biosorption and Bioremediation*, *35*, 317–327.

Bohrer, R. A. (1994). Food products affected by biotechnology. *University of Pittsburgh Law Review*, *55*, 653–679.

Bueckert, D. (1996). Study says cow hormone may raise cancer risk for humans. *Canadian Press* (January 23).

Busch, L. (1991). Biotechnology: Consumer concerns about risks and values. *Food Technology*, *45*, 96–101.

Canadian Biotechnology (1997). Website, at http://www.biotech.ca/fbc.

Clark, J. R., & Lee, D. R. (1997). Too safe to be safe: Some implications of long- and short-run rescue laffer curves. *Eastern Economic Journal*, *23*, 127–137.

Coffee, J. C. (1986). Shareholders versus managers: The strain in the corporate web. *Michigan Law Review*, *85*, 1–109.

Creative Research International (1996). *Environmental applications of biotechnology: Focus groups. Report to Environment Canada*. Ottawa: Environment Canada.

Day, K. (1996). Genetics research begets questions; Biotech industry seeks ethics advice to deal with complex issues. *Washington Post* (May 8), A1.

Day, P. (1994). Agriculture: Business interest. *Canadian Press* (March 9).

Douthitt, R. A. (1995). Consumer risk perception and recombinant bovine growth hormone: The case for labeling dairy products made from untreated herd milk. *Journal of Public Policy and Marketing*, *14*, 328–330.

Ekos Research Associates Inc. (1996). *Focus groups on agri-food: Applications of biotechnology*. Report prepared for the Office of Consumer Affairs. Ottawa: Industry Canada.

Estruch, J. J., Carozzi, N. B., Desai, N., Duck, N. B., & Warren, G. W. (1997). Transgenic plants: An emerging approach to pest control. *Nature Biotechnology*, *15*, 137–141.

Gordon, M. (1997). Suffering of the lambs. *New Scientist*, *154*, 16–17.

Goudey, J., & Nath, D. (1997). *Canadian biotech '97: Coming of age*. Toronto: Ernst & Young.

Grobe, D., & Douthitt, R.A. (1995). Consumer acceptance of recombinant bovine growth hormone: Interplay between beliefs and perceived risks. *Journal. of Consumer Affairs*, *29*, 128–143.

Hall, C. T. (1996). Biotechnology revolution – 20 years later Genentech's founders pioneered it all. *San Francisco Chronicle* (May 28), B1.

Hanley, C. J. (1996). Indigenous peoples resist worldwide gene study; ethics: Native Americans, Australian aborigines and others worry about victimization by science and commercial interests of diversity project. *Los Angeles Times* (Bulldog Edition), July 7, p. 8.

I'm modified, buy me (1998). *The Economist*, January 17, p. 60.

Ibrahim, Y. M. (1996). Genetic soybeans alarm Europeans. *New York Times* (November 7), D1.

Karcas, I. (1996). Greens rally against invasion of the killer corncobs. *The Independent* (June 30), p. 12.

Kendall, D. (1986). Agriculture enters new technological revolution. Associated Press, December 28.

Lee, R. J., Van Orden, D. R., Corn, M., & Crump, K. S. (1992). Exposure to airborne asbestos in buildings. *Regulatory Toxicology and Pharmacology*, *16*, 93–107.

Loaharanu, P. (1996). Irradiation as a cold pasteurization process for foods. *Veterinary Parasitology*, *64*, 71–82.

MacIntosh, J. G. (1994). *Legal and institutional barriers to financing innovative enterprise in Canada*. Monograph prepared for the Government and Competitiveness Project. Kingston, Canada: Queen's University, School of Policy Studies. Discussion Paper 94-10.

MacIntosh, J. G., & Cummings, D. (1996). *Consumer controversy and the funding of biotechnology research*. Toronto: University of Toronto. Working Paper.

Marwick, C. (1997). Put human cloning on hold, say bioethicists. *Journal of the American Medical Association*, *278*, 13–14.

McCann, H. (1995). But some producers, consumers remain wary of technology. *Detroit News* (May 1), p. 1.

Neale, G. (1996). Activists step up their fight over test tube food. *Daily Telegraph* (December 9), p. 6.

Nilsson, J., Jonasson, P., & Samuelsson, E. (1996). Integrated production of human insulin and its C-peptide. *Journal of Biotechnology*, *48*, 241–250.

Nooitgedagt, A. J. (1997). Food irradiation with ionizing radiation: An overview. *Tijdschrift voor DiergeneesKunde*, *122*, 158–163.

Nordlee, J. A., Taylor, S. I., Townsend, J. A., Thomas, L. A., & Bush, R. K. (1996). Identification of a Brazil-nut allergen in transgenic soybeans. *New England Journal of Medicine*, *334*, 688–692.

O'Neill, B. (1997). Swords drawn in the battle for knowledge. *The Guardian* (January 2), p. 17.

Optima Consultants (1994). *Understanding the consumer interest in the new biotechnology*. Ottawa: Industry Canada.

Pan, A., Yang, M., Tie, F., Li, L., Chen, Z., & Ru, B. (1994). Expression of mouse metallothionein-I geneconfers cadmium resistance in transgenic tobacco plants. *Plant Molecular Biology*, *24*, 341–351.

Patrick, A. W., & Williams, G. (1993). L'insuline humaine le point d'une controverse. *Louvain Medicin*, *112*, 337–343.

Peltzman, S. (1975). The effects of automobile safety regulation. *Journal of Political Economy*, *83*, 677–725.

Perry, R. (1995). Use common sense in food safety rules. *Dallas Morning News* (August 6), p. 6J.

rbST Task Force (1995). *Review of the potential impact of recombinant bovine somatotropin (rbST) in Canada: Full report*. Ottawa: Agriculture Canada.

Ronk, R. J., Maryanski, J. H., & Thompson, P. (1990). Governmental regulatory issues – Reducing uncertainties in food safety. *HortScience*, *25*, 1482–1484.

Schlumberger, H. D., & Stradler, P. (1997). Modern pharmaceutical biotechnology: Situation worldwide and in Germany, 1995. *Arzneimittelforschung*, *47*, 106–110.

Sparks, P., & Shepherd, R. (1994). Public perceptions of the potential hazards associated with food production and food consumption: An empirical study. *Risk Analysis*, *14*, 799–806.

Wade, N. (1998). Senate plans to weigh ban on cloning. *New York Times* (February 10), p. A16.

Weiss, R. (1996). New variety of corn reaps controversy; EU Ministers balking at gene-altered crop. *Washington Post* (October 8), p. A1.

Yom, H. C., & Bremel R. D. (1996). Genetic engineering of milk composition: Modification of milk components in lactating transgenic animals. *American Journal of Clinical Nutrition, 58* (Suppl.), 299S-306S.

THE AUTHORS

Jeffrey G. MacIntosh is Professor of Law at the Faculty of Law, University of Toronto, 78 Queen's Park, Toronto, Canada, M5S 2C5. E-mail: j.macintosh@utoronto.ca. Douglas J. Cumming is a LL.B./Ph.D. student at the Institute for Policy Analysis, University of Toronto, 140 St. George Street, Toronto, Canada, M5S 3G6. E-mail: dcumming@chass.utoronto.ca.

Marc Legault and Alison McDermott
in collaboration with
Bartha M. Knoppers and Derek Ireland

Appendix A: Integration Document Biotechnology, the Consumer, and the Canadian Marketplace

BACKGROUND AND PURPOSE OF THE RESEARCH PROGRAM

What Do We Mean by Biotechnology?

The term biotechnology is used in a variety of contexts and reflects a range of activities, processes, and applications. Biotechnology is widely applied in the agricultural, environmental, fisheries, forestry, health, and mining industries to develop products or applications with special characteristics. The Canadian Environmental Protection Act (CEPA) defines biotechnology as:

> The application of science and engineering in the direct use of living organisms or parts or products of living organisms in their natural or modified forms.

This definition is sufficiently broad to encompass both traditional biotechnology and new techniques of genetic manipulation to develop novel varieties and traits.

Traditional biotechnology involves, for example, using yeast to make beer or selectively breeding dogs to create new varieties of breeds. New techniques, such as those provided through recombinant-DNA technology or genetic engineering, allow researchers to more precisely enhance desired traits in plants or animals and to develop new varieties of micro-organisms, plants and, to some extent, animals.

The term "biotechnology," used throughout this paper, refers to a wide variety of applications, many of which have little in common other than their reliance on an expanding body of basic knowledge of science at the molecular level. In fact, one participant at the Symposium argued that the term "product of biotechnology" has little or no meaning.[1]

Journal of Consumer Policy **21**, 1998.
Bartha M. Knoppers & Alan D. Mathios (eds), Biotechnology and the Consumer.

Use of biotechnology in food production and in medicine has generated heated debate among industry, government, the scientific community, and consumer and environmental groups.

What Is the Background to the Current Debate on Biotechnology?

Over the past few years, governments, the private sector, and consumer groups have conducted surveys and focus groups to gauge public awareness, views, and concerns about biotechnology. The findings reveal some appreciation of the potential benefits of biotechnology, as well as information gaps and concerns about potential risks and the ability of government and industry to manage these risks in the public interest.

For instance, a 1994 consumer survey indicated that about 60 percent of Canadians were uncertain or undecided about genetic engineering, while the remainder were almost equally divided between "true believers" and "avid opponents." Previous research also underlined that consumers react differently to various applications depending on how they perceive the application will benefit them and society generally and on how the benefits and longer-term risks will be shared between consumers and industry. Subsequent focus group studies supported these findings and inferred that people react this way for a number of reasons:

- because they lack information and understanding of biotechnology;
- because some biotechnology industries perhaps do not fully understand the needs, preferences, and concerns of the final consumer;
- because some consumers do not fully trust the ability of government and industry to manage this complex technology in the public interest.

Much of the recent controversy has focused on ag-biotechnology applications. However, Caulfield (this volume) states that "there are those who believe the recent rapid increases in our genetic knowledge coupled with subtle social pressure, including commercial forces, could result in a *consumer led* 'new eugenics' movement."

Purpose of the Office of Consumer Affairs Research Program

In 1995, the Office of Consumer Affairs (OCA) assumed responsibility under the federal government's National Biotechnology Strategy to conduct surveys, focus groups, and other research to better understand consumer attitudes toward biotechnology.

After an in-depth review of existing surveys and following focus groups and consultations with partners inside and outside government, the Office decided that the next step should involve the development of an improved analytical framework to guide future work on consumer attitudes and public confidence regarding biotechnology.

The approach of using analytical research to encourage discussion among stakeholders (consumers, government, and industry) stems from the view that good public policy research and subsequent policy development may be able to promote understanding and consensus across stakeholders. Policy discussions should start from a solid understanding of the underlying issues. For the OCA work on biotechnology and the consumer, these issues might include the following: how consumers influence the economy and societal values by participating in policy making and by bringing to their purchasing decisions ethical, social, and environmental concerns; how consumers and other market participants behave and interact with one another in the marketplace; how companies compete and are structured; and, how science-based industries with complex market transactions use consumer and other information.

The Annex provides further information on the OCA research program.

This integration paper attempts to bring together the major findings of the expert papers put together in this volume and to draw from them the most important policy and strategic implications for government, industry, consumers, and others. The paper consists of three main sections on the roles and responsibilities of (a) consumers, (b) producers, and (c) government regarding consumer information, awareness, and confidence. A short concluding chapter summarizes the major findings from the research and the Symposium.

This is the second version of this paper. The first was completed early in September 1997 and distributed to all participants who attended the Symposium on *Biotechnology and the Consumer* hosted by the Office of Consumer Affairs. Other interested people who were not able to attend the Symposium also received the earlier version. The first draft served as a reference point and the central discussion paper at the two day Symposium held in Ottawa on September 24 and 25. The Symposium was attended by 150 people representing industry, consumer and environmental groups, and other non-governmental organizations (NGOs), provincial governments, and various federal government departments and agencies.

This second version incorporates major points made in the Symposium discussions and comments received from participants and others in the six weeks following the Symposium. The Office of Consumer Affairs would like to thank all of our colleagues who attended the Symposium, commented on the first draft, and contributed to our work in other ways.

CONSUMER APPROACHES AND PERSPECTIVES

Introduction to the Consumer Challenge

The consumer brings many roles, responsibilities, and perspectives to a new technology such as biotechnology. In her paper in this volume, Einsiedel highlights the challenge for proponents of biotechnology as follows:

> In the past, public reactions to products generally took place in the postmarketing phase of development. Increasingly, however, the products of innovative technologies are undergoing scrutiny by publics in earlier stages of the technological development cycle. This appears to be the case with biotechnology. The coalescing of the impacts of various social changes and movements over the last several decades, including consumerism, environmentalism, health activism, and the visibility of public interest campaigns relating to a variety of technologies . . . have pushed forward the timelines for consideration of market reactions to technological products.

Consumer perspectives of biotechnology are being shaped by the rapid development of another enabling technology – information technology, including the information highway and publishing technologies. These advances have significantly reduced the costs of producing and disseminating information (as well as misinformation) and made information about competing products and technologies more accessible to the busy consumer. They have contributed to an era of informed consumerism, where individuals increasingly have the tools, information, and time to pursue their own interests in the marketplace. These factors, in turn, change the manner in which firms compete to satisfy consumer preferences.

Today's consumer is more educated, better informed, and increasingly able to determine and express individual preferences and choice, and to challenge existing practices and arrangements. This is reflected in surveys of Canadians regarding the consumer interest in biotechnology and its many applications:

> They . . . expect government to communicate balanced and understandable information on biotechnology. Through this, the Canadian public may achieve a higher level of comfort and self-confidence to make intelligent choices (Optima Consultants, 1994).

However, a better informed and educated consumer does not guarantee a broad acceptance of the products. The survey by Optima Consultants (1994) indicated that only 51 percent agree that a high-technology society is important to improving Canada's quality of life.

Similarly, the 1995 CROP Environics survey of Canadian consumers, quoted in Ireland (1998), shows the following consumer response to purchasing high-technology goods and services:

- less than 10% are eager to embrace the new goods and services;
- bring varying degrees of enthusiasm;
- bring varying degrees of reluctance;
- about 10% resist and distrust technology.

Finally, while past surveys and focus groups suggest that attitudes towards science and technology may be positively correlated with the respondent's level of education (e.g., Sheehy, Legault, & Ireland, this volume), more recent surveys (including one in July 1997 by the Environics Research Group) suggest there is little evidence that support for biotechnology and the current regulatory framework for this technology increases with education and income.

Interactions Between Consumers and Biotechnology

Consumers may have concerns about whether certain technologies are needed or acceptable. Many of the social and ethical concerns raised in the biotechnology debate are associated with processes and structures that are not specific to biotechnology (see, e.g., Envision Research, 1997). Consumers may base their opinions on their perception of factors such as

- the technology's ethical ramifications;
- its impacts on the environment;
- potential health risks;
- its effect on the structure of the agricultural industry and other sectors in Canada and developing countries;

- its role in reinforcing and accelerating other trends in society;
- how it may affect future generations;
- the extent to which the technology and its management by government and industry are trusted by concerned consumers.

Support for biotechnology may be based on individual consumers' perceptions of the role of the technology in

- providing new products, health treatments, and diagnostic techniques;
- providing existing products at lower prices and/or better quality;
- providing new jobs;
- facilitating economic growth and diversification for certain industries and geographical regions;
- increasing food production and improving the environment and health care.

In broad terms, consumers influence the biotechnology industry in two ways. As members of the public, consumers can participate in the processes which provide for public input on these matters. They also attempt to influence the industry through market mechanisms by bringing their concerns about a product into their purchasing decisions.

(i) Consumers as Citizens of a Nation State

Before delving more deeply into the marketplace role of consumers, it is worth commenting on the existing mechanisms that facilitate the more "public" role of consumers – as, e.g., citizens, voters, and participants in public interest groups. The degree of effectiveness of the public roles of consumers and their representatives can affect consumers' marketplace decisions. A public feeling of powerlessness, for example, may create frustration that manifests itself in a consumer backlash against certain technologies and products.

What may be the government's motivation for seeking public input in the area of biotechnology? Clearly, policy making should reflect public concerns, and citizen activism is a crucial part of any democratic process. Also, since consumers will be affected directly by many aspects of biotechnology, they are obvious sources of advice regarding consumer protection. By representing their own interests, for example through consumer groups, they can provide a balance to industry input in the policy-making process.

Another motivation behind consumer involvement is that it increases consumer awareness which, in turn, leads to better informed consumers. Leroux, Hirtle, and Fortin (this volume) in fact point out that "the majority of bodies studied pursue public educational and information dissemination objectives." Also, having consumers and their representatives participate in the policy-making process legitimizes the policy outcome in the eyes of the public and increases the overall acceptance of biotechnology.

Consultation approaches. Leroux et al. examine the strengths and weaknesses of several consultation mechanisms that have been used to involve citizens and groups who wish to take part in the decision-making processes relating to the social and ethical issues raised by biotechnology. They present the strengths and weaknesses of the mechanisms from various standpoints. Their findings are summarized in Table I.

The findings indicate that when selecting a public consultation mechanism, it is necessary to take into account not only the country's social, economic, political, and technological realities, but also, the characteristics of the target public and the accessibility of the mechanism under consideration.

International comparisons. In comparing the Canadian experience regarding public participation with that of other countries, Einsiedel finds the Canadian experience inadequate. She describes the result of participation of social groups in the policy formation process in the Netherlands as a "broad consensus . . . over the policy adopted, accompanied by a commitment to share responsibilities for outcomes". In her view, public involvement and participation have not reached this stage in Canada, and "public and consumer representation on biotechnology remains inadequate."

As for providing information to the public, every organization cited by Einsiedel as being involved in such activities is either expressly industry promotional or is connected with the biotechnology industry through funding arrangements. Whether justified or not, any perception of undue industry influence could hamper the development of a more integrated industry/consumer approach. It will also discourage acceptance of biotechnology by more cautious consumers.

Commentators, including Einsiedel, have suggested that the limited public representation on biotechnology issues in Canada is related

TABLE I
Strengths and Weaknesses of Several Consultation Mechanisms

Consultation Approach	Representatativeness of the Public	Quality of Information	Feasibility	Advantages	Disadvantages
Distribution of discussion paper and request for comments	+++	++	+++	Simple methodology; Detailed oral or written comments	Participation of an informed public; Fragmentary, incomplete information
Opinion poll	++++	+	++++	Accessible; Wide range of respondents	Static picture of public opinion; Limited, one-way information
Focus groups	+	++	++	Process conducive to discussion and exchange; Emergence of new, original opinions	Limited representativeness; Factors of time and place
Sequential consultation	+++	+++	++	Circulation of information good; Verification of positions at different stages of policy development	Greater participation by interest groups; Management of information gathered more complex
Consensus conference	++	++	++	Active participation by interested persons; Consensus on a given subject	Limited number of participants in a single event; Logistics
The Internet	?	?	+	Technology offering a variety of tools; Investment according to organization's resources	Limited accessibility for public at large; Lack of control of participants' identity

Legend: +poor / ++++very good / ?unknown

to the relatively low degree of public debate here, as compared with the U.S. and Europe. Protests in Europe about genetically engineered crops demonstrate the potentially explosive nature of the biotechnology issue, and MacIntosh and Cummings (this volume) document the potential impact that such controversy can have on research and development expenditures. The European protests highlight the importance of providing a means for a genuine public input into the policy process, before consumer and other public-interest controversies reach their peak.

The research suggests that improved public representation on biotechnology issues in Canada may help to validate and legitimize public policies. Canada can learn from the experiences of other countries, as well as our own recent efforts (e.g., by AgWest Biotech, the Consumers Association of Canada, and the Food Biotechnology Centre) to inform the public and build public confidence, and can apply these lessons to the next stage of the Canadian Biotechnology Strategy.

(ii) Consumers as Citizens in a Global Economy

Globalization of the world economy may limit public input on Canadian policies. Harmonization pressures at the national and international levels are one reason for this. The internationalization of most areas of public policy means that regulators are increasingly constrained by international agencies, other countries' national regulatory bodies, international interest groups, and the international rule-based dispute settlement processes. Pressures to harmonize to reduce costs and/or increase the speed of regulation can arise as large international firms "shop" globally for regulatory approval (Doern & Sheehy, in this volume).

Canada's difficulty in pursuing independent policies is exacerbated by trade pressures from foreign governments. For example, as a major exporter of agricultural products, Canada is sensitive to trade irritants with trading partners such as the United States, which recently demonstrated its willingness to take biotechnology trade disputes (as regards, e.g., bovine growth hormone in cattle and genetically modified soybeans) to the WTO and other multilateral fora. This phenomenon extends beyond biotechnology and is consistent with the kinds of limitations that globalization can impose on domestic policy discretion.

Chen and McDermott (this volume) examine the policies of some of Canada's significant trading partners (U.S., Japan, and the European Union) and note that all these countries actively promote biotechnology. They find the regulatory approaches similar, with regulatory oversight being more or less science and risk-based (more on this in the Government section below). However, in contrast to Canadian policies, European health and safety policies are concerned with the process by which foods are produced: There are separate regulations for genetically engineered foods, including the mandatory labelling of genetically modified foods. The Canadian approach which looks at the final product, matches the U.S. approach. Labelling of genetically modified products and the patentability of transgenic animals remain controversial issues that divide the U.S. and the EU.

Chen and McDermott conclude that, from an industrial-development perspective, Canada should consider harmonizing its regulatory system, to the greatest degree possible, with those of its trading partners. For example, new biotechnology firms need a large market to successfully commercialize and harmonization with trading partners can help to give Canadian firms the large market required for successful commercialization. It should be noted however that with the differences in regulatory approaches – between for example, the U.S. and the EU – harmonization is difficult, and smaller countries such as Canada must decide the jurisdiction with which it will harmonize. Until we see greater convergence in the regulatory approaches of the U.S. and EU, regulatory harmonization on a broader scale will be difficult.

Although globalization may be seen as taking power away from citizens by reducing effective public input, it can also give people greater sovereignty in their roles as consumers. With greater international competition, the quality and performance of products are being set in the marketplace by the people who buy the products – not only by national companies and regulatory agencies. And consumers are being rewarded with better product choices, often with higher quality and lower prices.

Thus, global forces provide two challenges for consumer citizens and their representatives.

First, consumer groups must become more knowledgeable about international developments, transmit this information to their members and the public in general, and develop alliances and partnerships with consumer and other associations in different countries. This would include working with multi-country groups, such as Consumers

International, which are actively pursuing an international agenda, including improved representation at the OECD and other international venues. This will help to ensure that the consumer voice is clearly heard in multilateral trade negotiations and other international fora.

Second, the limitations of the consumers' role in domestic policy formulation means that the consumers' role in the marketplace must be enhanced. In the case of biotechnology, this means consumers must be well informed. This subject is addressed in the next sections of this paper.

(iii) Consumers in the Marketplace

A free market provides choice to consumers and allows them to signal their preferences back to researchers, producers, and marketers. Economic literature suggests that two kinds of market failures can place these functions at risk (e.g., Averitt & Lande, 1997). The first, which operates on the supply side of the market, limits the range and quality of products available to the consumer. This type of market failure is not seen to be a significant problem in biotechnology which is intended to improve the quality of existing products and add many new product choices to the marketplace.

The second type of market failure operates on the demand side and results from misleading, incomplete, or imperfect information which reduces the consumer's ability to choose rationally among the products in the marketplace. The OCA and other research suggests there is strong potential for consumer information problems with respect to many biotechnology applications and the technology more generally.

Consumer acceptance of biotechnology depends, among other things, on the type of information that is available. Accordingly, information – its *usefulness*, *quality*, *accuracy*, *timeliness*, and *credibility* – is expected to play a pivotal role in consumer behaviour, influencing both public advocacy and the ability of consumers to make informed choices in the marketplace. Policy makers and producers have an interest in promoting information that supports these purposes. Yet, according to our findings, Canadian consumers remain largely uninformed (Sheehy et al.), and current information activities do not appear to be remedying the situation.

Information Factors That Complicate the Consumer's Role in the Marketplace

The following sections examine the potential for information problems associated with the nature of biotechnology products and the technology more generally.

The "credence nature" of most biotechnology applications. Biotechnology applications and biotechnology products have qualities that are not necessarily obvious to the consumer. For example, a genetically altered vegetable can be indistinguishable from a regular vegetable. This, as well as the complexity of the technology, makes consumers dependent on specialists to learn about the benefits and/or risks of the products. When consumers rely on outside information, the information source must be one that they find credible.

Einsiedel finds that the credibility of the source depends on its perceived *expertise* as well as a feeling of trust.

Information asymmetries resulting from the technology's complexity. Biotechnology involving genetic engineering is relatively new and complex. This creates a large gap between what the specialist knows and what the non-specialized consumer knows. This asymmetry creates two kinds of problems:

1. It opens the possibility that the information could mislead the consumer, which is why the credibility of the source is important.
2. Because the technology is complex, it is expensive (in terms of time, money, and effort) to communicate the information because it must be brought to a level that consumers can understand. This involves either reducing technical scientific data to information that can be used easily by consumers to make informed choices in the marketplace, or increasing the public's understanding of the technology.

Traditionally, government has the responsibility for increasing education levels with respect to science and technology. Government is also responsible for assessing risk and ensuring public safety. Producers can also provide information about product benefits and risks, through advertising and other forms of promotion, with the government implementing laws and regulations on misleading advertising, misrepresentations, labelling, and, where required, appropriate warning information. These roles are further explored in the sections on the Producer and Government roles.

Consumer incentives to seek out information. Hadfield and Thomson (this volume) analyse how information is acquired. Consumers assess the value of information and compare it with the "cost" of acquiring it. The cost of information can involve the time to process and learn from it and/or the search costs associated with retrieving it. Whether a consumer will actually seek out and acquire this information is determined by the cost relative to the perceived or expected value. As a result, consumers will acquire high-cost information only when they perceive it will yield high benefit. The scientific nature and complexity of biotechnology information make it expensive for consumers to acquire and absorb.

Efforts by the government and industry to reduce the cost of the information and increase its accessibility are well documented.[2] Thus, the costs of acquiring information have been subsidized by government and industry; however, consumers remain relatively uninformed about biotechnology. These two realities suggest that consumers do not perceive information about biotechnology as particularly valuable to their purchasing decisions in the marketplace. Accordingly, given current consumer beliefs and the biotech applications now available for purchase, it can be questioned whether additional efforts by government and industry to reduce information costs or increase its accessibility will succeed in addressing consumers' basic valuations of the usefulness of biotechnology information and thereby succeed in altering consumer behaviour and increasing consumer acceptance of biotechnology.

One reason why some consumers may not value new information is that they may already have strong beliefs that a product or technology is risky. If consumers already believe that a product is high risk, they will require substantial contrary information in order to adjust beliefs sufficiently to change their behaviour. This is perceived to be costly, and as a result learning is not likely to occur (Hadfield & Thomson).

Limitations to acquiring information. The research on risk evaluation supports the view that, up to a certain point, providing consumers with additional information can increase consumer well-being. That is, when individuals are given additional information, they tend to respond in a rational manner, by taking more or fewer precautionary actions depending on the risks involved. However, as they become increasingly informed, the additional increments of information have less impact on behaviour. This suggests that a person's ability to process risk information is limited and diminishes with the amount of additional information.

Similarly, research on the psychological assessment of risks is consistent with arguments on the limitations of consumers' ability to employ risk information usefully. If the information is too complex for people to absorb, they will resort to heuristics (such as buying time-proven products or those with well-known brands, or consuming only "naturally produced" foods) and other simple rules to facilitate decision making.

As a result, information policies should distinguish between the *processability* and *availability* of information.

> Any solution to the consumer information problems posed by biotechnology should build on what is known about the source of consumer information problems (not an absence of information per se but rather a limited and costly capacity to absorb and apply information) and about the ways in which consumers can exploit less costly and more manageable ways of obtaining and applying information (Hadfield & Thomson).

Similar comments were made during the Symposium, where participants noted that most of the information that consumers need is known and available. The bigger issue is how to best provide that information.

The role of the media as a source of consumer information. Proponents and critics of biotechnology either praise or criticize the role of the media – often depending on the content or impact of an article or news item. It is important therefore to understand the media's role, strengths, and limitations.

On the one hand, some argue that the media are one of the most important sources of consumer information because people tend to regard it as impartial and therefore credible. However, the print and other media may not be the best source of biotechnology information for consumers.

Strauss (this volume) describes some of the challenges the print media face in addressing a subject such as biotechnology. First, the media are always rushing toward a deadline and this can compromise a story's accuracy and clarity. Also, the definition of news being "about conflict, or progress, or disaster, or novelty, or human interest," where "things are made more newsworthy by their timeliness, their proximity to readers, and the prominence of those involved," dictates that news about "progress, slow change, of the kind which clearly takes place in biotechnology, doesn't get reported until something happens." Reporters face an additional "dramatic imperative," so that information "may sit around for ages as it awaits a dramatic framework." The push for stories of dramatic appeal may tend to create an anti-biotechnology bias, as potential risks are highlighted more than the benefits or may cause the stories to under-emphasize the real and useful information which the public can actually employ to understand biotechnology.

Finally, the media may suffer pressures from interest groups, particularly in small markets where a few large firms dominate advertising revenues. In all markets pressures exist to design the information so that it "makes a good story." In addition, although reporters are able to separate hype from facts, like consumers, they also depend to a great extent on specialists and their reporting may ultimately reflect the particular bias of sources. A predilection among science writers to consult experts can in itself create biases, so that journalists may tend to "become defenders and champions of what they legitimately should only report upon" (Strauss).

Conclusion

This section has only touched the surface of the more detailed discussions on consumer information needs, found in the research papers. Public scrutiny is starting early in the development of biotechnology. Different papers point to the need for better consumer involvement. Public involvement and participation in policy formulation are important and in the case of biotechnology in Canada must be increased for policies and the technology to be better accepted by consumers. As compared to the experience in other countries, the participation of the Canadian consumer is still seen as insufficient by many of the OCA authors and participants at the Symposium.

Factors such as globalization, which complicate consumer involvement, further underline the need to enhance the consumer's role in the marketplace. In the case of biotechnology, this means the consumer must be well-informed; and many papers indicate that better, more comprehensible, information to consumers is critical. The current situation is seen as unsatisfactory by many concerned consumers, as well as by NGOs and many business representatives. The information provided must be accessible and credible to the consumer, and the information sources themselves must be seen as credible and trustworthy. This is important in biotechnology where information is complex. Finally, new information must be highly valued by consumers to have any influence on their behaviour.

PRODUCER ROLES

From the consumer's perspective, producers conduct two essential roles with respect to biotechnology:

1. They conduct research and development (R&D), commercialize innovations, and compete with other companies – and against more traditional as well as other biotech products – for consumer dollars.
2. They provide information on specific applications and on biotechnology in general through advertising, point-of-sale information, general promotion and public-education endeavours, and by participating in government bodies and multi-stakeholder initiatives.

The following sections briefly discuss these two roles from the perspective of consumer choice and information.

Biotechnology and Commercialization

Commercialization is now taking place. As is the case with most "technological revolutions," the development of the first biotechnology products was more the result of scientific discovery and feasibility than of economic cost or demand considerations. This situation is rapidly changing, with the focus shifting to market demands and opportunities for profit. Biotechnology is now at the stage of commercializing a wide variety of products. These commercialization activities may well increase consumer welfare, as new products serve the different needs, preferences, and tastes of consumers, and potentially reduce prices.

However, the successful development of biotechnology applications still depends on their technical feasibility, profitability, environmental safety, and public health benefits, and their acceptance by both producers and consumers (Crampes & Hollander, this volume). Commercialization of biotechnology still has had few successes relative to other new technologies, up to the time of writing, although participants at the Symposium stressed that the number of biotechnology applications entering the market could increase dramatically over the next few years. Only one in 10 pharmaceutical products entering clinical trials is ultimately commercialized and, of those, only 30 percent generate sufficient revenues to cover the R&D costs, as reported in Crampes and Hollander.

Regulation is a key factor in a biotechnology firm's success (Caulfield) and many countries recognize the potential of biotechnology and support its development and diffusion.

Although industrial development is not the primary focus of this paper, the following sections, drawing from research by Crampes and Hollander and by Green (both in this volume), will briefly review the determinants of successful commercialization with a view to identifying whether this stage raises any issues related to competition policy and other marketplace, public interest, and regulatory concerns. As noted earlier in the text, the exercise of consumer sovereignty can be impeded by either failures on the supply side of the market which limit the product choices available to the consumer, or by information and related failures on the demand side which impair the consumer's ability to make informed choices among the product choices which are available. This part of the text focuses on market failures on the supply side and includes an examination of the scientific and commercial advances, the economic regulation of these industries, and the producer's roles with regard to information.

Commercial advances in the United States. Activities in the United States, the international leader with respect to developments in biotechnology, show that commercial advances are being made and new applications are coming to the market. The U.S. experience provides a good example of how the biotechnology industry can develop.

Whereas the U.S. federal government has been active in supporting basic scientific research, with particular emphasis on medical applications, it has done much less in terms of direct promotion to stimulate the development of the biotechnology industry as a whole and the commercialization of biotechnology applications. The American lead has much more to do with the dynamism of small private firms and their linkages with large organizations as well as the increasing linkages to university research and to venture capital financing (Crampes & Hollander). These linkages between firms result mainly from the difficulties of bringing together, under one roof, various activities such as research, development, production, commercialization, and regulatory approval. In order to understand these linkages it is useful to examine the biotechnology "industry."

What is the biotechnology "industry"? As noted earlier, biotechnology represents more a specific technology than a specific industry. Biotechnology applications are used as an enabling technology to improve products and processes in many different industries. This makes it difficult from an industrial organization perspective to characterize and examine its static and dynamic economic elements in terms of industrial structure, firm behaviour, and performance (Green).

Most current biotechnology companies are focused almost exclusively on R&D. For example, in the U.S., in 1991, the biotechnology industry invested 47 percent of its sales in R&D (compared to a figure of only 14 percent for the traditional pharmaceutical industry). Biotech companies often are small firms, with relatively little vertical integration. This is in contrast to the pharmaceutical industry which is dominated by large firms in which R&D is only one activity which is vertically integrated with other activities, such as production, marketing, and distribution (Green). Successful commercialization of biotechnology applications therefore depends on formal or informal linkages often involving vertical integration with other firms for production, marketing, and distribution.

How biotechnology companies compete in the marketplace. In general, biotechnology companies compete through high R&D expenditures, forming alliances and other linkages with larger companies, and being first to market new innovations – although one implication from the Green and the Crampes and Hollander studies is the dangers of being a "first mover" in biotechnology. A preferable strategy might be to be a follower and thus allow the first mover to absorb all of the public controversy.

Given the narrow definition of biotechnology firms as being producers first and foremost of R&D, the most important form of economic regulation with respect to biotechnology relates to the protection of intellectual property rights (Chen & McDermott; Crampes

& Hollander). The importance of property rights is to ensure that these firms have sufficient incentives to engage in R&D activity, as well as to attract venture capital and to encourage foreign companies to invest.

Usually, protection of intellectual property rights is in the form of patents. However, many biotechnology products are unpatentable because they do not meet the requirements of novelty or utility. In the U.S., inventors have the alternative of patenting the *process*, rather than the product, or of obtaining the special status of an orphan drug. This type of intellectual property protection is not available in Canada.

How does competition and competitiveness among biotechnology firms impact consumer welfare? On the one hand, intellectual property protection stimulates innovation in products, processes, and applications, which benefits consumers by providing greater product choice and lower prices in the marketplace. However patent protection can also result in monopoly behaviour leading to higher prices and less product diversity for the consumer. In addition, strategic alliances and vertical integration across firms can lead, at least theoretically, to anti-competitive behaviour. However, in general, informal and formal strategic alliances raise fewer competition issues than mergers and takeovers, and vertical integration raises fewer competition concerns than horizontal integration.

Among biotechnology firms, however, the danger of such behaviour does not appear to be high at the present time. First, the small size of most new biotechnology firms indicates that the concentration levels of firms within any industry are low, reducing the opportunity for monopoly pricing and other forms of anti-competitive behaviour.

Second, vertical integration among biotech firms through merger is slow to develop, and still depends on the benefits of full integration relative to establishing contractual relationships with distributors as is currently the case. These latter activities do not appear to present any negative impacts on competition.

Third, biotechnology often leads to greater product and process diversity. As mentioned earlier, product diversity is desirable for a market comprised of many different consumer needs and preferences. Process diversity provides producers who use the technology with more assurance against economic and technical risks by providing them with ways to respond to unexpected input price hikes, regulatory changes, and predatory agents who control certain essential inputs.

Commercialization in human genetics. Commercialization in human genetics (and some other medical care fields) raise quite different market and public interest issues than other areas of biotechnology. Accordingly, Timothy Caulfield was commissioned by OCA to examine the importance and effects of the ethical, social, and legal ramifications of genetic services in the health care system. There is concern that commercial pressures will result in premature entrance of genetic services into the marketplace. For example, the building of an expensive laboratory infrastructure creates a need for a market. What is the effect of industry creating for itself a market by determining the criteria regarding who will be offered a genetic test?

> Decisions about who should be considered an "at risk" individual, and therefore within the criteria to be offered a genetic test, should be based on sound scientific, psychological, and social information and not a desire to increase the potential market (Caulfield).

The most controversial fear associated with the commercialization of human genetics is that it could result in a consumer-led eugenics movement. "Will biotech companies, through their efforts to create a robust market for the emerging genetic services, affect our perception of normalcy?" (Caulfield). For example, if a producer creates a hormone that successfully treats dwarfism, it can decide to increase its potential market by promoting the perception that normal shortness is a disease.

Caulfield raises questions relating to potential costs to the health care system as well as liability:

- Will commercial pressures create artificially high expectations among patients?
- Will there be over-use of genetic services?
- Will health fears lead to inappropriate use of diagnostic tests?

Caulfield recommends further research into the benefits and possible negative ramifications of commercialization, with a view to "regulate the potentially adverse impact of market forces while still allowing the rapid dissemination of genetic innovations."

Producers as Providers of Information on Biotechnology

Producer incentives to provide quality information to consumers. The incentives for biotechnology producers to provide information to consumers about the benefits of their products, can depend on whether they are engaged in *product* innovation or *process* innovation (Crampes & Hollander).

Product innovation includes new products with particular characteristics (e.g., nutriceuticals), substitutes for existing products (e.g., unicellular proteins), and alternative uses of existing products (e.g., bio-ethanol developed from corn). In this case, the characteristics of the products are beneficial to consumers. Producers have incentives to provide information about the products so that the consumer can identify them and differentiate them from other products. Information can be provided through marketing and advertising programs.

Process innovation includes medical processes (for instance, diagnostic kits, vaccines), production enhancing applications (for example, hormones to speed growth of plants or animals), and cost-reducing innovations due to resistance to disease and pestilence. Process innovations have beneficial effects on producer productivity either through increased production and/or decreased costs of production. These may benefit consumers indirectly, depending on the extent to which the productivity gains are translated into lower market prices. Such gains will only reduce prices if the marketplace is a competitive one. Here producers have few incentives to provide information about the biotechnology processes used.

The positive aspects of producer information. Producers typically convey information to consumers through marketing and advertising efforts. The economic literature distinguishes two polar views on advertising: the *positive* view, explored here, and the *adverse* view, explored in the next section.

The positive view sees advertising as providing information which enables consumers to make rational choices. Under this view, advertising announces the existence of a product and describes its qualities, characteristics and use. It reduces consumers' learning costs and helps them choose between brands of products. Advertising can also reduce product differentiation associated with a lack of information about some products and has been identified as encouraging the production of high-quality goods. Seen in this context, producers of biotechnology applications have incentives to develop and communicate information about the beneficial aspects of products to consumers. "If these potential consumers could be informed about these product features at a low enough cost, demand for the product would increase enough to create profit opportunities for producers" (Mathios, this volume).

When product attributes are too complicated to convey through advertising, or if the attributes cannot be determined on purchase (what economists like to call "credence goods"), firms can use advertising as a "performance bond." Kleit (this volume) shows how pharmaceutical firms invest heavily in their brand names, promoting recognition and demonstrating to consumers a willingness to stand behind their products. "If firms renege on their implicit promise of quality, consumers punish them by refusing to buy their products again" (Kleit). As a result, clear and strong links between information on products and company names creates reputation effects which allay consumer misgivings about product safety and quality.

However, even scientific information can be effectively provided by producers in a way that consumers can understand, as is examined in Mathios' paper. Mathios shows how producers in the U.S. were effective in providing information on the links between diet and disease to promote healthier food products. Producers were able to educate the public using a variety of media to target specific markets and seemingly alter consumer behaviour to bring about dietary improvements. Government advertising, on the other hand, which was not product specific and tended to be print media oriented, failed to reach all sectors of the public. This example serves to show that producers can provide useful and accessible scientific information and increase consumer welfare. If consumer demand for information increases, there will be an incentive for producers to satisfy this demand.

Limitations to producer information. The adverse view perceives advertising as a means to mislead consumers, and as an instrument to "create" product differentiation which is not real and thus offers no real benefit to consumers. Seen from this perspective, advertising reduces product competition and introduces barriers to the entry of new products and companies. These potential disadvantages relate to the more conventional role for consumer protection – that is, the monitoring of truth in producer representations of their products. Since consumers cannot verify science-based claims directly, there exists a potential for deception, reducing the credibility of producer information:

Unless the market or government has the mechanisms to punish firms that lie, or consumers can verify information in some way, consumers would be expected to be sceptical of producer-provided information, limiting food producers' incentive to make claims (Mathios).

Research on advertising practices in the U.S. pharmaceutical industry is consistent with both the positive and the adverse views of advertisements. That is, advertising is used to provide information to consumers about new products but it can also generate barriers to the entry of new firms and products in order to protect market share (Kleit).

In the case of biotechnological advances in health care, Caulfield warns that if producers market their new products or genetic tests directly to the consumer, there is potential for abuse. For example, women fear breast cancer. Marketing campaigns can prey on this fear by first increasing their anxiety and then offering testing as a solution. Caulfield recommends continued regulation of advertising distributed by pharmaceutical companies in the field of genetics, allowing only direct-to-physician advertising as is now permitted under existing laws and regulations.

Conclusion

Commercialization of biotechnology products is now taking place, with limited success up to the present time (but with more success expected in the near future). Economic regulation is seen as a key factor in a biotechnology firm's success. Biotechnology firms largely conduct only R&D and thus the most important form of economic regulation for biotech companies is the protection of intellectual property rights.

In the commercialization of human genetics, there are still ethical, social, and legal issues that must be assessed. Further research into the benefits and possible negative ramifications is recommended.

Producers also have a role as providers of information. There are both positive and adverse views on producer advertising, and both are relevant to biotechnology. Past experience has shown that producers can sometimes provide information which is superior to that emanating from government.

Whether or not advertising is used to misinform consumers and/or enhance consumers' market power will depend in part on competition in product markets and on government monitoring and enforcement of relevant statutes in the areas of misrepresentation and misleading advertising. Research and experience to date suggest that markets and industries involving biotech applications could be characterized by fairly healthy company rivalry, suggesting that advertising may play more of a positive than a negative role in informing consumers about biotechnology applications.

However, even with strong company rivalry, producers will tend to accentuate the positive attributes of their products in their advertisements. This underlines the need for other information sources – such as government and consumer groups – to complement the efforts of industry and to ensure the entire story is told.

GOVERNMENT ROLES

Public Perceptions of the Government's Roles and Responsibilities

With respect to biotechnology in general and consumer interests more specifically, the government has taken on various roles, including:

- law maker and regulator
- ensuring public safety
- supporter of the basic science and R&D to provide consumers with new innovative products
- information provider
- organizer and manager of non-market mechanisms (for example, advisory bodies) to promote public understanding and the public interest
- representative of Canadian consumer, industry, socio-ethical, environmental, and other public interests at various international fora (OECD, WTO, NAFTA, various UN initiatives on Biodiversity, and the Human Genome project).

Table II summarizes the Optima Consultants (1994) survey of Canadian views about the role of government in biotechnology. Of the ten listed potential roles that governments could play, the survey found that the top six deal with issues of safety, ethics, and consumer information. In line with market economics, the public survey indicated comparatively little support for direct government involvement in financing and commercialization of biotechnology activities.

TABLE II
Level of Agreement with Statements Regarding Government's Role in Biotechnology

	Agree (percent)	Neutral (percent)	Disagree (percent)
Protect the safety of workers in biotechnology industries	87	8	5
Determine the safety of biotechnology products	87	8	4
Enforce regulations on activities of biotechnology	84	10	5
Consult the public on regulating biotechnology products and uses	81	13	5
Conduct a public information campaign about biotechnology	77	14	9
Assess the benefits of biotechnology	76	16	7
Be involved in the ethical aspects of biotechnology	75	16	8
Educate the public by offering seminars on biotechnology	74	16	9
Financially support biotechnology research in companies	37	33	29
Develop biotechnology products for commercial purposes	33	28	37

The federal government has responded to the need for safety of product and process innovations by developing a Canadian framework for regulating biotechnology. The survey also identified government roles for conducting a public information campaign, assessing benefits and risks and educating the public about biotechnology – all of which relate to the informational roles of government.

The following sections focus on the information and regulatory/ stewardship roles of government in relation to biotechnology and the consumer.

Information Role

In general, when governments are faced with consumer concerns and perceptions about risks, their role is (a) to provide impartial and balanced information about the risks to consumers, and (b) to ensure consumer involvement in risk assessment processes.

Also, as examined above, biotechnology is subject to market failures on the demand side, resulting from incomplete or inadequate information. The government as well has a role in working with other stakeholders to facilitate the market acceptance of appropriate biotechnology innovations. Accurate, up-to-date information in plain language is important to that success.

Government as a provider of information on new technologies. Two rationales exist for the government to provide information to the public. The first relates to the fact that it is not in the producer's interest to discuss the cost and risk aspects of their products. Second, government regulators may have informational advantages over consumers and producers in communicating risk information because of their regulatory functions in collecting and examining risk data. Providing consumers with risk information holds the potential to yield superior market outcomes compared to direct regulation. This is because consumers are a diverse group, and amongst themselves, will display different preferences for risks. A single regulatory rule cannot fully reflect this diversity.

Information role when risks are unclear. Although there are strategies for increasing consumer understanding of measurable risk, it is less clear what an appropriate policy response would be when dealing with the risk uncertainty that arises when scientists cannot fully understand or anticipate the secondary effects of genetic engineering. Even when risks are not identifiable or measurable, the government still has a role in providing information.

In light of the scientific uncertainty, the fact that there are no known scientific risks is not a basis for concluding that there is no need to inform consumers about potential risks (Hadfield & Thomson).

An unknown or unmeasurable risk should not be treated as no risk at all.

Consumers' complete information needs. Consumers benefit by acquiring information that corrects their mistakes about product safety. But even when risks are known and measurable, their information needs go beyond science-based risk assessment. Consumers also have ethical beliefs, or, in the case of food products, they may have special dietary preferences. Thus, although consumers benefit from acquiring information about risks, that information can still be incomplete. Consumers' idiosyncrasies and different information needs must be taken into account.

Allergic response is one form of idiosyncrasy. Another is individualized approaches to nutrition and diet, such as combining diets which recommend eating foods from one group – such as vegetables – only in combination with foods from specified other groups, such as grains. Yet another is represented by the array of ethical preferences with respect to food (Hadfield & Thomson).

Potential limitations to the government's role as information provider. There are also potential limitations to government provision of information to the public, particularly when it is the major source of information. When technologies and markets change rapidly, it is difficult for government officials to stay *au courant*, resulting in incomplete or incorrect information. There is also concern that government agencies can be influenced by special interest groups, affecting the information that is provided to the consumer. One solution to this problem is an increased role for markets in providing information.

Towards a marketplace solution to information about biotechnology. Labelling is one way to address consumers' various information needs while harnessing the incentives offered by producer-provided information described in the previous chapter. There is already a movement towards this in Europe, where some countries are looking at mandatory labelling of products derived through genetic engineering. The notion of government encouraging the labelling of biotechnology products, or mandating it by law, has received support in Canadian public opinion surveys. It is argued that labelling serves the goal of consumer protection from a consumer-centered perspective and allows consumers to match their transactions with their intentions.

The benefits are seen to be threefold.

First, a label guarantees the identification of biotechnology applications in the market, increasing consumer awareness. This, in turn, provides incentives for consumers to seek out and learn about biotechnology, according to their personal needs, preferences for risks and benefits, and value systems. This creates an increased demand for information that the marketplace can satisfy, leading to better informed consumers. The effectiveness of producer provided information is examined in the chapter in this volume by Mathios.

Second, labels give producers strong incentives to meet this demand for information, particularly when consumers are inclined to perceive high risks.

> Intuitively, if producers are faced with the prospects that consumers will, through lack of information, avoid biotechnology products, they will devote resources to providing information. More importantly, they will devote resources to understanding the complex nature of consumer beliefs, information-processing capacities, and so on, and devising effective methods for achieving the goal of conveying useful information that consumers can understand and apply (Hadfield & Thomson).

Third, labels would better ensure that purchasing decisions are based on informed consent – which, through time, would have the effect of turning involuntary into voluntary risks. Research suggests that consumers prefer voluntary risks over involuntary risks, even when objectively the actual degree of risk is about the same. Therefore, in this manner, labelling could, through time, enhance consumer confidence and acceptance of biotechnology.

However, no studies have yet been done on the compliance, administrative, and other costs, the possible international implications, and whether similar benefits could be obtained through voluntary means that are less costly than mandatory labelling. As well, whether voluntary or mandatory, this approach may be more relevant to some types of biotechnology applications such as processed foods which contain a large biotech component, than to others where there may be only a trace of a genetically modified substance. As noted by one symposium participant, it will become rare to find any processed food that does not, at some stage, include ingredients that are produced with biotechnology applications, with the result that almost all foods would require a label. A more efficient solution could be to label those foods that are completely free of biotechnology applications.

Regulatory Role: Ensuring Safety

As stressed at the beginning of the paper, consumers and the general public place priority on the regulatory or stewardship role of government with respect to health, safety, and protecting the environment. This section explores the consumer interest within the context of the principles underlying the federal regulatory framework for biotechnology. More specifically, it will attempt to describe consumer perspectives and approaches to risks and how these compare with the approaches used within the Canadian regulatory framework for biotechnology.

The federal government framework to regulate biotechnology. The federal government framework to regulate biotechnology emerged from the Canadian Biotechnology Strategy (formerly the National Biotechnology Strategy), which has promoted the development of biotechnology in Canada since 1983. The framework was designed to satisfy the following set of *six principles*:

1. Maintain Canada's high standards for the protection of human health and the environment.
2. Build upon existing legislation and institutions, clarifying jurisdictional responsibilities, and avoiding duplication.
3. Develop guidelines, standards, codes of practice and monitoring capabilities for pre-release assessment of the risks associated with release to the environment.
4. Develop a sound scientific database, upon which risk assessments and evaluations of products can be made.
5. Promote development and enforcement of Canadian regulations in an open and consultative manner, in harmony with national priorities and international approaches.
6. Foster a favourable climate for development, accelerating innovation and adoption of sustainable Canadian biotechnology products and processes.

Consumer perspectives on the regulatory system tend to support the broad objectives of maintaining the same level of human health and environmental standards that Canada currently enjoys. For instance, the poll by Optima Consultants (1994) indicated that consumers were anxious about the safety of the environment with 81 percent agreeing that land, air, and water are more contaminated than ever before, and that government should increase its regulation of the environment (62 percent). Although consumers support regulation of biotechnology to promote health and environmental safety, some question its capacity to address long-term risks and hidden hazards (Ekos Research, 1996).

> Even with pure motives and strict controls, participants felt that no one really knows what will happen. They point to things that have gone wrong in the past – citing Thalidomide, radiation, and DDT (Creative Research, 1996).

None of the things cited above involved biotechnology, but because biotechnology uses the same regulatory structures as other products and technologies, participants were quite justified in citing problems outside biotechnology. This issue is further explored later in this section.

An analysis of the regulatory structure should therefore examine how adequately the design captures the advantages and limitations of scientific knowledge in order to promote a high level of human health and environmental safety.

Doern & Sheehy analyse the strengths and weaknesses of the regulatory framework. They find it is nested within the existing statutes, laws, and departmental authorities of Health Canada, Environment Canada, Fisheries & Oceans, and the Canadian Food Inspection Agency.

> The biotechnology regulatory framework was deliberately based on a decision not to create a separate new regulatory agency and hence in a very real sense its creators knew that it was embedding the biotechnology framework in a set of existing regulators (Doern & Sheehy).

Doern and Sheehy note that, as a result, the regulatory framework for biotechnology maintains both the strengths and weaknesses of the existing regulatory frameworks and preserves the political-economic and organizational cultures of each departmental authority.

Thus, while building on what is already in place is efficient, the regulation of biotechnology is as a consequence influenced by factors which have contributed to a decline in public trust in existing regulatory agencies:

> First, regulatory agencies have been perceived to have been overly influenced by industry; second, agencies are seen to be inappropriately biased in favor of promoting particular technologies; third, agencies have been known to mismanage particular health or environmental activities; fourth, government officials and experts have been known to disagree amongst themselves (Einsiedel).

These factors do not relate solely to the framework for regulating biotechnology. However, given that the framework is based on existing regulatory bodies, these broader concerns do influence how the public perceives the regulatory processes associated with this technology, and can thus heighten public anxiety about certain biotechnology applications.

All federal legislation must comply with the "equivalency" clause as stated in the Canadian Environmental Protection Act (CEPA) regarding environmental assessment standards. Thus CEPA can be seen as the ultimate back-up legislation which ensures a minimal environmental and notification standard across government (Doern & Sheehy). However, the CEPA basis for regulation is "substances" while the other departments regulate on the basis of "products."

> Regulating *products* is different from regulating *substances* and there are widely varying traditions and needs regarding regulatory flexibility and how its is managed (Doern & Sheehy).

To further complicate the matter, the *Standing Committee on Environment and Sustainable Development*, as quoted by Doern and Sheehy, states that

> the concept of equivalence or harmonization [between federal regulatory departments], does not mean that the same precautionary tests and data are required for all products, but that appropriately comprehensive and stringent standards shall apply to all products, whichever legislation those products fall under.

This complexity of the statutory base for regulating biotechnology can cause two problems. First, with respect to the differentiation between "product" and "substance," it may become increasingly difficult to reconcile these differences once the caseload of biotechnology substances or products increases. It is generally assumed that consumer acceptance of biotechnology will increase as more products move through the regulatory process and are marketed, thus increasing the consumer's direct experience with such applications.

However, as the caseload increases, conflicts and tensions within the regulatory system could tend to undermine consumer confidence in both the technology and the regulatory framework. Particularly given recent government cutbacks and regulatory controversies, questions could arise about how and whether government can cope with this new complex technology.

Doern and Sheehy find that the complexity of the statutory base for regulating biotechnology can cause problems. Any conflicts or tensions within the regulatory system could undermine consumer confidence in both the technology and the regulatory framework. Furthermore, to gain consumer support, the regulatory system must be able to communicate its processes to the public.

> Because it is so central to their commercial or research interests, biotechnology producers know first-hand what the regulatory road map looks like . . . In real terms, however, the case handling road map is not objectively very clear or self-evident for those other interests which may be more occasional participants in the regulatory process or whose own incentive systems make it difficult for them to know the regulatory regime as well as business interests do (Doern & Sheehy).

To prevent further erosion of trust in the government's capacity to make science-based decisions, it is important that regulatory decisions be clear. All affected parties must be able to find out how a particular regulatory decision was reached, and be comfortable with – if not necessarily in agreement with – both the process and the decision.

Several authors, including Doern and Sheehy, note that the

government's dual mandates, to both regulate (Principle 5) and promote (Principle 6) the technology, have given rise to public concerns that this may compromise its ability to protect human health and environmental safety (Creative Research, 1996; Hadfield & Thomson, this volume; Wohl, this volume). One concern is that these dual mandates appear to compromise the government's ability to provide accurate and credible information on the true costs and risks of this technology. Doern and Sheehy note that the government-led biotechnology strategies are seen by environmentalists as "lobby from within," giving insufficient weight to environmental and human health issues. Included in the comments of the Canadian Institute for Environmental Law and Policy on the research is the following statement:

> Given the evidence of a lack of social consensus within Canadian society it seems profoundly inappropriate that the government of Canada focus its efforts on the development of means to facilitate the acceptance of biotechnology products by Canadian consumers. This is especially true in the context of the review of the (Canadian) Biotechnology Strategy, which is one of the government's primary vehicles through which public funds are committed to support the biotechnology industry.

Evaluating risk. The fourth principle of the federal biotechnology regulatory framework acknowledges the limitations of scientific understandings of biotechnology. It proposes to develop a sound scientific basis for the assessment of risks, with explicit mention of developing the capacity to pre-assess risks associated with release into the environment. Some product risks can be measured by a quantitative assessment of the degree of harm and frequency of occurrence. But there is also "risk uncertainty" which encompasses risks not known to science or which are difficult to assess accurately.

> . . . biotechnology, at the forefront of scientific development, presents the scientists who work in the field and the regulators who must determine appropriate policy in the field with extraordinary information needs of their own. Those who produce and those who must regulate the science involved are themselves at places uncertain about the match between what they expect or intend and what they will realize . . . just as consumers must figure out how to take account of and act in spite of uncertainty, so must the regulators (Hadfield & Thomson)

When regulators are faced with risk uncertainty, they approach the limitations of scientific understandings by sponsoring additional scientific inquiry by independent researchers.

Regarding measurable or *quantitative* risks, the current "risk-based" approach to regulation allows for some degree of risk and seeks to manage or confine it to "acceptable levels." However consumer assessments of risk often diverge from scientific assessments. Canada's existing regulatory approach is "science-based," which means that objective scientific standards, rather than consumer perceptions of risk, are the basis of regulatory decisions.

From the point of view of scientists, consumers' decisions may not seem rational, but their risk assessment includes other *qualititative* factors, factors which are not included in the science-based assessment of risk; see Wohl's chapter in this volume as well as the comments given by the Symposium rapporteur, Bryne Purchase (this volume). These qualitative factors can be considered an equally legitimate basis for regulatory decisions. Acknowledging and in some manner incorporating these factors into the government's management of this technology may contribute to bringing consumer perspectives more into line with scientific assessments.

Consumers' risk assessment also depends on their trust in science and trust in the information source. The issue of trust and credibility of information sources is explored in the first section of this paper.

However, a single regulatory rule may be inadequate if consumers display a wide range of preferences for benefits and risks. Crampes & Hollander point out that if information is properly provided, and if public and professional organizations can identify products derived from biotechnology (with labels, point-of-sale information, or some other means), then the marketplace may be the preferred place to achieve the appropriate balance between natural and genetically engineered products.

The risks posed by biotechnology may be considered less acceptable than risks from other activities if people feel that they do not have a choice, or if they feel that the distribution of risks and benefits between producers and consumers is not fair.

There are other considerations which increase the rationale for individual consumer choice. Wohl cites survey and focus group results which indicate that consumers are concerned about the moral and ethical implications of the use of biotechnology, and additional non-health implications, such as the distribution of benefits.

Regulation and ethical concerns. Ethical concerns are not directly reflected in the six principles of the federal framework for the regulation of biotechnology (Doern & Sheehy). It is not obvious from the literature and the Symposium discussions how such considerations can be directly incorporated into the current regulatory framework, without placing additional burdens on regulators and compromising the benefits of international trade in these products.

Ethicists and others have argued for parallel structures, such as (a) the development of ethical frameworks (to be embodied perhaps in voluntary codes of practise for industry and the scientific community), and (b) the establishment of one or more advisory bodies, to ensure that ethical values and social norms play an appropriate role in the development of this technology – from the basic science, R&D, and product testing through patenting, commercialization, and marketing. For example, a well designed ethical framework which is effectively applied to the basic science, research, and product development would act as an early warning system which would perhaps prevent ethically questionable biotechnology applications from being brought to the stage of regulatory approval and commercialization. This would both serve the public interest and prevent industry from allocating scarce resources to ethically and commercially questionable products and processes.

These are some of the challenges to be addressed in the Renewal of the Canadian Biotechnology Strategy (CBS), and perhaps to be given priority in the terms of reference for a more broadly based national advisory body on biotechnology which could result from CBS Renewal.

Conclusion

This section has reviewed the information and regulatory roles of the government. Consumers have a wide range of needs and concerns, and therefore providing consumers with information may be preferable to more regulation, which cannot encompass differing consumers' needs. One approach being pursued by the EU and proposed by some NGOs in Canada is the labelling of certain biotechnology products, especially food products, thus allowing consumers to make choices according to their personal preference and risk tolerance. Labelling may also be an effective means of harnessing producer incentives to meet information needs. However, as noted above, a number of issues need to be further assessed in developing government policy and corporate strategies on labelling.

This section has also reviewed the federal regulatory framework for biotechnology. The fact that the regulation of biotechnology is currently divided among four departments makes case handling and compliance activities complex. Potential conflicts within the regulatory system, as well as lack of transparency of case handling procedures, could serve to undermine consumer support for the regulatory system. However, consumers also recognize the importance of the role of regulatory agencies in protecting them from harm and promoting a clean environment. Producers as well recognize the importance of product approvals from a sound and trusted regulatory framework for sending signals of quality to their customers and for marketing their products in Canada and other countries.

Governments will continue to devote significant energy and resources to ensuring that products marketed in Canada meet minimum thresholds for health, safety, and the environment. However, in the absence of absolute guarantees of no-risk (a "state of grace" which is impossible to achieve without virtually infinite resources), governments may also need to explore how consumers:

- can be provided with better information and choice;
- can participate more fully in policy-making at every stage; and
- can be made more aware of the risks they may be having to accept in consuming certain products, so that they more fully understand the implications of these product choices (thus turning involuntary into voluntary risks).

CONCLUDING REMARKS

Biotechnology is an enabling technology which can make existing industries more dynamic and competitive. It is also new, rapidly developing, and complex, raising consumer concerns about the safety and ethics of its applications. As biotechnology continues to develop, the government has a role to play in working with industry, consumer groups, and other NGOs to build consumer trust that will permit the market success of potentially beneficial new products or processes.

While important efforts have been made in the recent past, public input in policy making is still insufficient. Consumer concerns should be reflected in policy outcomes. Consumer input will also serve to legitimize policy decisions in the eyes of the public.

Greater information to the public remains a laudable means of promoting public confidence about this technology. Information must be credible in terms of content and source, accessible, and provided in a manner that the non-specialized consumer can understand. The research indicates that the industry has been active in preparing and disseminating positive information about biotechnology.

The low level of consumer awareness about biotechnology may relate more closely to the limited value consumers place on the usefulness of biotechnology information to increase their effectiveness in the marketplace and their quality of life. The lack of explicit consumer choice diminishes consumers' interest in this technology, since they cannot distinguish products in the marketplace and hence see less need to learn about biotechnology.

Risk assessment needs to take into account consumer perspectives of risk, which include factors beyond science-based assessments. One suggested approach would be for the regulatory system for biotechnology to incorporate the consumer perspectives on risk. Another approach, which would better address diverse consumer needs, is the promotion of participation and choice, with improved risk and benefit information. This would provide consumers with the opportunity to assess products according to their risk, socio-ethical, or other preferences and to choose products accordingly. This could then be supplemented by alternatives to regulation such as a broadly based advisory body and the establishment and use of ethical frameworks by industry, government, and the scientific community.

In addition to the known, scientific assessments of risk, the notion of potential or unknown risks remains a challenge for consumers and regulators alike.

Regulatory responsibility for biotechnology is divided among four government departments. The way in which health and safety regulatory decisions are reached must be clear to all affected parties in order to maintain public trust in the government's ability to make science-based decisions. The government's dual mandate of both regulator and promoter of biotechnology, also affects public perceptions of trust.

More must be done to address social and ethical concerns. Government could minimize the potential negative impact on industry and consumers by taking some direct responsibility for addressing these issues – through, for example, an advisory body and other consultative groups. Government could also facilitate the establishment of ethical frameworks to be embodied in voluntary codes of practise implemented by industry and the scientific community.

Earning and keeping public trust will be a major challenge facing government, industry, and consumer groups in the Renewal of the Canadian Biotechnology Strategy. Canadians have changed both as citizens and as consumers. We have become a "harder sell," and trust must now be earned, not taken for granted.

The OCA research and the Symposium underline the complexity of biotechnology – in terms of its various uses, the issues raised by different applications, and how we as individual Canadian consumers and citizens respond to it.

> Shakespeare's Hamlet says, "There is nothing either good or bad but thinking makes it so." If this is true, then one is invited to reflect on where our thoughts come from. Why would we think the science or applications of biotechnology either good or bad? In this regard there is an old saying that, "What you see depends on where you sit." In all of this complex reality we tend to see what we focus on. And what we focus on is dictated very much by our prior knowledge and experience of the world. In short, our attitudes and perspectives will be shaped by our interests and our professional expertise (Purchase, in this volume).

This complexity and diversity – of uses, issues, and consumer preferences and concerns – makes it difficult to build understanding and consensus around government policies and corporate strategies regarding biotechnology. There was, however, one area of consensus among most if not all participants at the Symposium – that it is time to move forward with this technology through:

- multi-stakeholder consultations, involvement, and partnerships;
- finalizing, vetting, and publicizing key policies of interest to concerned consumers;
- preparing the information already available on this technology for broader distribution;
- sorting out the responsibilities of government, industry, and public interest groups in providing consumer information.

Moving forward also means addressing and consulting on the new issues which are now emerging, such as:

- genetic testing and mapping;
- xenotransplantation (use of genetically engineered animals for organ production for human transplants, e.g., "Dolly");
- biodiversity and biosafety;
- international rules and disciplines to influence, extend, or replace national rules;
- access to foreign markets;

in addition to the broader socio-ethical agenda now being explored under CBS Renewal.

In short, many participants stated that this is not the time for more research and generation of new information. As noted by one Symposium attendee: "Biotechnology has been studied to death." Rather, it is time to capitalize on the information and mechanisms now in place – or expected to be established under CBS Renewal – in order to improve consumer access to information that they need to make their own choices about biotechnology and its many applications. It is hoped that this report and the OCA research will make a modest contribution to moving the biotechnology agenda forward in the months and years to come.

ACKNOWLEDGMENTS

This paper is the result of the work of many individuals and organizations. The Office of Consumer Affairs appreciates the participation and contributions of the following:

The External Policy Advisory Committee:
Canadian Federation of Agriculture
Industrial Biotechnology Association of Canada
National Biotechnology Advisory Committee (NBAC)
Consumers' Association of Canada (CAC)
Federation nationale des associations de consommateurs du Quebec (FNACQ)
Canadian Institute of Biotechnology (CIB)
Canadian Institute for Environmental Law and Policy (CIELAP)
Dr. Richard Goldbloom, Grace Health Centre, Dalhousie University
Environmentally Sound Packaging Coalition of Canada
B.C. Biotech Alliance
Jim Hutch, Hutchtech Consulting Inc.
National Institute of Nutrition
Life Sciences and Health Care, Royal Bank of Canada
Technical Food and Consumer Products Manufacturers of Canada
Dr. Thomas Hudson, Montreal General Hospital

The Internal Policy Advisory Committee:
Chemicals and Bio-Industries Branch, Industry Canada
Federal Science & Technology, Industry Canada

Competition Bureau, Industry Canada
Intellectual Property Policy Directorate, Industry Canada
Health Protection Branch, Health Canada
Science Branch, Canadian Forest Service, Natural Resources Canada
Science & Technology Section, Department of Foreign Affairs and International Trade
National Research Council of Canada
Treasury Board Secretariat
Biotechnology Strategies, Agriculture and Agri-Food Canada
Corporate Governance Branch, Industry Canada
Environmental technology Advancement Directorate, Environment Canada
Aquaculture & Oceans, Science Branch, Department of Fisheries and Oceans
Economic and Regional Development Secretariat, Privy Council Office
Western Economic Diversification Canada

NOTES

[1] It was noted that one restaurant in the U.S. is now advertising that their restaurants use no "products of biotechnology." The participant asked how this term is being defined. "Does this term cover cattle and chicken which are fed 'Roundup Read' soybean and BT corn? Are these products of biotechnology? The cooking oil for the french fries comes from soybean or canola which is 'herbicide tolerant.' Are these fries a product of biotechnology? The fructose in the Coca Cola syrup comes from pooled corn which could include BT corn. Is the coke a product of biotechnology?"

[2] For example, the Canadian Institute of Biotechnology (CIB), under contract to the Office of Consumer Affairs and the Canadian Food Inspection Agency, prepared the report "About biotechnology: The communication experience in Canada."

ANNEX

Background on the Preparation of the Research and the Integration Document

The research agenda was developed from information gathered over the course of the Optima, Decima, and CROP surveys, follow-up focus groups, and general discussions with stakeholders at various biotechnology events. As such, although it did not represent an exhaustive list of the wide variety of economic issues, nor of the ethical and social aspects of biotechnology, the papers were intended to address an important subset of the main consumer and marketplace concerns on biotechnology.

To ensure that the analysis of the expert papers on which this document is based is rigorous and accessible to all biotechnology stakeholders, the Office of Consumer

Affairs engaged a General Editor (Dr. Bartha Maria Knoppers who is Professor of Law at the Faculty of Law, Université de Montréal) to oversee the preparation of papers; and established two Project Advisory Committees – one comprised of interdepartmental biotechnology stakeholders and the other comprised of non-governmental organizations with an interest in biotechnology issues.

The General Editor served as an intellectual guide for the preparation of manuscripts. In this capacity, she:

- fleshed out in a detailed manner the themes to be addressed in each paper;
- identified potential authors for the preparation of manuscripts;
- oversaw the thoroughness and quality of papers;
- added credibility to the process by ensuring that the ideas and notions developed and advanced in the expert papers were at arm's length from all stakeholders;
- oversaw the preparation of the first draft and reviewed the second draft of this integration paper which draws on the findings of the expert papers and the results from the Symposium.

As many of the consumer themes that are identified in the expert papers deal with issues related to the particular functions of government departments involved in biotechnology – be they regulatory, policy development, public awareness, or industrial development – the Office deemed it essential to structure an Internal Project Advisory Committee (IPAC) committee that would permit the variety of functional views and expertise to be reflected in the research effort.

The authors of the papers submitted detailed outlines of the methodology, scope, and findings of their papers which went to all IPAC members for their review and commentary. Once comments were incorporated into revised outlines, the authors started writing their manuscripts. All papers were then circulated to IPAC members for their review and commentary. Authors were required to consider all comments and views put forward by the IPAC.

To ensure that all biotechnology stakeholders, particularly the consumer and industry NGOs, had the opportunity to contribute their views and commentaries on the expert papers, the Office of Consumer Affairs also established an External Project Advisory Committee (EPAC). This committee ensured that the research ideas and potential policy recommendations were more broadly communicated to the NGOs and other groups. The activities of this group parallelled the activities of the IPAC with respect to the review of draft copies and commentaries.

It is hoped that the research contained in this document and the individual papers will contribute to developing more appropriate structures, defining roles and activities, and promoting collaboration among non-governmental biotechnology stakeholders to address the varied views and concerns of consumers in order to support:

1. Future survey, focus group, policy analysis, and other socio-economic research which is more policy oriented, deeper probing, better informed, and better directed towards (a) addressing the fundamental risk/benefit/information issues and other concerns of consumers and other users, and (b) meeting the needs of government regulators and policy makers and of the private sector.
2. Future policy formulation and, where required, regulatory changes which have broad understanding and support among stakeholders within and outside government.
3. The establishment of a strong, sustainable consensus among government, the scientific and academic communities, industry, consumer groups, and other organizations concerning:
    - the kinds of biotechnology information needed by consumers and their repre-

sentatives to make informed decisions in the marketplace and to contribute knowledgeably and effectively to policy and regulatory development in the biotechnology field;
- the advantages and disadvantages of the various alternatives available to provide that information – for example, labelling, point of sale information, industry advertising and promotion, and more general information campaigns on the part of government, industry, and NGOs;
- the respective roles and activities of different government, industry, consumer groups, and other stakeholders in providing information which addresses consumer concerns on biotechnology.

4. The development of a greater understanding and consensus across stakeholder groups regarding the role of product markets – relative to other instruments and processes (public opinion monitoring, public and industry consultations, policy formulation, the marketplace for ideas) – in addressing the many questions and concerns raised by biotechnology and the consumer, including:
   - information asymmetries (for example, regarding health and safety and other forms of traditional market failure);
   - social, cultural, and ethical issues;
   - the role of the consumer in supporting the development of biotechnology and related applications, services, and industries;
   - mechanisms for consumer consultation and participation.

Many of the problems faced by biotechnology are shared by other science-based activities. It is hoped that the lessons learned from this research will assist in the development of government policy and corporate strategies in other industries and markets which are based on the commercial applications of advanced science.

REFERENCES

Averitt, N. W., & Lande, H. (1997). Consumer sovereignty: A unified theory of antitrust and consumer protection law. *Antitrust Law Journal, 65*, 713–756.

Caulfield, T. (1998). The commercialization of human genetics: A discussion of issues relevant to the Canadian consumer. *This volume*. Also in *Journal of Consumer Policy, 21*.

Chen, Z., & McDermott, A. (1998). International comparisons of biotechnology policies. *This volume*. Also in *Journal of Consumer Policy, 21*.

Crampes, C., & Hollander, A. (1998). Biotechnological innovation and industrial performance. *This volume*.

Creative Research International (1996). *Environmental applications of biotechnology: Focus groups. Final report*. Ottawa: Industry Canada.

Doern, G. B., & Sheehy, H. (1998). The federal biotechnology regulatory system: A commentary on an institutional work in progress. *This volume*.

Einsiedel, E. F. (1998). The market for credible information in biotechnology. *This volume*. Also in *Journal of Consumer Policy, 21*.

Ekos Research Associates Inc. (1996). *Focus groups on agri-food applications of biotechnology. Summary report*. Ottawa: Industry Canada.

Envision Research (1997). *Socioethical implications of biotechnology*. Report prepared for Western Economic Diversification. Ottawa: Western Economic Diversification.

Green, C. (1998). The industrial economics of biotechnology. *This volume*.

Hadfield, G. K., & Thomson, D. (1998). An information-based approach to labelling biotechnology. *This volume*. Also in *Journal of Consumer Policy, 21*.

Ireland, D. (1998). Competition policy, intellectual property, and the consumer. In: R. B. Anderson & N. T. Gallini (Eds.), *Competition policy and intellectual property rights in the knowledge-based economy*, pp. 293–338. Ottawa: Industry Canada Research Series, Vol. 9.

Kleit, A. N. (1998).Using advertising to generate information and signals for product quality: Lessons for biotechnology markets in Canada from pharmaceutical markets in the United States. *This volume*.

Leroux, T., Hirtle, M., & Fortin, L. N. (1998). A comparative study of mechanisms developed abroad for examining the ethical and social issues in biotechnology. *This volume*. Also in *Journal of Consumer Policy, 21*.

McIntosh, J. G., & Cumming, D. (1998). Consumer controversy and the funding of biotechnology research. *This volume*.

Mathios, A. M. (1998). Economic perspectives on the dissemination of science-based information to consumers. *This volume*.

Optima Consultants (1994). *Understanding the consumer interest in the new biotechnology industry*. Ottawa: Industry Canada.

Purchase, B. (1998). Rapporteur's remarks: Symposium on Biotechnology and the Consumer. *This volume*.

Sheehy, H., Legault, M., & Ireland, D. (1998). Consumers and biotechnology: A synopsis of survey and focus group research. *This volume*.

Strauss, S. (1998). Biotechnology and the media. *This volume*.

Wohl, J. B. (1998). Consumers' decision-making and risk perceptions regarding foods produced with biotechnology. *This volume*. Also in *Journal of Consumer Policy, 21*.

THE AUTHORS

Marc Legault is a Senior Policy Analyst, International Investment and Services Policy, Industry Canada, 235 Queen Street, Room 502B, Ottawa, Ontario, K1A 0H5, Canada. At the time of writing he was with the Office of Consumer Affairs, Industry Canada. Alison McDermott is an economist at Industry Canada. Her address is: 1801 Hollis St, P.O. Box 940, Station "M", 4th Floor, Halifax, Nova Scotia B3J 2V9. E-mail: mcdermott.alison@ic.gc.ca. Bartha M. Knoppers is Professor of Law at the Faculty of Law, Université de Montréal, P.O. Box 6128, Succursale Centre-ville, Montréal, Québec, H3C 3J7, Canada. Derek Ireland was previously Director of Consumer Research and Analysis. The paper reflects neither the views nor the policies of the Office of Consumer Affairs, Industry Canada, or the Government of Canada.

Bryne Purchase
Appendix B: Rapporteur's Remarks Symposium on Biotechnology and the

INTRODUCTION

The following is a reconstruction of remarks I made as the rapporteur at the Symposium "Biotechnology and the Consumer."

Complex Reality

I was chosen for this assignment because of my relative innocence in matters related to biotechnology. Much of what I have learned has been at this Symposium from the various speakers' remarks and the interventions from the floor. I have also learned a great deal from reading the background papers a few days before. Throughout, I have been struck by the complexity of the issues surrounding this technology. There are, for example, many different types of high level expertise involved. And it is not just a matter of a remarkable science and technology with a wide variety of applications. The issues include technical matters related to communications and education, law and economics, and political, ethical, and moral choices. And there are also many different interests involved, most of whom are represented at this Symposium. Each expertise, and each interest, brings a different and necessary perspective to the table.

Origin of Our Thoughts?

Shakespeare's Hamlet says, "There is nothing either good or bad but thinking makes it so." If this is true, then one is invited to reflect on where our thoughts come from. Why would we think the science or applications of biotechnology either good or bad? In this regard there is an old saying that, "What you see depends on where you sit." In all of this complex reality we tend to see what we focus on. And what we focus on is dictated very much by our prior knowledge and experience of the world. In short, our attitudes

Journal of Consumer Policy **21**, 1998.
Bartha M. Knoppers & Alan D. Mathios (eds), Biotechnology and the Consumer.

and perspectives will be shaped by our interests and our professional expertise.

Moreover, throughout the Symposium we were very forcefully reminded of the fact that the way we pose questions very often determines the answers we will get. That is, our questions will highlight the part of the complex reality that we choose to illuminate. As a result a limited and biased set of questions might lead us to facilitate in law and economics a technology that was not as desirable as some alternative course. Therefore, it is extremely important that all of the relevant types of expertise and interests be represented *early* in the consultative process so that the public debate about biotechnology, or any other, not be unfairly biased from the very outset. Otherwise, the alternative technologies, if there are any, may never be given a reasonable chance to become the mainstream reality.

Three Perspectives

I have taken the liberty of organizing my remarks around three different perspectives: the consumer/citizen perspective, the industry perspective, and the government perspective. These are not always competing perspectives, although sometimes they are. My objective is to outline very briefly some of the highlights of each perspective. Doing so is, I believe, a first step, although an essential one, in moving toward, if not a consensual perspective, a common understanding of each separate perspective.

THE CONSUMER PERSPECTIVE

Benefits

First and foremost it is important to remember in a Symposium about consumers that the consumer benefits most when there are products available that provide utility beyond the price that consumers must pay. And numerous speakers outlined many examples of biotechnology applications that yield real value in the way of improving human health and prolonging and improving the quality of human life. There are also a number of applications in the area of animal health. Some would value this in its own right. Others would value it for the potential contribution to increasing productivity in the agri-food business, where there are also applications in many other areas that bring the prospect of improved food quality and lower cost. Similarly there are applications in the fisheries, in forestry, and even in environmental clean-up, all of which hold out the promise of significant benefit to consumers.

What is particularly exciting is that while the number of products currently on the market is in the order of a hundred or so, there are a huge number of potential new products under development and in the pipeline. So that we may well be simply on the threshold of an explosion of bio-engineered products in the marketplace. Now is surely the time therefore to address some of the fundamental concerns raised by biotechnology.

Risks and Uncertainty

On the other side of potential benefits to consumers, there are downside risks. There are risks to our health and safety. And, perhaps even more importantly, many see risks to our values and beliefs. In this latter respect we can easily see, for example, the threat to our society from human cloning. But there are potential applications where there will be differing perspectives; and we were reminded that we often have no "standard of proof" by which to evaluate arguments about potential threats to our social fabric.

With respect to risk itself we were told as well that there is a difference between the expert evaluation and the public perception of risk. At the same time our regulatory structure for health and safety is based on a scientific conception of risk. Moreover as global trade expands and regulations are harmonized, it is almost surely to be done on the basis of science, the only common language and basis of human understanding available for such an undertaking. Therefore, we are dealing with a problem that will not be easily resolved, or indeed go away.

In respect to risk, the public is heavily influenced by the "dread" factor in certain adverse outcomes. It matters to public attitudes, as well, whether or not the risk was entered into voluntarily and whether there are strategies that make the risk controllable. The public cares also about the distribution of the benefits from risk taking, as well as the distribution of the risk itself. And while there is no such thing as a zero risk, there is often a great deal of scientific and social uncertainty involved in biotechnologies. That is to say, we simply do not have enough information to make scientific statements about risks. That consumers must make choices anyway, perhaps makes them vulnerable to the risk of misinformation and the potential for irrationally forgoing benefits. This, as an example from another technology, may be the case with respect to the irradiation of food which appears simply unacceptable to the consuming public, notwithstanding its very great benefits.

Consumer Information

For consumers to make rational choices they need to be informed. Numerous speakers talked about the multiple types of information that would be relevant to rational consumer choice. But not all of this information is available in the marketplace. On the other hand we heard the results of consumer research that reveal that some consumers often do not use or want the information given. There are extremely high costs associated with getting and interpreting information on biotechnology products; and there are low expected benefits.

Biotechnology products are often "credence" goods where it is not possible for the consumer to know their efficacy, either before or after consuming the product. Given this, and the often high level of uncertainty, consumers typically substitute a simple decision rule. In other cases, however, the public must rely on "gatekeepers" who will make these decisions for them. This is the case for example in health care and particularly as the health insurance system in both Canada and the United States moves toward "managed care" and away from simply open-ended access to diagnostics or treatments.

Rational Choice Model?

A number of speakers questioned even the idea that consumers make rational choices along the lines typically modeled by economists. Instead it was suggested that consumer choice is driven more by emotion, in particular the powerful emotions of love and fear, than by rational calculation. This is particularly challenging to an economist's perspective on reality, since rational choice is the basis for their entire understanding of market failure and the efficiency rationale for government regulation. In this regard it may be that communications experts and marketers have a better appreciation for the actual triggers of consumer choice. On the one hand, this represents a powerful ability to inform and communicate with the public, and it could be a useful ally of public information programs. But, on the other hand, there are some who fear the potential for manipulation.

Heterogeneity Problem

There is, of course, no single consumer perspective and no single consumer. There is a wide variety of people with differing needs, tastes, and values that make up the consuming public. There are also widely differing knowledge levels and differing capabilities to access and process information. For the most part, however, the distribution of consumer opinion leaves at least 60 percent of the population in an undecided category. These "undecided" are prepared to be convinced one way or another. Whereas at one extreme of the distribution of public opinion there are those who see benefits no matter what; and, at the other extreme, are those who are opposed no matter what. In general the opinion surveys appear to reveal that where there are big pay-offs (as potentially in health care) then there is a greater tolerance for risk.

Multiple Points of Influence and No Influence

Some speakers and the background reading argued that consumers have multiple points of influence on biotechnology development. There are, for example, non-market actions that can be pursued through the political process and in influencing the funding policies of public granting agencies; there are also actions that can be taken in the marketplace either in terms of the individual choice to consume, or through collective choice, for example, by joining consumer boycotts. And indeed, potential public controversy was on the list of considerations of biotechnology firms and their financiers in their decisions about what to research and finance.

At the same time there were several interventions to the effect that some groups felt they had no influence, and that the industry perspective predominated even the development and organization of this very Symposium. These groups made a strong case for early involvement and for tendering of research contracts.

THE INDUSTRY PERSPECTIVE

Not an Industry

Perhaps the first and most important point to make is that the biotechnology industry is not an industry in the classic sense of the term. It is instead an enabling technology with many applications across a wide range of industries. It is therefore potentially more significant economically than indicated by the small number of biotechnology firms now operating in Canada and the few research jobs they represent. Biotechnology affects the potential future competitiveness of many of Canada's largest and most traditional industries such as forestry, fishing, agriculture, and agri-food. Biotechnology is likely to play an important role in the success of some of the newest high technology industries including pharmaceuticals, health diagnostics, and environmental clean-up.

Science Based Competition and University Linkages

Biotechnology firms compete, to the degree that they operate in the same broad industry category, on the basis of science based innovation. Their purpose is principally to invent new products. And in this regard they are highly risky ventures. In fact these firms are perhaps akin to the wildcat operator in the oil industry or the prospector in the mining industry. Moreover, what perhaps distinguishes this technology from other advanced technologies is the extraordinary degree to which these firms have raised private financing in North American equity markets to search for new products. This is not the model found in Europe or Japan where large enterprises are direcly funding the research themselves. And, incidentally, in reflection of the enormous financial risk in this field, there has not been a large pay-off so far, although the financial news may soon brighten as more products reach commercial stages.

Biotechnology firms are often founded by university researchers and maintain very close links to universities and basic research. Indeed some have expressed concern about the impact that commercial imperatives may have on the traditional independence of the university. It is obviously a matter to be addressed by the governance structure of those institutions.

Industry Structure and Its Evolution

Because there is no biotechnology industry, it is surely inappropriate to speak of an industry structure. However, there appear to be few barriers to entry into biotechnology research, but considerable barriers to survival. On these grounds it is contended that there are none of the traditional market power concerns. Instead the background economic analysis tended to focus more on the vertical contractual relationships now forming not only between universities and biotech research firms but also between these firms and immensely larger and more commercial enterprises in the traditional health care and agri-business industries.

There will perhaps be greater cause for concern when there are more commercial products and the research protocols have become more routine. At that point, the large agri-business, chemical and pharmaceutical companies may bring to bear their large existing market share, their superior marketing expertise, their production capabilities, and their access to capital. And, also at this point, the now common various arms-length contracts now common will be replaced by outright acquisition. It is also possible that a more heavily regulated industry will accelerate the process of vertical integration and industrial concentration. The Symposium heard that it already takes thirteen years and $500 million from the initial research until a new drug hits the market.

Industry Cooperative Action

Populated as it is with many small and research oriented enterprises pursuing many different applications, the chances for strong industry cooperative action would seem limited. Yet, as the Symposium heard, there are a number of biotechnology associations and these are increasingly working together for a common cause. This collective action is aimed not only at governments, but also over the heads of governments directly at the consumer.

Private Enterprise as an Information Provider

Much consumer information typically comes from the media. However, the Symposium heard of the systemic weaknesses of the newspaper industry for example in developing reliable information on biotechnology and the consumer. Moreover, currently most of the information about biotechnology will be found in the financial section of the paper. But the Symposium also reviewed the role of private enterprise as a supplier of information to consumers in the marketplace. Industry acknowledges the crucial role of consumer acceptance, and it is deeply concerned about the potential for misinformation and unnecessarily frightening consumers who are not well informed.

There was some discussion as well on the strong natural incentives for industry to inform consumers of new biotechnology products (for example in health care). Some evidence was presented on food health claims by private enterprise in the United States to the effect that it is very effective in changing the population's eating behavior for the better. And that it appeared much more effective than simply government information programs. It was also found that producers are trustworthy in their claims. But at the same time it was noted that there was a lack of incentive to inform consumers of cost reducing process innovations.

THE GOVERNMENT PERSPECTIVE

Multiple Objectives and Approaches

Governments, of course, always have more than one agenda, simply because they must represent more than one perspective. For example, they wish to maximize the consumer benefits from biotechnology; at the same time, however, they wish to protect the community's health and safety. But they are also clearly interested in helping to create jobs and wealth in the economy. Needless to say governments are also interested in maintaining the social and political fabric of the nation. Therefore they are unlikely to wish to put political institutions to the great strain of highly divisive ethical or moral debates. And in recent years, governments have sought to do all these things in the context of expenditure reduction.

These diverse goals are clearly not always antithetical. For example, the creation of jobs and wealth is usually not at the excessive expense of consumers. There are gains from trade, to both sides of a fair exchange. Otherwise it would not be voluntary. But sometimes the goals can be conflicting. For example, the government's capacity to maintain "fairness" in the marketplace may be compromised by its reduced ability to do basic science as a result of budget cutbacks.

The result may be a diminished capacity to protect consumer health and safety. Or the promotion of industry interests may come at the expense of consumers where it involves a restriction of competition or an artificial narrowing of choice.

The Symposium also heard about some potential alternative options to government as the provider of both a regulatory framework and consumer information, while still ensuring that the applications of biotechnology may be maximized to the net benefit of both the individual consumer and society in general. The government acts simply as a catalyst to the collective actions of others as providers of information or regulation.

Trust as a Generic Issue

Perhaps the most frequently repeated observation was that the government was perceived to have a declining level of public trust (although the public trust in industry is not very substantial either). Even NGO's can develop problems of public trust as the Canadian blood scandal illustrates. Whatever the approach taken by government, industry, or consumer groups, the issue of trust must be addressed directly. On this score, the Symposium heard that, as a rule, people trust people, not facts. In short, information alone is not credible unless coming from a credible source. Moreover, trust needs to be earned; acting in a trustworthy manner is therefore a long term investment.

These issues go to the heart of the design and productivity of democratic institutions. The Symposium heard repeatedly about the desirability of designing institutions that were equitable in their accessibility, transparent in their proceedings, and independent from any special interest. These are qualities that must clearly be reflected in any institutional design for dealing with biotechnology issues or any other matters of great public importance.

Government as Regulator, Information Provider, and Catalyst

The Symposium was provided with a brief overview of the existing regulatory framework. It was noted that currently biotechnology is not regulated as a technology. Rather, each application has a regulatory framework reflecting traditional areas of concern for example in agriculture and agri-food, health care, fisheries, and the environment. Some coordination is achieved within the context of the Environmental

Protection Act in terms of notification and assessment. But there was nonetheless a call for more coordination. At the same time, however, the existing framework, given the many different applications, has a basic logic. There is a need to distinguish between applications. Indeed, there was a call to prioritize concerns so that relatively trivial matters need not be endlessly debated or reviewed. In the longer term, however, the technology itself may need to be regulated as certain applications become the subject of fundamental ethical questions that apply across the entire spectrum of applications.

The government could still be involved in providing generic information on biotechnology. But obviously the extraordinary range of applications means that others must provide information as well. Several industry and consumer groups spoke of their initiatives in this regard. There was also a paper on the potential use of voluntary codes. These codes have many positive features relative to traditional command and control regulations, most importantly their flexibility. However, the large number of small biotechnology firms raises some important questions about the efficacy of voluntary codes and a number of interveners expressed a desire to know more about the details of their application in this instance.

The question of mandatory labeling was raised numerous times. Strong arguments for and against were presented. While it is difficult to deny the basic consumer right to know, there is the fundamental question of its practical application. For example, basic marking information is essentially useless in a context of a multiplicity of products with some bio-engineered component. However, there are examples of the use of voluntary codes with respect to consumer information in the labeling of novel foods and the matter of "informed consent" in medicine. These may serve as models.

International Realities

The Symposium was reminded of the fact that whatever decisions are made domestically, sovereignty is constrained by the need to participate in global commerce. There are constant pressures to harmonize. Canada is closer to the United States in the existing structure of biotechnology regulation. Regulation differs from that practised in the European Union and there are labeling and patent problems already emerging between the great trading blocks. While the international institutional infrastructure to deal with these issues is still nascent, it is nonetheless emerging and evolving and can be expected to continue to do so.

Next Steps

It was clear from different interventions and speakers throughout the two days that the following were considered essential next steps:

- building a consultative infrastructure
- increasing the amount of grassroots, direct citizen input on a regional basis
- constructing a national information library on biotechnology
- developing a cross-cutting national advisory body.

CONCLUSIONS

The Symposium represents a well planned and highly useful first step in developing a national approach to this important and often controversial technology. The Symposium organizers and planners are to be highly praised. But this is also clearly a work in progress. And we have heard that many, if not all, of the issues raised in these two days will be dealt with by the advisory group developing the renewed National Biotechnology Strategy. It will be exciting and no doubt rewarding for all those involved. To which I can only add, Bonne Chance!

THE AUTHOR

Bryne Purchase is Chair, Executive and Professional Programs at the School of Policy Studies, Queen's University, Kingston, Ontario, Canada K7L 3N6. E-mail: bpstrategy@igs.net.

MIX
Papier aus verantwortungsvollen Quellen
Paper from responsible sources
FSC® C105338

If you have any concerns about our products,
you can contact us on
ProductSafety@springernature.com

In case Publisher is established outside the EU,
the EU authorized representative is:
Springer Nature Customer Service Center GmbH
Europaplatz 3, 69115 Heidelberg, Germany

Printed by Libri Plureos GmbH
in Hamburg, Germany